INTRODUCTION TO GEOGRAPHIC INFORMATION SYSTEMS

Second Edition

INTRODUCTION TO GEOGRAPHIC INFORMATION SYSTEMS

Kang-tsung Chang

University of Idaho

Boston   Burr Ridge, IL   Dubuque, IA   Madison, WI   New York   San Francisco   St. Louis
Bangkok   Bogotá   Caracas   Kuala Lumpur   Lisbon   London   Madrid   Mexico City
Milan   Montreal   New Delhi   Santiago   Seoul   Singapore   Sydney   Taipei   Toronto

Higher Education

INTRODUCTION TO GEOGRAPHIC INFORMATION SYSTEMS, SECOND EDITION

Published by McGraw-Hill, a business unit of The McGraw-Hill Companies, Inc., 1221 Avenue of the Americas, New York, NY 10020. Copyright © 2004, 2002 by The McGraw-Hill Companies, Inc. All rights reserved. No part of this publication may be reproduced or distributed in any form or by any means, or stored in a database or retrieval system, without the prior written consent of The McGraw-Hill Companies, Inc., including, but not limited to, in any network or other electronic storage or transmission, or broadcast for distance learning.

Some ancillaries, including electronic and print components, may not be available to customers outside the United States.

This book is printed on acid-free paper.

1 2 3 4 5 6 7 8 9 0 QPF/QPF 0 9 8 7 6 5 4 3

ISBN 0–07–252811–7

Publisher: *Margaret J. Kemp*
Sponsoring editor: *Thomas C. Lyon*
Developmental editor: *Lisa A. Leibold*
Executive marketing manager: *Lisa L. Gottschalk*
Senior project manager: *Jayne Klein*
Senior production supervisor: *Laura Fuller*
Media technology producer: *Renee Russian*
Designer: *David W. Hash*
Cover/interior designer: *Nathan Bahls*
Cover images: *© The Living Earth, Inc./© Corbis*
Compositor: *GAC—Indianapolis*
Typeface: *10/12 Times Roman*
Printer: *Quebecor World Fairfield, PA*

Library of Congress Cataloging-in-Publication Data

Chang, Kang-tsung.
 Introduction to geographic information systems / Kang-tsung Chang. — 2nd ed.
  p. cm.
 Includes bibliographical references and index.
 ISBN 0–07–252811–7 (alk. paper)
 1. Geographic information systems. I. Title.

G70.212.C4735   2004
910'.285—dc 21   2003005261
                                              CIP

www.mhhe.com

BRIEF CONTENTS

CONTENTS

CHAPTER 3

THE VECTOR DATA MODEL *40*

CHAPTER 4

VECTOR DATA INPUT *64*

CHAPTER 5

SPATIAL DATA EDITING 92

CHAPTER 6

ATTRIBUTE DATA INPUT AND
MANAGEMENT 115

RASTER DATA   *133*

DATA DISPLAY AND CARTOGRAPHY   *153*

DATA EXPLORATION   *180*

CHAPTER 10

VECTOR DATA ANALYSIS   *206*

CHAPTER 11

RASTER DATA ANALYSIS   *224*

CHAPTER 12

TERRAIN MAPPING AND ANALYSIS   245

CHAPTER 13

SPATIAL INTERPOLATION   274

CHAPTER 14

GIS MODELS AND MODELING *305*

CHAPTER 15

REGIONS *328*

CHAPTER 16

NETWORK AND DYNAMIC SEGMENTATION   *349*

PREFACE

THE IMPORTANCE OF GIS

A geographic information system (GIS) is a computer system for capturing, storing, querying, analyzing, and displaying geographically referenced data. Geographically referenced data describe both the location and characteristics of spatial features such as roads, land parcels, and forests. The ability of GIS to handle and process both location and attribute data distinguishes GIS from other information systems and establishes GIS as a technology necessary for a wide variety of applications.

Since the 1970s GIS has been important in natural resource management including land-use planning, natural hazard assessment, wildlife habitat analysis, and timber management. More recently, GIS has been used for crime analysis, emergency planning, land records management, market analysis, and transportation planning. And integration of GIS with other technologies has found applications in precision farming, interactive mapping on the Internet, and location-based services.

GIS is no longer a tool for specialists. In recent years, developments of the user-friendly interface, powerful and affordable computer hardware and software, and public digital data have brought GIS to mainstream use. It is not unusual nowadays to find students from more than 20 academic departments in an introductory GIS class.

The design of a GIS involves fundamental concepts from fields such as geography, cartography, spatial analysis, mathematics, and database management. To capture GIS data, for example, we must consider the spatial referencing system, a concept long established in cartography. A spatial referencing system, either geographic or projected, determines how the locations of spatial features are measured. The design of a GIS is also influenced by the constantly evolving computer technology. For instance, a recent development that allows location-based data to be stored and managed in a single database has led to a new GIS data model that differs from the traditional model of storing location data and attribute data in a split system. To be proficient in GIS, we must learn the basic concepts as well as new technologies.

ENHANCEMENTS TO THE SECOND EDITION

This book is intended to provide students in a first or second GIS course with a solid foundation in both concepts and practice. I have retained the same comprehensive coverage of GIS topics as featured in the first edition. This edition has five segments. Chapters 1 to 7 explain the fundamentals of GIS including coordinate systems, data models, data input, spatial data editing, and attribute data management. Chapters 8 and 9 include data display and data exploration. Chapters 10 and 11 examine the basic tools for GIS analysis and their applications. Chapters 12 and 13 cover terrain mapping and analysis, and spatial interpolation. Chapters 14 to 16 deal with GIS models and modeling, regions, and network and dynamic segmentation.

For this edition I have expanded discussions on GIS applications (Chapter 1), datums (Chapter 2), map types (Chapter 8), descriptive statistics and dynamic graphics (Chapter 9), viewshed analysis and watershed analysis (Chapter 12), elements and comparison of spatial interpolation methods (Chapter 13), and geographic regions (Chapter 15). I have added new materials on ESRI's geodatabase model (Chapter 3), the U.S. National Standard for Spatial Data Accuracy (Chapter 3), geocoding of street addresses (Chapter 4), LIDAR and ASTER data (Chapter 7), areal interpolation (Chapter 10), and ESRI's geometric network (Chapter 16). I have completely revised Chapter 14 by including elements of GIS modeling, multicriteria evaluation and its methods, linear and logistic regression, and an in-depth coverage of process models.

This edition contains many new references from a large variety of disciplines; students and professors should find *An Introduction to Geographic Information Systems,* Second Edition, to be up-to-date and resourceful. Boxes are again used throughout the book to provide additional information and worked examples of important computing algorithms (mainly in Chapters 12 and 13). Boxes are also used to cover software-specific materials. To improve the quality of visual aids, this edition includes over 70 revised figures and 25 new figures.

COMBINING CONCEPTS WITH APPLICATIONS

This second edition continues to emphasize the practice of GIS. I have included problem-solving tasks in the Applications section of each chapter, complete with data sets and instructions. The number of tasks totals 75, with 2 to 7 tasks in each chapter. The instructions for performing the tasks have been rewritten to correlate to *ArcGIS 8.2.* For those who prefer to use *ArcView 3.2,* the tasks and instructions are available at the McGraw-Hill website at http://www.mhhe.com/changgis2e. The

hands-on experience to be gained from the Applications sections is intended to complement the discussions in the text and to enhance the working knowledge of GIS functions. We can learn GIS concepts from readings, but GIS users can most fully grasp concepts behind the menus and icons through practice. Moreover, a regular requirement in the current GIS job market is to be skilled in using GIS packages.

Most exercises in this edition use *ArcView 8.2* and its extensions of Spatial Analyst, 3D Analyst, and Geostatistical Analyst. *ArcInfo 8.2,* especially its *ArcInfo Workstation,* is used only for tasks that cannot be performed in *ArcView 8.2,* such as topological editing, geometric transformation, regions, network analysis, and dynamic segmentation.

The website for the second edition, located at www.mhhe.com/changgis2e, contains a password-protected instructor's manual. Contact your McGraw-Hill sales representative for a password.

ACKNOWLEDGMENTS

I would like to thank the following reviewers who provided many helpful comments on the first edition:

ShunFu Hu
University of Wisconsin at Oshkosh

Jon T. Kilpinen
Valparaiso University

C. K. Leung
California State University, Fresno

Scott M. Pearson
Mars Hill College

Peter Siska
Stephen F. Austin University

I would also like to thank 102 reviewers who commented on the individual chapters of the first edition. I have read their reviews carefully and incorporated many comments into the revision. Of course, I take full responsibility for the book.

I wish to express my thanks to the University of Idaho for granting me a sabbatical leave for the fall semester of 2002 to work on this edition. At McGraw-Hill, Tom Lyon, Lisa Leibold, Marge Kemp, David Hash, and Jayne Klein deserve much credit for their guidance and assistance during various stages of the project.

I dedicate this book to my parents, Ching-tu and Yeh Chang, who trusted me and supported my graduate study in the United States many years ago. Finally, I am grateful to the support of my family.

Kang-tsung Chang

INTRODUCTION

1.1 WHAT IS A GIS?

A **geographic information system (GIS)** is a computer system for capturing, storing, querying, analyzing, and displaying geographically referenced data. Also called geospatial data, **geographically referenced data** are data that describe both the location and characteristics of spatial features such as roads, land parcels, and vegetation stands on the Earth's surface. The ability of GIS to handle and process geographically referenced data distinguishes GIS from other information systems and establishes GIS as a technology important to a wide variety of applications.

1.1.1 GIS Applications

Since the beginning, GIS has been important in natural resource management including land-use planning, natural hazard assessment, wildlife habitat analysis, riparian zone monitoring, and timber management. Here are some examples on the Internet:

- The U.S. Geological Survey has three case studies at their website showing how GIS was used to select potential well sites, to analyze the effect of an earthquake on the response time of fire and rescue squads, and to simulate 3-D views of an area before and after mining development (**http://www.usgs. gov/research/gis/title.html/**).
- The U.S. Forest Service uses GIS and other computer technologies to map forest fires and to model fire behavior (**http://www.fs. fed.us/fire/fire_new/tools_tech/**).
- The U.S. Department of Housing and Urban Development provides an interactive mapping service that can combine information on community development and housing programs with environmental data from the U.S. Environmental Protection Agency (**http://www.hud.gov/offices/cio/ emaps/index.cfm**).

In more recent years GIS has been used for crime analysis, emergency planning, land records management, market analysis, and transportation applications. Here are some examples on the Internet:

- The National Institute of Justice uses GIS to map crime and hotspots and to analyze the spatial patterns of crime by location and time (**http://www.ojp.usdoj.gov/nij/maps/**).
- The Federal Emergency Management Agency is building a flood insurance rate map database and is planning to use GIS to link the database to physical features (**http:// www.fema.gov/mit/tsd/mm_main.htm**).
- Larimer County, Colorado, provides public access to the county's land records in a GIS database, which includes information on parcel location, owner name and address, sale, value, and legal description (**http:// www.larimer.org/**).
- The Hertz Corporation offers the NeverLost in-car navigation system that can guide a driver with turn-by-turn directions to any destination in their GIS database (**http:// www.hertz.com/**).

Integration of GIS with global positioning system (GPS) and the Internet has also introduced new and exciting applications such as precision farming, interactive mapping, and location-based services (Box 1.1).

1.1.2 Components of a GIS

Like any other information technology, GIS requires the following four components to work with geographically referenced data:

- **Computer System.** The computer system includes the computer and the operating system to run GIS. Typically the choices are PCs that use the Windows operating system (e.g., Windows NT, Windows 2000, Windows XP) or workstations that use the UNIX operating system. Additional equipment may include monitors for display, digitizers and scanners for spatial data input, and printers and plotters for hard-copy data display.
- **GIS Software.** The GIS software includes the program and the user interface for driving the hardware. Common user interfaces in GIS are menus, graphical icons, and command lines.
- **Brainware.** Equally important as the computer hardware and software, the brainware refers to the purpose and objectives, and provides the reason and justification for using GIS.
- **Infrastructure.** The infrastructure refers to the necessary physical, organizational, administrative, and cultural environments that support GIS operations. The infrastructure includes requisite skills, data

Box 1.1   Application Examples of Integrating GIS with Other Technologies

Over the years, new applications have been found by integrating GIS with other technologies. Among the examples are precision farming, interactive mapping on the Internet, and location-based services. Precision farming refers to site-specific farming activities such as herbicide or fertilizer application. Technologies for precision farming include GPS to mark the location of weeds or plots, GIS to store and process data and to prepare maps, and advanced computer-controlled sprayers to specifically target areas with crops for chemical application.

Interactive mapping lets Internet users select map layers for display and make their own maps. Websites set up for interactive mapping use GIS to provide a database and the functionalities of pan, zoom, and query. The USGS maintains a website for interactive mapping using the National Atlas of the United States (**http://www.nationalatlas.gov/**). The U.S. Census

Bureau has its On-Line Mapping Resources website, where Internet users can map public geographic data of anywhere in the United States (**http://www. census.gov/geo/www/maps/CP_OnLineMapping. htm**). The National Association of Realtors has an interactive mapping service that can assist potential homebuyers in finding listings that meet their criteria and viewing these listings on a street map (**http:// www.realtor.com/**).

A new application that has been heavily promoted by GIS companies is called location-based services (LBS). This application lets users deliver and receive information that is related to a particular location through wireless telecommunication. For example, an LBS user may use a Web-enabled mobile phone to be located and then to access, through mobile location services, information that is relevant to the user's location such as nearby ATMs and restaurants.

standards, data clearinghouses, and general organizational patterns.

1.2 A BRIEF HISTORY OF GIS

GIS is not new. Since the late 1960s computers have been used to store and process geographically referenced data. Early examples of GIS-related work from the late 1960s and 1970s include the following:

- Computer mapping at the University of Edinburgh, the Harvard Laboratory for Computer Graphics, and the Experimental Cartography Unit (Coppock 1988; Chrisman 1988; Rhind 1988).
- Canada Land Inventory and the subsequent development of the Canada Geographic Information System (Tomlinson 1984).

- Publication of Ian McHarg's *Design with Nature* and its inclusion of the map overlay method for suitability analysis (McHarg 1969).
- Introduction of an urban street network with topology in the U.S. Census Bureau's DIME (Dual Independent Map Encoding) system (Broome and Meixler 1990).

For many years, though, GIS has been considered to be too difficult, expensive, and proprietary. The advent of the graphical user interface (GUI), powerful and affordable hardware and software, and public digital data has broadened the range of GIS applications and brought GIS to mainstream use in the 1990s. Box 1.2 shows a list of GIS software producers and their main products.

According to a published survey (Crockett 1997), ESRI Inc. and Intergraph Corp. have dominated the market for GIS software. These two

Box 1.2   A List of GIS Software Producers and Their Main Products

The following is a list of GIS software producers and their main products as of 2002:

- Environmental Systems Research Institute (ESRI) Inc. (**http://www.esri.com/**): **ArcGIS, ArcView 3.x**
- Autodesk Inc. (**http://www3.autodesk.com/**): **Autodesk Map**
- Land Management Information Center at Minnesota Planning (**http://www.mnplan. state.mn.us/EPPL7**): **EPPL7**
- Baylor University, Texas (**http://www3. baylor.edu/grass/**): **GRASS**
- Clark Labs (**http://www.clarklabs.org/**): **IDRISI**
- International Institute for Aerospace Survey and Earth Sciences, the Netherlands (**http://www. itc.nl/ilwis/**): **ILWIS**

- Manifold.net (**http://www.manifold.net/**): **Manifold System**
- MapInfo Corporation (**http://www.mapinfo. com/**): **MapInfo**
- ThinkSpace Inc. (**http://www.keigansystems. com/**): **MFworks**
- Intergraph Corporation (**http://www.inter graph.com/**): **MGE, GeoMedia**
- Bentley Systems, Inc. (**http://www2.bentley. com/**): **Microstation**
- PCI Geomatics (**http://www.pcigeomatics. com/**): **PAMAP**
- TYDAC Inc. (**http://www.tydac.ch/**): **SPANS**
- Caliper Corporation (**http://www.caliper.com/**): **TransCAD, Maptitude**
- Northwood Technologies Limited (**http://wnp. marconi.com/**): **Vertical Mapper**

companies also led the GIS industry in software revenues in 2000 according to a report in the December 2001 issue of *GEOWorld.*

The main software product from ESRI Inc. is ArcGIS, a scalable system with ArcView, ArcEditor, and ArcInfo. All three versions of the system operate on the Windows platforms and share the same applications and extensions, but they differ in their capabilities. ArcView has data integration, query, display, and analysis capabilities. ArcEditor has the additional functionality for data editing. ArcInfo consists of ArcInfo Desktop and ArcInfo Workstation. ArcInfo Desktop has more data conversion and analysis capabilities than ArcEditor, and ArcInfo Workstation is a command-line application, basically the same as ARC/INFO 7.x. Intergraph Corp. has two main products: GeoMedia and MGE. GeoMedia is a desktop GIS package for data integration and visualization and is compatible with standard Windows development tools. MGE, which operates on the Windows operating systems or UNIX, consists of a series of products

for data production and analysis. Data can be migrated from GeoMedia to MGE, and vice versa.

Smaller GIS companies often deploy strategic partnerships to strengthen their roles on the GIS market (Box 1.3). GRASS is unique among GIS software packages because it is free. Originally developed by the U.S. Army Construction Engineering Research Laboratories, GRASS is currently maintained in both the United States (Baylor University) and Germany (University of Hannover). Some GIS packages are targeted at certain user groups. TransCAD, for example, is a package designed for use by transportation professionals. Microsoft, Oracle, and IBM have also entered the GIS industry. Microsoft MapPoint targets business analysts who need to analyze and map geographic data (**http:// www.microsoft.com/**). Oracle Spatial can store, access, and manage spatial data in Oracle's relational database management system (**http://www.oracle.com/**). Through purchase of Informix, IBM offers Spatial DataBlade, an extension that allows location-based data to be stored in

Box 1.3   Strategic Partnerships in the GIS Industry

A strategic partnership is designed to increase the competitiveness of the companies involved on the GIS market (Wilson 2001). A partnership may involve marketing a package, providing data exchange functionalities, jointly developing a new module or extension, or licensing a new technology. The following is a list of such partnerships:

- MapInfo and Vertical Mapper, MapInfo and SPANS
- PC Geomatics and ILWIS

- GeoMedia and Mfworks
- ESRI and ERDAS (a raster imaging/image processing company)

The GIS industry has also witnessed acquisitions of companies by other companies. Examples include Atlas GIS by ESRI, Vision* Solutions by Autodesk, ETAK by TeleAtlas, and SmallWorld by GE. SmallWorld was once a GIS company, but GE-SmallWorld (**http://www.gepower.com/networksolutions/**) now concentrates on utility distribution operation software.

IBM's relational database (**http://www-3.ibm. com/software/data/informix/blades/spatial/**).

Along with the proliferation of GIS activities, numerous GIS textbooks have been published (see Appendix A), and several journals and trade magazines are now devoted to GIS and GIS applications (see Appendix B).

1.3 GEOGRAPHICALLY REFERENCED DATA

Geographically referenced data separate GIS from other information systems. Therefore, before discussing GIS operations, we must understand the nature of geographically referenced data. Take the example of roads. To describe a road, we refer to its location (i.e., where it is) and its characteristics (e.g., length, name, speed limit, and direction), as shown in Figure 1.1. The location, also called geometry or shape, represents spatial data, whereas the characteristics are attribute data. Thus a road, like any geographically referenced data, has the two components of spatial data and attribute data.

1.3.1 Spatial Data

Spatial data describe the locations of spatial features, which may be discrete or continuous. **Dis-crete features** are individually distinguishable features that do not exist between observations. Discrete features include points (e.g., wells), lines (e.g., roads), and areas (e.g., land-use types). **Continuous features** are features that exist spatially between observations. Examples of continuous feature are elevation and precipitation. A GIS represents these spatial features on the Earth's surface as map features on a plane surface. This transformation involves two main issues: the spatial referencing system and the data model.

The locations of spatial features on the Earth's surface are based on the geographic coordinate system with longitude and latitude values, whereas the locations of map features are based on a plane coordinate system with x-, y-coordinates. **Projection** is the process that can transform the Earth's spherical surface to a plane surface and bridge the two spatial reference systems. But because the transformation always involves some distortion, hundreds of plane coordinate systems that have been developed to preserve certain spatial properties are in use. To align (register) with one another spatially for GIS operations, map layers must be based on the same coordinate system. A basic understanding of projection and coordinate systems is therefore crucial to users of spatial data.

The data model defines how spatial features are represented in a GIS (Figure 1.2). The **vector**

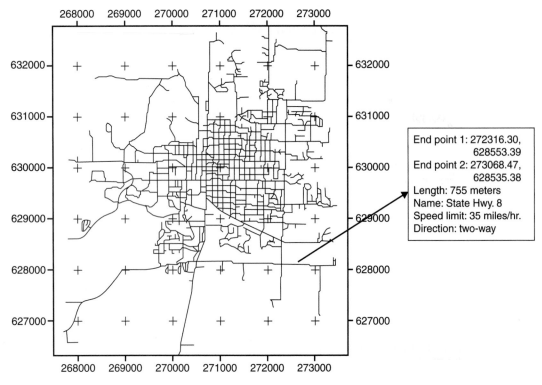

Figure 1.1
An example of geographically referenced data. The street network on the left is based on a plane coordinate system. The box on the right describes a street segment in terms of the *x*- and *y*-coordinates of its end points and the attributes.

data model uses points and their *x*-, *y*-coordinates to construct spatial features of points, lines, and areas. The **raster data model** uses a grid and grid cells to represent the spatial variation of a feature. The two data models differ in concept: vector data are ideal for representing discrete features; raster data are better suited for representing continuous features. They also differ in data structure: raster data have a simple data structure with rows and columns and fixed cell locations; vector data may be topological or nontopological, and simple or higher-level (Figure 1.3).

Topology, as used in GIS, expresses explicitly the spatial relationships between features, such as two lines meeting perfectly at a point and a directed line having an explicit left and right side. Topological or topology-based data are useful for detecting and correcting digitizing errors in geographic data sets and are necessary for some GIS analyses, whereas nontopological data display faster. To distinguish between them, ESRI Inc. uses the term **coverage** for topological data and **shapefile** for nontopological data. And the Ordnance Survey in Great Britain offers topological and nontopological data separately to accommodate end users with different needs.

Higher-level data are built on simple vector data of points, lines, and polygons. The **triangulated irregular network (TIN),** which approximates the terrain with a set of nonoverlapping

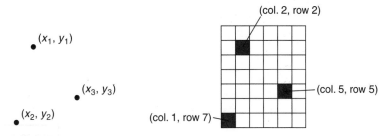

Figure 1.2
The vector data model uses *x*-, *y*-coordinates to represent point features on the left, and the raster data model uses cells in a grid to represent point features on the right.

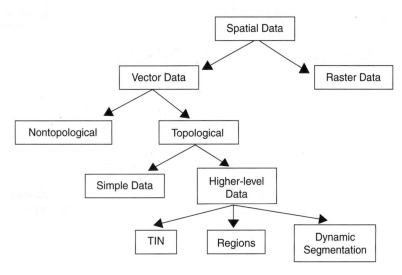

Figure 1.3
Data models for spatial data.

triangles, is made of points and edges (lines). The **regions** data model, which allows regions to overlap and to have spatially disjoint components, is built on lines and polygons. And **dynamic segmentation,** which combines plane coordinates with linear measures, is built on top of a network (a connected line coverage). Higher-level data are useful in GIS because they can handle more complex spatial relationships.

A recent trend in GIS is the adoption of the object-oriented data model. The **object-oriented**

data model uses objects to organize spatial data. An object has a set of built-in properties and can perform operations upon requests. To represent spatial features, objects may take the form of points, multipoints (a collection of points), polylines (a set of line segments), or polygons (including disjoint polygons). As discussed in Section 1.3.3, the object-oriented data model differs from the traditional vector data model because the geometries of objects are stored as attribute data.

1.3.2 Attribute Data

Attribute data describe the characteristics of spatial features. For raster data, each cell has a value that corresponds to the attribute of the spatial feature at that location. A cell is tightly bound to its cell value. For vector data, the amount of attribute data to be associated with a spatial feature can vary significantly. A road segment may have the attributes of only length and speed limit, whereas a soil polygon may have dozens of physical and chemical properties, interpretations, and performance data. How to join spatial and attribute data is therefore important in the case of vector data.

1.3.3 Joining Spatial and Attribute Data

There are two models for joining spatial and attribute data. Until recently, most commercial GIS packages store attribute data separately from spatial data in a split data system, often called the **georelational data model.** Spatial data are stored in graphic files, attribute data are stored in tables, and they are linked through the feature IDs. The object-oriented data model, on the other hand, stores spatial data as an attribute (e.g., geometry) along with other attribute data in a single table. The object-oriented data model eliminates the complexity of coordinating and synchronizing two sets of data files in a split data system and brings GIS closer to other nonspatial information systems.

Whether spatial and attribute data are stored in a split system or a single system, the relational database model is the norm for data management in GIS. A **relational database** is a collection of tables (relations). The connection between tables is made through a key, a common field whose values can uniquely identify a record in a table. The feature ID serves as the key in the georelational data model, and the object ID provides the key in the object-oriented model.

A relational database is efficient and flexible for data search, data retrieval, data editing, and creation of tabular reports. Each table in the database can be prepared, maintained, and edited separately from other tables. And the tables can remain separate until a query or an analysis requires that attribute data from different tables be linked or joined together.

With GIS increasingly becoming part of an organization's much larger information system, attribute data needed for a GIS project are likely to come from an enterprisewide database. If spatial data distinguish GIS as a special type of information system, attribute data tend to connect GIS with other information management systems. This unique combination enhances the role of GIS in an organization.

1.4 GIS Operations

Although GIS activities no longer follow a set sequence, to explain what GIS users do, we can group GIS activities into spatial data input, attribute data management, data display, data exploration, data analysis, and GIS modeling (Figure 1.4). This section provides an overview of GIS operations. It also serves as a preview of the chapters to follow by highlighting each group of GIS activities.

1.4.1 Spatial Data Input

The most expensive part of a GIS project is data acquisition. Two basic options for data acquisition are (1) use existing data and (2) create new data. Digital data clearinghouses have become commonplace on the Internet in recent years. The strategy for a GIS user is to look at what exists in the public domain before deciding to either buy data from private companies or create new data.

New digital spatial data can be created from satellite images, GPS data, field surveys, street addresses, and text files with x-, y-coordinates. But paper maps remain the dominant data source. Manual digitizing or scanning can convert paper maps into digital format. A newly digitized map typically requires editing and geometric transformation. Editing removes digitizing errors, which may relate to the location of spatial data such as missing polygons and distorted lines, or the topology such as dangling arcs and unclosed polygons. Geometric transformation converts a newly digitized map, which has the same physical dimension

Spatial data input	1. Data entry: use existing data, create new data 2. Data editing 3. Geometric transformation 4. Projection and reprojection
Attribute data management	1. Data entry and verification 2. Database management
Data display	1. Use of maps, charts, and tables
Data exploration	1. Attribute data query 2. Spatial data query 3. Geographic visualization
Data analysis	1. Vector data analysis: buffering, overlay, distance measurement, map manipulation 2. Raster data analysis: local, neighborhood, zonal, global 3. Terrain mapping and analysis 4. Spatial interpolation: global, local 5. Regions-based analysis 6. Network analysis, and dynamic segmentation
GIS modeling	1. Binary models 2. Index models 3. Regression models 4. Process models

Figure 1.4
A classification of GIS activities.

as its source map, into a real-world coordinate system. Geometric transformation relies on a set of control points and often requires adjustment of the control points to reduce the amount of transformation error to an acceptable level.

1.4.2 Attribute Data Management

To complete a GIS database, we must enter, verify, and manage attribute data. Similar to spatial data, attribute data entry and verification also involves digitizing and editing. Attribute data are usually managed in a relational database. Because a relational database consists of separate tables, some of which may even reside at remote locations, the tables must be designed to facilitate data input, search, retrieval, manipulation, and output. Two basic elements in the design of a relational database are the key and the type of data relationship:

the key establishes a connection between corresponding records in two tables, and the type of data relationship dictates how the tables are actually joined or linked.

1.4.3 Data Display

Because maps are most effective in communicating spatial information, mapmaking is a routine GIS operation for data visualization, query, analysis, and presentation.

A map has a number of elements: title, subtitle, body, legend, north arrow, scale, acknowledgment, neatline, and border. These elements work together to bring spatial information to the map user. The first step in data display is to assemble map elements. Windows-based GIS packages have simplified this process by providing choices for each map element through menus and palettes. The second step is map design. Map design is a creative process that cannot be easily replaced by default templates and computer codes. The mapmaker must experiment with map symbols, type variations, and color schemes. A well-designed map can help the mapmaker communicate spatial information to the map user. A poorly designed map can confuse the map user and even distort the information intended by the mapmaker.

1.4.4 Data Exploration

Data exploration is data-centered query and analysis. It allows the user to explore the general trends in the data, to take a close look at data subsets, and to focus on possible relationships between data sets. Data exploration can be a GIS operation by itself or, more often, a precursor to formal data analysis.

Effective data exploration requires interactive and dynamically linked visual tools. A Windows-based GIS package is ideal for data exploration. We can display maps, graphs, and tables in multiple but dynamically linked windows so that, when we select a data subset from a table, it automatically highlights the corresponding features in a graph and a map. This kind of interactivity increases our capacity for information processing and synthesis.

The importance of maps in GIS operations has broadened the scope of data exploration. **Geographic visualization** refers to the use of maps for setting up a context for visual information processing. Although functionally and technically similar to data-centered exploration, geographic visualization is map-based and can take advantage of map-based tools such as data classification, data aggregation, and map comparison.

1.4.5 Data Analysis

Figure 1.4 classifies data analysis into six groups. The classification shows the importance of the data model to data analysis in GIS. The basic data models are vector and raster, each having its own set of common analytical tools. For vector data, these tools include map overlay, map manipulation, buffering, and distance measurement. Map overlay, recognized by many as the most important GIS tool, combines spatial data and attribute data from different maps to create the output (Figure 1.5). Map manipulation tools perform dissolving,

clipping, merging, splitting, erasing, and other tasks. Buffering measures straight-line distances from map features and creates buffer zones. Distance measurement, as the term suggests, measures distances between map features.

Common tools for raster data include local, neighborhood, zonal, and global operations. A local operation involves individual cells; a neighborhood operation, a specified neighborhood; a zonal operation, a group of connected cells; and a global operation, an entire grid. These operations may apply to a single grid or multiple grids. As examples, a local operation involving a single grid can convert the measurement unit of a slope map from percent to degrees, and a local operation involving multiple grids can compute the average of the input grids on a cell-by-cell basis. In both cases, a mathematical function relates the input to the output.

A raster operation involving multiple grids (Figure 1.6) is conceptually similar to a vector-based map overlay operation (Figure 1.5). But the procedures differ: the raster operation can take ad-

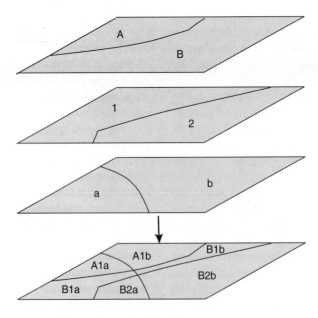

Figure 1.5
Vector-based map overlay combines spatial data and attribute data from different maps to create the output.

vantage of the fixed cell locations, whereas the vector operation must deal with the intersection of polygon boundaries. In other words, the raster operation is computationally more efficient than the vector operation. This difference shows the importance of the data model in GIS analysis.

The terrain has been the object for mapping and analysis for hundreds of years. Mapping techniques such as contouring, profiling, hill shading, hypsometric tinting, and perspective views are useful for visualizing the land surface. Topographic measures including slope, aspect, and surface curvature are important for studies of timber management, soil erosion, hydrologic modeling, wildlife habitat suitability, and many other fields. Terrain analysis also includes viewshed analysis to determine areas of the land surface that are visible from an observation point (e.g., a forest lookout station) and watershed analysis to derive flow direction, stream networks, and watershed boundaries. Terrain mapping and analysis can use elevation grids (raster data) or TINs (vector data) as the primary data source. But the computing algorithm depends on the data type.

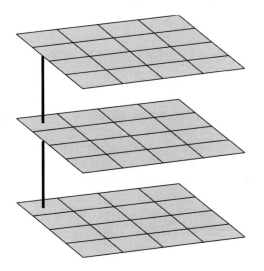

Figure 1.6
Raster data analysis with multiple grids can take advantage of the fixed cell locations. Geometric intersection is not an issue with raster data.

A statistical surface based on data such as precipitation or water table does not exist physically but can be visualized in a way similar to the land surface. **Spatial interpolation** refers to the process of using points with known values to construct a statistical surface. Because a statistical surface is normally represented as a grid, we may consider spatial interpolation as a process of converting point data into raster data. Spatial interpolation can be global or local. A global method uses every known point available to estimate an unknown value in a grid cell. Trend surface and regression are two well-known global methods. A local method uses a sample of known points for estimation. GIS applications often use such local methods as Thiessen polygons, density estimation, inverse distance weighted, splines, and kriging.

Higher-level vector data offer additional analytical tools that are not possible with simple points, lines, and polygons. Regions are more efficient than simple polygons for storing and managing geographically referenced data that overlap or have spatially disjoint components. Because some spatial features (e.g., ecological systems and census units) form a nested hierarchical structure, regions are potentially useful for analyzing those features in terms of the effect of spatial scale. Regions also allow map overlay and data query to be performed in a single operation with multiple input maps (up to 32 in ArcInfo Workstation). With simple polygons, data query must follow map overlay, which is limited to two maps at a time.

A network is a connected line coverage such as a road or stream network with the appropriate attributes for the flow of objects. Network analysis includes shortest path analysis for finding the path between points on the network with the minimum cumulative cost in time or distance, allocation for analyzing the spatial distribution of resources, and location-allocation for analyzing the spatial distribution of the supply and demand. The dynamic segmentation model combines a network and a linear measurement system. Dynamic segmentation makes it possible to link events that are linearly measured to a network for analysis. Examples include linking accident locations and speed limits to a road network and linking fishery habitat conditions to a stream network.

1.4.6 GIS Models and Modeling

A model is a simplified representation of a phenomenon or a system, and **GIS modeling** refers to the use of a GIS and its functionalities in building a model with geographically referenced data (i.e., a spatially explicit model). Often discipline specific, GIS models can be as diverse as analytical tools in a GIS. Figure 1.4 lists four general types of GIS models: binary, index, regression, and process models.

Binary and index models are similar in that both rely on map overlay, either vector- or raster-based, to combine geographically referenced data from different maps for multicriteria evaluation. But they differ in the evaluation procedure and the output. A binary model uses a query statement to produce a binary (true or false) map that separates map features that satisfy the criteria from those that do not. An index model is much more elaborate in its evaluation procedure. Weighted linear combination is the most common method for building an index model: it computes the index value by summing the weighted criterion values. Other methods provide alternatives in terms of criterion weights, data standardization, and data aggregation. Regardless of the modeling procedure, the result of an index model is a ranked map such as a land suitability map.

A regression model also uses map overlay to gather data for analyzing the statistical relationship between a dependent variable and independent variables. Linear regression is applicable when both the dependent and independent variables are numeric. Logistic regression is used when the dependent variable is categorical and the independent variables are categorical, numeric, or both. Predicting the dependent variable is the primary purpose of a regression model.

A process model integrates existing knowledge about the environmental processes in the real world and quantifies the processes with a set of relationships and equations. Modules and submodels are often needed to cover different components of a process model. Some of these modules may use mathematical equations derived from empirical data, whereas others may use equations derived from laws in physics. In building a process model, a GIS is usually delegated to the role of performing modeling-related tasks such as data visualization, database management, and data exploration. Examples of process models include those for predicting soil erosion, nonpoint source pollution, water quality and quantity, sediment, and agricultural chemical yield.

1.5 ORGANIZATION OF THIS BOOK

Chapter 2 discusses the concepts of projection, projection parameters, coordinate systems, and reprojection (projecting from one coordinate system to another). Common map projections and coordinate systems including the transverse Mercator, the Lambert conformal conic, the Universal Transverse Mercator (UTM) grid system, and the State Plane Coordinate (SPC) system are discussed in the chapter along with their parameters.

Chapters 3 through 6 cover different aspects of vector data. Chapter 3 introduces the vector data model with topics on topology, nontopological data, higher-level objects, the object-oriented data model, location accuracy, and spatial data accuracy standards. Chapter 4 describes use of existing digital data and creation of new data from satellite images, GPS readings, street addresses, and paper maps. The chapter also discusses metadata (information about data), data conversion, and geometric transformation. After explaining topological errors and location errors, Chapter 5 illustrates different tools for correcting digitizing errors in a newly digitized map or an outdated map. Chapter 6 examines attribute data input and management including the relational database model and its design, relational database examples, and attribute data entry and verification.

Chapter 7 considers the raster data model and a large variety of raster data such as remotely sensed data, digital elevation models, digital orthophotos, digital raster graphs, scanned maps, and graphic files. The chapter also discusses raster data structure and compression, geometric transformation, conversion between vector and raster data, and data integration.

Chapter 8 focuses on data display and cartography. Topics in the chapter include map symbols, color, text and text placement, guidelines for map design, and production of soft-copy and hard-copy maps. Chapter 9 covers data exploration. After reviewing exploratory data analysis and dynamic graphs, the chapter describes query methods using attribute data, spatial data, or both to take advantage of the capability of a Windows-based GIS in dynamically linking maps, graphs, and tables. Tools for map-based geographic visualization are also illustrated.

Chapters 10 and 11 cover the basic tools for GIS analysis and their applications. Chapter 10 includes the vector-based tools of buffering, map overlay, distance measurement, and map manipulation. Slivers, error propagation, and areal interpolation are also discussed in the context of map overlay. Chapter 11 covers local, neighborhood, zonal, and global operations using raster data from a single grid or multiple grids. Global operations include physical distance measures, cost distance measures, and spatial autocorrelation.

Chapter 12 examines terrain mapping and analysis using elevation grids and TINs. Terrain mapping includes contouring, profiling, hill shading, hypsometric tinting, and perspective view. Terrain analysis includes slope, aspect, surface curvature, viewshed, and watershed. Worked examples are provided for the computing algorithms, and a comparison is made between elevation grids and TINs for terrain applications. Chapter 13 discusses the statistical surface, elements of spatial interpolation, global methods, and local methods. Throughout the chapter, a small data set and worked examples are used to explain the interpolation algorithms. The chapter concludes with quantitative measures for comparing the interpolation methods.

Chapter 14 gives an overview of GIS models and modeling. Following an introduction to type of model, the modeling process, and the role of GIS in modeling, the chapter describes the basics of building binary, index, regression, and process models. The chapter offers many examples from different disciplines for each type of model.

Chapter 15 covers regions by first explaining geographic regions, implementation of geographic regions using the regions data model, and the importance of incorporating spatial scale in GIS analysis. The chapter then describes the procedure for creating and managing regions and regions-based analysis. Chapter 16 covers network and dynamic segmentation. The chapter explains the elements of a network and network analysis in the first part and the dynamic segmentation model and its applications in the second part.

1.6 CONCEPTS AND PRACTICE

Each chapter in this book has two main sections. The first section covers a set of topics and concepts; the second section covers applications with three to seven problem-solving tasks. Each chapter also has boxes to provide either additional relevant information or worked examples for the computing algorithms.

This book stresses both concept and practice. GIS concepts explain the purpose and objectives of GIS operations and the interrelationship among GIS operations. A basic understanding of map projection, for example, explains why we must project map layers into a common coordinate system before using them together and why we must input numerous projection parameters. Our knowledge of map projection is long lasting: the knowledge will neither change with the technology nor become outdated with new versions of a GIS package.

For most GIS users, GIS is a problem-solving tool (Wright et al. 1997). To apply the tool correctly and efficiently, we must become proficient in using the tool. One constant complaint with command-driven GIS packages is that they are too difficult to learn. The graphical user interface has improved the human-computer interaction but it still requires us to sort out a multitude of menus, buttons, and tools and to know how and when to use them.

Practice, which is a regular feature in mathematics and statistics textbooks, is really the only way for us to become proficient in using GIS. Practice can also help us grasp GIS concepts. For instance, the root mean square (RMS) error, a measure of the goodness of geometric transformation as well as spatial interpolation, is difficult to understand mathematically. But after a couple of runs of geometric transformation, RMS starts to make more sense because we can see how the error measure changes each time with a different set of control points.

Practice sections in a GIS textbook require data sets and GIS software. Many data sets used in this book are from GIS classes taught at the University of Idaho over the past 16 years. Most exercises in the practice sections use ArcView 8.2 and the extensions of Spatial Analyst, 3D Analyst, and Geostatistical Analyst. ArcInfo 8.2, especially its ArcInfo Workstation, is used only for tasks that cannot be performed in ArcView 8.2, including topological editing, geometric transformation, regions, network analysis, and dynamic segmentation.

KEY CONCEPTS AND TERMS

Attribute data: Data that describe the characteristics of spatial features.

Continuous features: Spatial features that exist between observations.

Coverage: An ESRI data format for topological vector data.

Data exploration: Data-centered query and analysis.

Discrete features: Spatial features that do not exist between observations, form separate entities, and are individually distinguishable.

Dynamic segmentation: A data model that is built on arcs of a line coverage and allows the use of real-world coordinates with linear measures.

Geographic information system (GIS):   A computer system for capturing, storing, querying, analyzing, and displaying geographically referenced data.

Geographic visualization: Use of maps for data exploration and visual information processing.

Geographically referenced data: Data that describe both the location and characteristics of spatial features on the Earth's surface.

Georelational data model: A GIS data model that stores spatial data and attribute data in two separate but related file systems.

GIS modeling: The process of using GIS in building models with spatial data.

Object-oriented data model: A data model that uses objects to organize spatial data. An object is an entity such as a land parcel that has a set of properties and can perform operations on request.

Projection: The process of transforming from the spherical geographic grid to a plane coordinate system.

Raster data model: A spatial data model that uses a grid and cells to represent the spatial variation of a feature.

Regions: Higher-level vector data that can have spatially disjoint components and can overlap one another.

Relational database: A collection of tables, which can be connected to each other by attributes whose values can uniquely identify a record in a table.

Shapefile: An ESRI data format for nontopological vector data.

Spatial data: Data that describe the geometry of spatial features.

Spatial interpolation: A process of using control points with known values to estimate values at other points.

Topology: A subfield of mathematics that is applied in GIS to ensure that the spatial relationships between features are expressed explicitly.

Triangulated irregular network (TIN): A data model that approximates the terrain with a set of nonoverlapping triangles.

Vector data model: A spatial data model that uses points and their x-, y-coordinates to construct spatial features of points, lines, and areas.

APPLICATIONS: INTRODUCTION

This applications section covers two tasks. Task 1 introduces the applications of ArcCatalog, Arc-Map, and ArcToolbox in ArcGIS 8.2 and the Spatial Analyst extension. Task 2 introduces ArcInfo Workstation and its use.

ArcGIS uses a single, scalable architecture and user interface. ArcGIS has three versions: Arc-View, ArcEditor, and ArcInfo. ArcView is the simplest version and has fewer capabilities than the other two versions. All three versions use the same applications of ArcCatalog, ArcMap, and ArcTool-box. But only ArcInfo has the additional ArcInfo Workstation, a command-driven application similar to ARC/INFO version 7.x. It should be noted that ArcView 8.2 is different from ArcView 3.x, a stand-alone desktop package that uses extensions and utility programs to extend its capabilities.

You can tell which version of ArcGIS you are using by looking at the title of an application. For example, the title of ArcCatalog may appear as Ar-cCatalog—ArcView or ArcCatalog—ArcInfo, depending on whether you are using ArcView or ArcInfo. You can also tell which version you are using by starting ArcToolbox. ArcView offers three types of tools: Conversion Tools, Data Management Tools, and My Tools. ArcInfo, on the other hand, has an additional suite of tools called Analysis Tools.

To check the availability of Spatial Analyst, 3D Analyst, and GeoStatistical Analyst, which should be available to all three versions of ArcGIS, you can do the following: close all ArcGIS applications, click the Start menu, point to Programs (All Programs on Windows XP), point to ArcGIS, and select Desktop Administrator. Double-click Availability in the next menu to display the available licenses.

Task 1: Introduction to ArcCatalog, ArcMap, and ArcToolbox in ArcGIS

What you need: *emidalat*, an elevation grid; *emidastrm.shp*, a stream shapefile.

Task 1 introduces the applications of ArcCatalog, ArcMap, and ArcToolbox in ArcGIS. You will use ArcCatalog to manage the data sets, Arc-Map to display the data sets and to derive a slope map from *emidalat*, and ArcToolbox to convert *emidastrm.shp* to a geodatabase feature class.

1. Start ArcCatalog. ArcCatalog lets you set up connections to your data sources, which may reside in a folder on a local disk or a database on the network. For Task 1, you will make connection to the folder containing the Chapter 1 database (e.g., chap1). Click the Connect to Folder button. Navigate to the chap1 folder and click OK. The chap1 folder now appears in the Catalog tree. Click the plus sign next to the folder to view the datasets in the folder.

2. Click *emidalat* in the Catalog tree. Click the Preview tab to view the elevation grid. Click the Metadata tab. The narrative shows that *emidalat* is a projected raster dataset in ESRI grid format. (*Emidalat* is a floating-point grid and has no attribute table.)

3. Click *emidastrm.shp* in the Catalog tree. You can preview the geography of *emidastrm.shp* as well as its table by selecting Table in the Preview's dropdown list.

4. ArcCatalog allows you to perform data management tasks. Right-click *emidastrm.shp* in the Catalog tree. A menu, which is called the context menu of *emidastrm.shp*, appears with selections

including Copy, Delete, Rename, Create Layer, Export, and Properties. For example, if you want to copy *emidastrm.shp* to a different folder, you can select Copy in its context menu, and right-click the folder where you want to copy *emidastrm.shp* to and select Paste.

5. Start ArcMap by clicking the Launch ArcMap button in ArcCatalog. An alternative is to start ArcMap by using the Programs menu. ArcMap is the main application for data display, data query, and data output. Several things are important to know about ArcMap. First, associated with ArcMap is a tool bar, which includes such tools as Zoom in, Zoom out, Pan, Select Elements, and Identify. When you hold the mouse point over a tool (a graphic icon), a short message called ToolTip appears in a floating box to tell you the function of the tool. Second, ArcMap has two views: Data View and Layout View. (The buttons for the two views are located at the bottom of the view window.) Data View is for viewing data, whereas Layout View is for viewing the map product for printing and plotting. Third, ArcMap organizes data sets into data frames. Each data frame contains data sets made of shapefiles, grids, or coverages, and, to spatially register with one another, these data sets must be based on the same coordinate system. If one or more of the data sets in a data frame do not have projection information, ArcMap will display a warning message about the missing information and use the first available projection information for data display.

6. Click the Add Data button in ArcMap, navigate to the chap1 folder, and select *emidalat* and *emidastrm.shp* to add to the map. To select more than one data set to add, click the first data set and then click other data sets while holding down the Control key. An alternative to the Add Data button is the drag-and-drop method. You can add a data set in ArcMap by dragging it from the

Catalog tree and dropping it in ArcMap's view window.

7. *Emidalat* and *emidastrm* are now added to the Table of Contents under Layers. A message appears informing you about one or more layers with missing spatial reference information. *Emidastrm.shp* does not have the projection information, although it is based on the UTM coordinate system, same as *emidalat*. You will learn in Chapter 2 how to define a coordinate system.

8. ArcMap places the last data set added to view at the top of the Table of Contents. You can change the order by dragging a data set to a new position and dropping it.

9. The generic name for a data frame is Layers. You can go through the following steps to change the name from Layers to Task 1: right-click Layers in the Table of Contents, select Properties from Layers' context menu, click the General tab, and enter Task 1 for the Name.

10. This step is to change the symbol for *emidastrm*. Click the symbol for *emidastrm* in the Table of Contents to open the Symbol Selector dialog. You can either select a preset symbol for river or make up your own symbol for *emidastrm* by specifying the color, width, and properties of the symbol.

11. This step is to display *emidalat* in the elevation zones of <900, 900–1000, 1000–1100, 1100–1200, 1200–1300, and >1300 meters. Right-click *emidalat* and select Properties from the context menu. Click the Symbology tab. Click Classified in the Show frame. Change the number of classes to 6, and click the Classify button. The Method dropdown list in the Classification dialog offers six methods: Manual, Equal Interval, Defined Interval, Quantile, Natural Breaks (Jenks), and Standard Deviation. Select Manual. There are two ways to set the break values for the elevation zones manually. To use the first method, you will check the box to snap

breaks to data values and then click the first break line and drag it to the intended value of 900. You will do the same to set the break lines at 1000, 1100, 1200, 1300, and 1337. To use the second method, you will click the first cell in the Break Values frame and enter 900. You will click and enter 1000, 1100, 1200, and 1300 for the next four cells. If the break value you entered is changed to a different value, reenter it. To unselect 1337, which is the highest value, click the empty space below the cell. After entering the break values, click OK to dismiss the Classification dialog.

12. You can change the color scheme for *emidalat* by using the Color Ramp dropdown list in the Layer Properties dialog. Sometimes it is easier to select a color scheme using words instead of graphic views. In that case, you can right-click inside the Color Ramp box and uncheck Graphic View. The Color Ramp dropdown list now shows White to Black, Yellow to Red, etc. Select Elevation #1. Click OK and dismiss the Layer Properties dialog.

13. Several extensions are available in ArcMap, including Spatial Analyst, 3D Analyst, and Geostatiscal Analyst. For example, you can use Spatial Analyst to derive a slope map from *emidalat*. Select Extensions from the Tools menu and check the box for Spatial Analyst. Then select Toolbars from the View menu in ArcMap. Make sure that the box for Spatial Analyst is checked. After docking the Spatial Analyst toolbar, click the Spatial Analyst dropdown arrow. Select Surface Analysis and then Slope. In the Slope dialog, make sure the input surface is *emidalat*, the output measurement is degree, the Z factor is 1, the output cell size is 30 (meters), and the output raster is temporary. Click OK. *Slope of emidalat* is now added to the Table of Contents.

14. You can save Task 1 as a map document before exiting ArcMap. Select Save As from the File menu in ArcMap. Navigate to the

chap1 folder, enter Task1 for the file name, and click Save. ArcMap automatically adds the extension .mxd to Task1. Data sets displayed in Task 1 are now saved with Task1.mxd. To re-open Task1.mxd, Task1.mxd must reside in the same folder as the data sets it references.

15. To exit ArcMap, simply select Exit from the File menu. To make sure that Task1.mxd is saved correctly, launch ArcMap. Select Open from the File menu, navigate to the chap1 folder, and double-click Task1.

16. The rest of Task 1 is to use ArcToolbox to convert *emidastrm.shp* to a geodatabase feature class. First, you must create a personal geodatabase. Right-click the chap1 folder in the Catalog tree, point to New, and select Personal Geodatabase. Rename the new personal geodatabase *emida.mdb*. If the extension .mdb does not appear, select Options from the Tools menu and uncheck the box next to Hide file extensions.

17. You can start ArcToolbox by clicking the Launch ArcToolbox button in ArcCatalog or by using the Programs menu. Click the plus sign next to Conversion Tools. Click the plus sign next to Export from Shapefile, and double-click on Shapefile to Geodatabase Wizard. In the first panel, select *emidastrm.shp* for the input shapefile by using the browse button or the drag-and-drop method. Click Next. In the second panel, choose *emida.mdb* as the geodatabase. In the third panel, make sure that you want to create a stand-alone feature class and click Next. Opt to accept the default parameters in the fourth panel and click Next. Click Finish in the Summary dialog.

18. Click the plus sign next to *emida.mdb* in the Catalog tree, and the feature class *emidastrm* appears. Click *emidastrm* to preview it. *Emidastrm* looks the same as *emidastrm.shp*.

19. To exit ArcToolbox, select Exit from the Tools menu. Likewise, select Exit from the File menu to exit ArcCatalog.

Task 2: Introduction to ArcInfo Workstation

What you need: *emidalat*, an elevation grid; *breakstrm*, a stream coverage.

The practice sections of this book may require use of ArcInfo Workstation in ArcGIS. The command-driven ArcInfo Workstation organizes commands into modules such as Arc, ArcPlot, ArcEdit, and Tables. Arc is the start-up module; ArcPlot displays coverages, grids, and TINs; ArcEdit includes editing functions; and Tables works with tables or INFO files. While working in ArcInfo Workstation, you need to move between modules, depending on the task you do.

Task 2 uses ArcInfo Workstation to display *emidalat* and *breakstrm*. *Breakstrm* shows the same streams as *emidastrm.shp* in Task 1 except that *breakstrm* is a coverage.

1. Use the Programs menu to start ArcInfo Workstation. After the Arc prompt appears, use the WORKSPACE (or W) command to change the directory to the chap1 folder, where *emidalat* and *breakstrm* reside.

2. First, you need to go to Arcplot and set up the environment to display *emidalat* and *breakstrm*:

Arc: arcplot

Arcplot: display 9999 /*define the computer monitor as the display device

Arcplot: mapextent emidalat /*define the map extent of *emidalat*

Note: To explain what a command does, this book includes the explanation following /.*

3. The next command display *emidalat* by using an item (value for this example) and a lookup table (emidalat.lut):

Arcplot: gridpaint emidalat value emidalat.lut /*display *emidalat* by elevation zone

Arcplot: arclines breakstrm 3 /*display *breakstrm* in green

Note: A look-up table like emidalat.lut specifies the symbol for each elevation zone.

4. Use QUIT to exit Arcplot and Arc:

Arcplot: quit

Arc: quit

REFERENCES

Broome, F. R., and D. B. Meixler. 1990. The TIGER Data Base Structure. *Cartography and Geographic Information Systems* 17: 39–47.

Chrisman, N. 1988. The Risks of Software Innovation: A Case Study of the Harvard Lab. *The American Cartographer* 15: 291–300.

Coppock, J. T. 1988. The Analogue to Digital Revolution: A View from an Unreconstructed Geographer. *The American Cartographer* 15: 263–75.

Crockett, M. 1997. GIS Companies Race for Market Share. *GIS World* 10 (11): 54–57.

McHarg, I. L. 1969. *Design with Nature.* New York: Natural History Press.

Rhind, D. 1988. Personality as a Factor in the Development of a Discipline: The Example of Computer-Assisted Cartography. *The American Cartographer* 15: 277–89.

Tomlinson, R. F. 1984. Geographic Information Systems: The New Frontier. *The Operational Geographer* 5:31–35.

Wilson, J.D. 2001. Mergers and Acquistions Remake the Competitive Landscape. *GEOWorld* 14(11): 30-33.

Wright, D. J., M. F. Goodchild, and J. D. Proctor. 1997. Demystifying the Persistent Ambiguity of GIS as "Tool" versus "Science." *Annals of the Association of American Geographers* 87: 346–62.

CHAPTER

2

MAP PROJECTIONS AND COORDINATE SYSTEMS

2.1 INTRODUCTION

GIS users work with map features on a two-dimensional or plane surface. These map features represent spatial features on the Earth's surface. The locations of map features are based on a plane coordinate system expressed in x- and y-coordinates, whereas the locations of spatial features are based on the geographic coordinate system expressed in longitude and latitude values. To bridge the two systems, **projection** transforms the spherical surface of the Earth to a map projection, which provides the basis for a plane coordinate system.

A basic principle in GIS is that map layers to be used together must have the same coordinate system. But hundreds of coordinate systems used in mapmaking and by different GIS data producers complicate the practice of this principle. For example, Figure 2.1 shows the road maps of Idaho and Montana downloaded from the Internet. The road maps obviously cannot be used together because they have different coordinate systems, which result in the overlapping of the road networks. To make them usable, these two road maps must be converted to a common coordinate system.

Increasingly, GIS users download digital maps from the Internet, or get them from government agencies and private companies. Some digital maps are measured in longitude and latitude values, whereas others are in different plane coordinate systems from the one intended for the GIS project. Invariably, these digital maps must be processed before they can be used together.

Processing in this case implies projection and reprojection. Projection converts digital maps from longitude and latitude values to plane coordinates, and **reprojection** converts from one plane coordinate system to another. Typically, projection and reprojection are among the initial tasks performed in a GIS project.

This chapter is divided into the following four sections. Section 2.2 describes the geographic coordinate system. Section 2.3 discusses projection and common map projections. Section 2.4 covers commonly used coordinate systems. Section 2.5 discusses how to work with coordinate systems in a GIS package.

2.2 GEOGRAPHIC COORDINATE SYSTEM

The **geographic coordinate system** is the location reference system for spatial features on the Earth's surface (Figure 2.2). The geographic coordinate system consists of meridians and parallels. **Meridians** are lines of longitude. Running north and south, meridians are used for measuring location in the E–W direction. Using the meridian passing through Greenwich, England, as the prime meridian or 0°, one can measure the longitude value of a point on the Earth's surface as 0° to 180° east or west of the prime meridian. **Parallels** are lines of latitude. Running east and west, parallels are used for measuring location in the N–S direction. Using the equator as 0° latitude, one can measure the latitude value of a point as 0° to 90° north or south of the equator.

Although applied to the spherical Earth's surface, the geographic coordinate system is similar to a plane coordinate system in terms of notation. The origin of the geographic coordinate system is where the prime meridian meets the equator. Thus, longitude values are similar to x values in a coordinate system and latitude values are similar to y values. In GIS, it is conventional to enter longitude and latitude values with positive or negative signs. Latitude values are positive if north of the equator, and negative if south of the equator. Longitude values are positive in the eastern hemisphere and negative in the western hemisphere.

Longitude and latitude values may be measured in the **degrees-minutes-seconds (DMS),** or **decimal degrees (DD),** system, in which 1 degree equals 60 minutes and 1 minute equals 60 seconds. This allows conversion between the two systems. For example, a latitude value of 45°52'30" would be equal to 45.875° (45 + 52/60 + 30/3600).

2.3 MAP PROJECTIONS

Projection transforms the spherical Earth's surface to a two-dimensional or plane surface (Robinson et al. 1995; Dent 1999). The outcome of this transformation is a **map projection:** a systematic construction of lines on a plane surface representing the geographic coordinate system. Besides allowing the use of paper maps and digital maps, a map projection enables map users to work with plane coordinates, rather than spherical geographic coordinates, which are more complex for measurements (Box

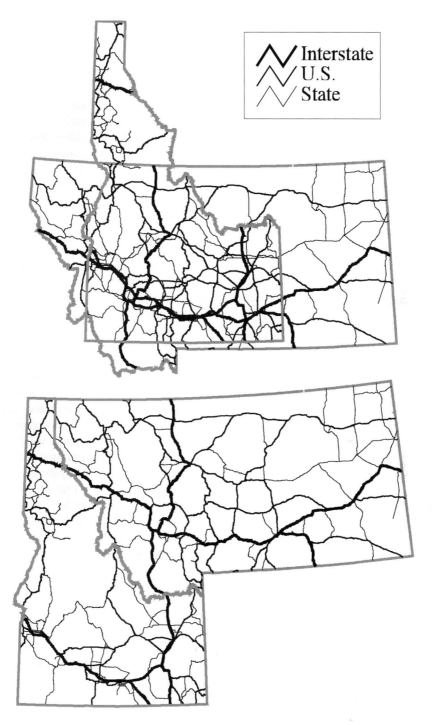

Figure 2.1
The top map shows two road coverages of Idaho and Montana based on different coordinate systems. The bottom map shows the road coverages based on the same coordinate system.

Figure 2.2
The geographic coordinate system.

2.1). But the transformation from the Earth's surface to a flat surface always involves distortion and no map projection is perfect. This is why hundreds of map projections have been developed for mapmaking (Maling 1992; Snyder 1993). Every map projection preserves certain spatial properties while sacrificing other properties.

Cartographers group map projections into four classes by their preserved properties: conformal, equal area or equivalent, equidistant, and azimuthal or true direction. A **conformal projection** preserves local angles and shapes. An **equivalent projection** represents areas in correct relative size. An **equidistant projection** maintains consistency of scale along certain lines. And an **azimuthal projection** retains certain accurate directions. The conformal and equivalent properties are mutually exclusive. Otherwise a map projection can have more than one preserved property such as conformal and azimuthal. The conformal and equivalent properties are global properties, meaning that they apply to the entire map projection. The equidistant

and azimuthal properties are local properties and may be true only from or to the center of the map projection. The preserved property of a map projection is usually included in its name such as the Lambert conformal conic or Albers equal-area conic.

The preserved property of a map projection is particularly important for small-scale mapping such as mapping of the world or a continent on a letter size paper. For example, to give the correct impression of population densities, a population map of the world should be based on an equivalent projection, rather than a conformal projection.

Cartographers often use a geometric object to illustrate how to construct a map projection (Figure 2.3). For example, by placing a cylinder tangent to a lighted globe, one can draw a projection by tracing the lines of longitude and latitude onto the cylinder. The cylinder in this case is the projection surface, and the globe is called the **reference globe.** Other common projection surfaces include a cone and a plane. A map projection is called a

Box 2.1 **How to Measure Distances on the Earth's Spherical Surface**

The equation for measuring distances on a plane coordinate system is:

$$D = \sqrt{(x_1 - x_2)^2 + (y_1 - y_2)^2}$$

where x_i and y_i are the coordinates of point i.

This equation, however, cannot be used for measuring distances between two points on the Earth's surface. Because meridians converge at the poles, the length of 1-degree latitude does not remain constant but gradually decreases from the equator to 0 at the pole. The standard and simplest method for calculating the shortest spherical distance between two points on the Earth's surface uses the equation:

$$\cos d = \sin a \sin b + \cos a \cos b \cos c$$

where d is the angular distance between points A and B in degrees, a is the latitude of A, b is the latitude of B, and c is the difference in longitude between A and B. To convert d to a linear distance measure, one can multiply d by the length of 1 degree at the equator, which is 111.32 kilometers or 69.17 miles. This method is accurate unless d is very close to zero (Snyder 1987).

Most commercial data producers deliver spatial data in geographic coordinates so that they can be used with any map projection the end user needs to work with. But more GIS users are using spatial data in geographic coordinates directly for data display and even simple analysis. Distance measurements from such spatial data are usually derived from the shortest spherical distance between points.

cylindrical projection if it can be constructed using a cylinder, a **conic projection** if using a cone, and an **azimuthal projection** if using a plane.

The use of a geometric object can also help explain two other projection concepts: case and aspect. For a conic projection, the cone can be placed so that it is tangent to the globe or intersects the globe (Figure 2.3). The first is the simple case, which results in one line of tangency, and the second is the secant case, which results in two lines of tangency. A cylindrical projection behaves the same way as a conic projection in terms of case. An azimuthal projection, in contrast, has a point of tangency in the simple case and a line of tangency in the secant case. Aspect describes the placement of a geometric object relative to a globe. A plane, for example, may be tangent at any point on a globe. A polar aspect refers to tangency at the pole, an equatorial aspect at the equator, and an oblique aspect anywhere between the equator and the pole (Figure 2.4).

The concept of case relates directly to the **standard line** in map projections, which refers to the line of tangency between the projection surface and the reference globe. For cylindrical and conic

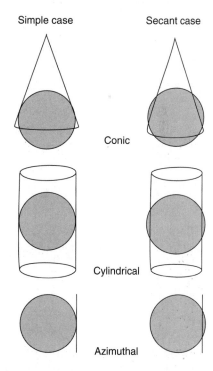

Figure 2.3
Use of a geometric object to construct a map projection.

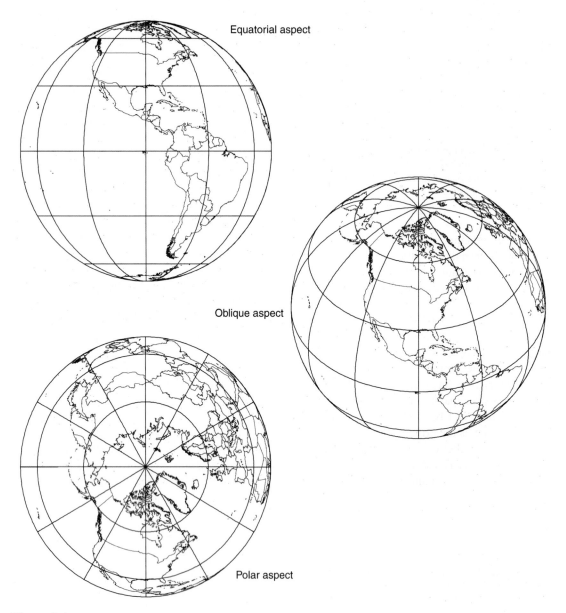

Figure 2.4
Aspect and map projection.

projections the simple case has one standard line whereas the secant case has two standard lines. The standard line is called the **standard parallel** if it follows a parallel, and the **standard meridian** if it follows a meridian. There is no projection dis-

tortion along the standard line because the line is the same as on the reference globe.

A common measure of projection distortion is scale, which is defined as the ratio of a distance on a map (or globe) to its corresponding ground dis-

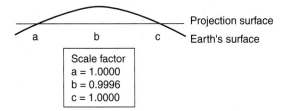

Figure 2.5
The central meridian in this secant case transverse Mercator projection has a scale factor of 0.9996. The two standard lines on either side of the central meridian have a scale factor of 1.

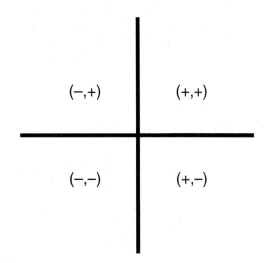

Figure 2.6
The central parallel and the central meridian divides a map projection into four quadrants. Points within the NE quadrant have positive x- and y-coordinates, points within the NW quadrant have negative x-coordinates and positive y-coordinates, points within the SE quadrant have positive x-coordinates and negative y-coordinates, and points within the SW quadrant have negative x- and y-coordinates.

tance. The **principal scale,** or the scale of the reference globe, can therefore be derived from the ratio of the globe's radius to the Earth's radius (3963 miles or 6378 kilometers). For example, if the globe's radius is 12 inches, then the principal scale is 1: 20,924,640 (1: 3963 × 5280). The local scale applies to a map projection. Unlike the principal scale, the local scale varies across a map projection. The **scale factor** is the normalized local scale, which is defined as the ratio of the local scale to the principal scale. The standard line has the scale factor of 1 because the standard line has the same scale as the principal scale. This is why the standard parallel is sometimes called the latitude of true scale. The scale factor becomes either less than 1 or greater than 1 away from the standard line.

The standard line should not be confused with the central line. Whereas the standard line dictates the distribution pattern of projection distortion, the **central lines** (the central parallel and meridian) define the center or the origin of a map projection. The central parallel, sometimes called the latitude of origin, often differs from the standard parallel. Likewise, the central meridian often differs from the standard meridian. A good example showing the difference between the central meridian and the standard line is the transverse Mercator projection. Normally a secant projection, a transverse Mercator projection is defined by its central meridian and two standard lines on either side. The standard line has a scale factor of 1, and the central meridian has a scale factor of less than 1 (Figure 2.5).

When a map projection is used as the basis of a coordinate system, the center of the map projection, as defined by the central parallel and the central meridian, becomes the origin of the coordinate system and divides the coordinate system into four quadrants. The x-, y-coordinates of a point are either positive or negative, depending on where the point is located (Figure 2.6). To avoid having negative coordinates, GIS users can assign x-, y-coordinate values to the origin of the map projection. The **false easting** is the assigned x-coordinate value and the **false northing** is the assigned y-coordinate value. Essentially, the false easting and false northing move the origin of the coordinate system to its SW corner so that all points will fall within the NE quadrant and have positive coordinates.

Finally, to help users choose from among hundreds of map projections, cartographers sometimes group them by how well they can be used to map the world, a hemisphere, a continent, a country, or a region.

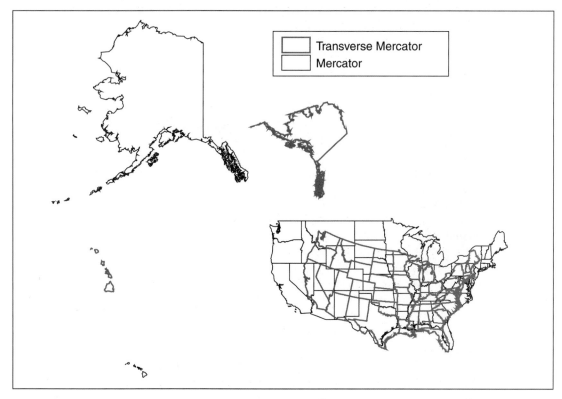

Figure 2.7
The Mercator and transverse Mercator projection of the United States. The central meridian is 90°W, and the latitude of true scale is the equator.

2.3.1 Commonly Used Map Projections

2.3.1.1 Transverse Mercator

The **transverse Mercator projection** is a variation of the Mercator projection, probably the best-
known projection for mapping the world (Figure
2.7). Instead of using the standard parallel as in the
case of the Mercator projection, the transverse
Mercator projection uses the standard meridian. As
discussed later, the transverse Mercator is the basis
for two common coordinate systems. The definition of the projection requires the following parameters: scale factor at central meridian, longitude
of central meridian, latitude of origin (or central
parallel), false easting, and false northing.

2.3.1.2 Lambert Conformal Conic

The **Lambert conformal conic projection** is a
good choice for mapping a midlatitude area of
greater east-west than north-south extent, such as
the conterminous United States or the state of
Montana (Figure 2.8). Typically used as a secant
projection, the projection is defined by the parameters of the first and second standard parallels,
central meridian, latitude of projection's origin,
false easting, and false northing.

2.3.1.3 Albers Equal-Area Conic

The Albers equal-area conic projection uses the
same parameters as the Lambert conformal conic
projection. The two projections in fact look

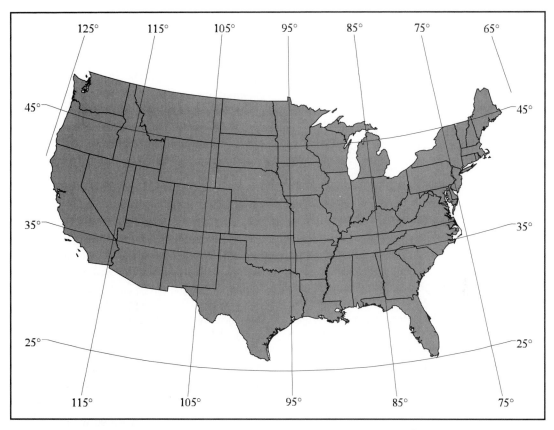

Figure 2.8
The Lambert conformal conic projection of the conterminous United States. The central meridian is 96°W, the two standard parallels are 33°N and 45°N, and the latitude of projection's origin is 39°N.

similar except that one is equal area and the other is conformal.

2.3.1.4 Equidistant Conic

The equidistant conic projection is also called the simple conic projection. The projection preserves the distance property along all meridians and one or two standard parallels. It uses the same parameters as the Lambert conformal conic projection.

2.3.2 Datum

So far the discussion of map projections has emphasized the use of a reference globe and assumed the Earth to be a perfect sphere. The Earth is wider along the equator than between the poles. To map spatial features more accurately, we need to work with two closely related parameters in projection: spheroid and datum.

A **spheroid** is a model that approximates the shape and size of the Earth. A much more accurate model than a sphere, a spheroid has its major axis (*a*) along the equator and its minor axis (*b*) connecting the poles (Figure 2.9). A parameter called the flattening (*f*), defined by $(a - b) / a$, measures the difference between the two axes of a spheroid. A spheroid is also called an **ellipsoid,** an ellipse rotated about its minor axis.

A **datum** is a mathematical model of the Earth, which serves as the reference or base for

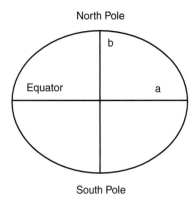

Figure 2.9
The semimajor axis is represented by a and the semiminor axis is represented by b.

calculating the geographic coordinates of a location (Burkard 1984; Moffitt and Bossler 1998). The definition of a datum consists of an origin, the parameters of the spheroid selected for the computations, and the separation of the spheroid and the Earth at the origin. Many countries have developed their own datums for local surveys. Among these local datums are the European Datum, the Australian Geodetic Datum, the Tokyo Datum, the Indian Datum (for India and several adjacent countries), and the Hu Tsu Shan Datum (for Taiwan).

Until the late 1980s, **Clarke 1866,** a ground-measured spheroid, was the standard spheroid for mapping in the United States. Clarke 1866's semimajor axis (equatorial radius) and semiminor axis (polar radius) measure 6,378,206.4 meters (3962.96 miles) and 6,356,583.8 meters (3949.21 miles) respectively. **NAD27** (North American Datum of 1927) is a local datum based on the Clarke 1866 spheroid, with its origin at Meades Ranch in Kansas. In 1986, the National Geodetic Survey (NGS) introduced **NAD83** (North American Datum of 1983), an Earth-centered (also called geo-centered) datum based on the **GRS80** (Geodetic Reference System 1980) spheroid. GRS80's semi-major axis measures 6,378,137.0 meters (3962.94 miles) and semiminor axis, 6,356,752.3 meters (3949.65 miles), respectively. In the case of

GRS80, the shape and size of the Earth was determined through measurements made by Doppler satellite observations.

Conversion from NAD27 to NAD83 can result in a substantial horizontal shift (Figure 2.10). Positions of points can change between 10 and 100 meters in the conterminous United States, more than 200 meters in Alaska, and in excess of 400 meters in Hawaii. For example, for the Ozette quadrangle map from the Olympic Peninsula in Washington, the shift is 98 meters to the east and 26 meters to the north. The horizontal shift in this case is 101.4 meters ($\sqrt{98^2 + 26^2}$). Until the switch from NAD27 to NAD83 is complete, GIS users must keep watchful eyes on the datum because digital maps based on the same coordinate system but different datums will not register correctly.

WGS84 (World Geodetic System 1984) is a reference system or datum established by the National Imagery and Mapping Agency (NIMA) of the U.S. Department of Defense (Kumar 1993). WGS84 agrees with GRS80 in terms of measures of the semimajor and semiminor axes. But WGS84 has a set of primary and secondary parameters. The primary parameters define the shape and size of the Earth, whereas the secondary parameters allow referencing of local datums used in different countries (National Imagery and Mapping Agency 2000). WGS84 is the datum for global positioning systems (GPS). The satellites used by GPS send their positions in WGS84 coordinates and all calculations internal to GPS receivers are based on WGS84. Conversion between WGS84 and other datums has become a routine operation in GIS projects (Box 2.2).

Although many GIS users in the United States are still migrating from NAD27 to NAD83, other reference systems that are more accurate than NAD83 have been developed for local surveys. In the late 1980s, the NGS began a program to establish the High Accuracy Reference Network (HARN) on a state-by-state basis, using GPS technology. In 1994, the NGS started a network of Continuously Operating Reference Stations (CORS). The positional difference of a control

Box 2.2 **Conversion between Datums**

Conversion between datums involves transformation and computation of geographic coordinates. Free software packages can be downloaded from the Internet for datum conversion. For example, Nadcon is a software package developed by the National Geodetic Survey (NGS) for conversion between NAD27 and NAD83. Nadcon can be downloaded at the NGS website (**http://www.ngs.noaa.gov/TOOLS/Nadcon/**

Nadcon.html) or at the Topographic Engineering Center of the U.S. Army Corps of Engineers website (**http://crunch.tec.army.mil/software/corpscon/corpscon.html**). GPS receivers usually have the options to read coordinates in different datums besides WGS84. Many GIS packages offer a large number of datums and spheroids to accommodate users from different countries.

point may be up to a meter between NAD83 and HARN but less than 10 centimeters between HARN and CORS (Snay and Soler 2000).

2.4 PLANE COORDINATE SYSTEMS

Because a plane coordinate system is based on a map projection, the terms of coordinate system and map projection are sometimes used interchangeably. Plane coordinate systems are designed for detailed calculations and positioning and are typically used in large-scale mapping such as at a scale of 1:24,000 or larger. Accuracy in a feature's absolute position and its relative position to other features is more important to a coordinate system than the preserved property of a map projection. To maintain the level of accuracy desired for measurements, a coordinate system is often divided into different zones, with each zone having a separate map projection.

Three coordinate systems are commonly used in the United States: the Universal Transverse Mercator (UTM) grid system, the Universal Polar Stereographic (UPS) grid system, and the State Plane Coordinate (SPC) system. As a group, coordinates of these common systems are often called real-world coordinates. The Public Land Survey System (PLSS) is also included in this section, although it is a land partitioning system and not a co-

ordinate system. Additional readings on these systems can be found in Robinson et al. (1995) and Muehrcke et al. (2001).

2.4.1 The Universal Transverse Mercator (UTM) Grid System

Used worldwide, the **UTM** grid system divides the Earth's surface between 84°N and 80°S into 60 zones. Each zone covers 6° of longitude, and is numbered sequentially with zone 1 beginning at 180°W. Figure 2.11 shows the UTM zones in the conterminous United States.

Each of the 60 UTM zones is mapped onto a secant case transverse Mercator projection, with a scale factor of 0.9996 at the central meridian. The standard meridians are 180 kilometers to the east and the west of the central meridian (Figure 2.12). In the northern hemisphere, UTM coordinates are measured from a false origin located at the equator and 500,000 meters west of the UTM zone's central meridian. In the southern hemisphere, UTM coordinates are measured from a false origin located at 10,000,000 meters south of the equator and 500,000 meters west of the UTM zone's central meridian. The UTM grid system maintains the accuracy of at least one part in 2500 (i.e., distance measured over a 2500-meter course on the UTM grid system would be accurate within a meter of the true measure).

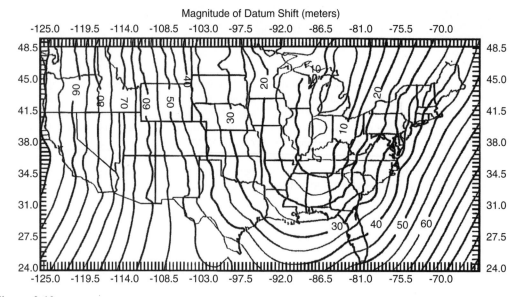

Figure 2.10
Magnitude of the horizontal shift from NAD27 to NAD83 in meters. See text for the definition of the horizontal shift.
(By permission of the National Geodetic Survey.)

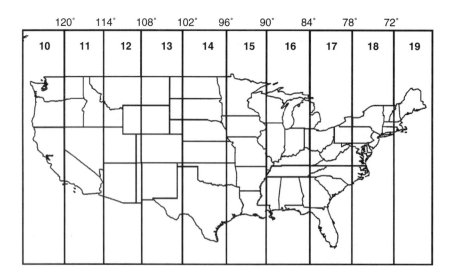

Figure 2.11
UTM zones in the conterminous United States.

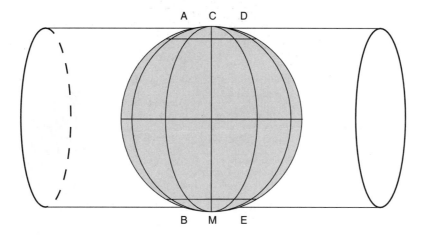

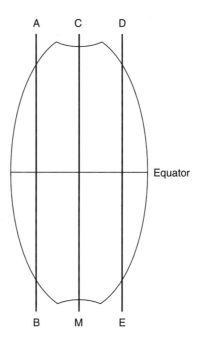

Figure 2.12

A UTM zone represents a secant case transverse Mercator projection. CM is the central meridian, and AB and DE are the standard meridians. The standard meridians are placed 180 kilometers west and east of the central meridian. Each UTM zone covers 6^0 of longitude and extends from 84^0N to 80^0S. The size and shape of the UTM zone are exaggerated for illustration purposes.

Because UTM coordinates are measured from a false origin, they are often very large numbers. For example, the NW corner of the Moscow East, Idaho, quadrangle map has the UTM coordinates of 500,000 and 5,177,164 meters. To preserve data precision for computations with coordinates, one can apply *x*-shift and *y*-shift values to all coordinate readings to reduce the number of digits. Therefore, if the *x*-shift value is set as −500,000 meters and the *y*-shift value as −5,170,000 meters for the previous quadrangle map, the coordinates for its NW corner will be changed to 0 and 7164 meters. Like false easting and false northing, *x*-shift and *y*-shift change the values of *x*-, *y*-coordinates in a digital map. Therefore, they must be documented along with the projection parameters in the metadata (information about data), especially if the map is to be shared with other users.

2.4.2 The Universal Polar Stereographic (UPS) Grid System

The **UPS grid system** covers the polar areas. The stereographic projection is centered on the pole and is used for dividing the polar area into a series of 100,000-meter squares, similar to the UTM grid system.

2.4.3 The State Plane Coordinate (SPC) System

The **SPC** system was developed in the 1930s to permanently record original land survey monument locations in the United States. To maintain the required accuracy of one part in 10,000 or less, a state may have two or more SPC zones. As examples, Oregon has the North and South SPC zones and Idaho has the West, Central, and East SPC zones (Figure 2.13). Each SPC zone is mapped onto a map projection. Zones that are elongated in the north-south direction (e.g., Idaho's SPC zones) use the transverse Mercator projection and zones that are elongated in the east-west direction (e.g., Oregon's SPC zones) use the Lambert conformal conic projection. Point locations within each SPC zone are measured from a false origin located to the southwest of the zone.

Figure 2.13
State Plane Coordinate systems for Washington, Oregon, and Idaho. The Lambert conformal conic projection is the basis for the Washington and Oregon coordinate systems, whereas the transverse Mercator projection is the basis for the Idaho coordinate system. The thinner lines are county boundaries.

Because of the switch from NAD27 to NAD83, there are SPC27 and SPC83. Besides the change of the datum, SPC83 has a few other changes. SPC83 coordinates are published in meters instead of feet. The states of Montana, Nebraska, and South Carolina have each replaced multiple zones by a single SPC zone. California has reduced SPC zones from seven to six. And Michigan has changed from transverse Mercator to Lambert conformal conic projections.

Some states in the United States have developed their own statewide coordinate system to maintain a desired level of accuracy. Idaho, for example, is divided into UTM zones 11 and 12 nearly in the center of the state. This division presents problems to GIS users in Idaho when their study area covers both zones. Because they must convert their database to a single zone so that maps in the database can register spatially, they cannot maintain the accuracy level designed for the UTM coordinate system. In 1994, the Idaho Geographic Information Advisory Committee adopted a statewide coordinate system (Box 2.3). The new system is still based on a transverse Mercator

Box 2.3 **Parameters for the Idaho Statewide Transverse Mercator Coordinate System**

Based on a transverse Mercator projection, the Idaho statewide coordinate system has the following technical parameters:

- Measurement unit: meter
- Central meridian: −114°

- Central meridian scale factor: 0.9996
- Datum: NAD27 (until NAD83 is adopted)
- Latitude of origin: 42°
- False northing: 100,000 meters
- False easting: 500,000 meters

Figure 2.14

An example of the Public Land Survey System (PLSS). The shaded survey township has the designation of T1S, R2E. T1S means that the survey township is south of the base line by one unit. R2E means that the survey township is east of the Boise (principal) meridian by 2 units. Each survey township is divided into 36 sections. Each section measures 1 mile by 1 mile and has a numeric designation.

projection but its central meridian passes through the center of the state (114°W). Changing the location of the central meridian allows the Idaho system to use one zone for the entire state.

2.4.4 The Public Land Survey System (PLSS)

The **PLSS** is a land partitioning system (Figure 2.14). Using the intersecting township and range lines, the system divides the lands mainly in the central and western states into 6 × 6 mile squares or townships. Each township is further partitioned

into 36 square-mile parcels of 640 acres, called sections. A recent development in GIS is to use the PLSS for creation of the land parcel layer. The Bureau of Land Management (BLM) has developed a **Geographic Coordinate Data Base (GCDB)** of the PLSS for the western United States (**http://www.blm.gov/gcdb/).** The GCDB contains longitude and latitude values and other descriptive information for section corners and monuments recorded in the PLSS. Legal descriptions of a parcel layer can then be entered using, for example, bearing and distance readings originating from section corners.

Box 2.4 Coordinate Systems in ArcGIS

ArcGIS divides coordinate systems into geographic and projected (plane). The user can define a coordinate system by selecting a predefined coordinate system, importing a coordinate system from an existing data set, or creating a new (custom) coordinate system. The parameters that are used to define a coordinate system are stored in a projection file. A projection file is provided for a predefined coordinate system. For a new coordinate system, a projection file can be named and saved for future use or for projecting other data sets.

The predefined geographic coordinate systems in ArcGIS have the main options of world, continent, and spheroid-based. WGS84 is one of the world projection files. Local datums are used for the continental projection files. For example, the Indian Datum and Tokyo Datum are available for the Asian continent. The spheroid-based options include Clarke1866 and GRS80. The predefined projected coordinate systems have the main options of world, continent, polar, national grids, UTM, State Plane, and Gauss Kruger (one type of the transverse Mercator projection mainly used in Russia and China). For example, the Mercator is one of the world projections, the Lambert conformal conic and Albers equal-area are among the continental projections, and the UPS is one of the polar projections.

A new coordinate system, either geographic or projected, is user-defined. The definition of a new geographic coordinate system must have a datum including a selected spheroid and its major and minor axes. The definition of a new projected coordinate system must include a datum and the parameters of the projection such as standard parallels and central meridian.

TABLE 2.1	The Grouping of Coordinate Systems in GIS and Examples	
	Predefined	Custom
Geographic	NAD27, NAD83	Datum transformation
Plane	UTM, State Plane	IDTM

2.5 WORKING WITH COORDINATE SYSTEMS IN GIS

Most GIS packages provide a wide variety of coordinate systems, datums, and spheroids. For example, Autodesk Map offers 3000 global systems, presumably 3000 combinations of coordinate system, datum, and spheroid. A constant challenge for the GIS user is how to work with coordinate systems. Table 2.1 shows a simple grouping of coordinate systems in a GIS package. Coordinate systems may be geographic or plane, and predefined or custom. The previous sections have already covered geographic and plane coordinate systems.

A predefined coordinate system, either geographic or plane, means that its parameters have already been defined. Examples of predefined coordinate systems include NAD27 (based on Clarke 1866) and Minnesota SPC83, North (based on a Lambert conformal conic projection). A GIS package usually provides a menu for the user to select a predefined coordinate system.

A custom plane coordinate system requires the user to specify its parameters including map projection parameters, datum, false easting, false northing, and so on. For example, to use the Idaho Statewide Transverse Mercator Coordinate System (IDTM, Box 2.3), one must enter its parameters. A custom geographic coordinate system typically involves a datum transformation, which recomputes longitude and latitude values from one system to another. A GIS package may offer several transformation methods such as three-parameter, seven-parameter, Molodensky, and abridged Molodensky. A good reference on datum transformation is a report from the National Imagery and Mapping Agency (2000), available on the Internet.

Like most GIS packages, ArcGIS provides a suite of tools to work with coordinate systems. Box 2.4 is a summary of coordinate systems in ArcGIS.

KEY CONCEPTS AND TERMS

Azimuthal projection: One type of map projection that retains certain accurate directions. Azimuthal also refers to one type of map projection that uses a plane as the projection surface.

Central lines: The central parallel and the central meridian. Together, they define the center or the origin of a map projection.

Clarke 1866 spheroid: A ground-measured spheroid, which is the basis for the North American Datum of 1927 (NAD27).

Conformal projection: One type of map projection that preserves local shapes.

Conic projection: One type of map projection that uses a cone as the projection surface.

Cylindrical projection: One type of map projection that uses a cylinder as the projection surface.

Datum: The basis for calculating the geographic coordinates of a location. A spheroid is a required input to the derivation of a datum.

Decimal degrees (DD) system: A measurement system for longitude and latitude values such as 42.5°.

Degrees-minutes-seconds (DMS) system: A measuring system for longitude and latitude values such as 42°30′00″, in which 1 degree equals 60 minutes and 1 minute equals 60 seconds.

Ellipsoid: A model that approximates the Earth. Also called *spheroid*.

Equidistant projection: One type of map projection that maintains consistency of scale for certain distances.

Equivalent projection: One type of map projection that represents areas in correct relative size.

False easting: A value applied to the origin of a map projection to change the *x*-coordinate readings.

False northing: A value applied to the origin of a map projection to change the *y*-coordinate readings.

Geographic Coordinate Data Base (GCDB): A database that is being developed by the U.S. Bureau of Land Management (BLM) to include longitude and latitude values and other descriptive information for section corners and monuments recorded in the PLSS.

Geographic coordinate system: The location reference system for spatial features on the Earth's surface.

GRS80 spheroid: A satellite-determined spheroid for the Geodetic Reference System 1980.

Lambert conformal conic projection: A common map projection, which is the basis for the SPC system for many states.

Map projection: A systematic construction of lines on a plane surface representing the geographic coordinate system.

Meridians: Lines of longitude on the geographic coordinate system for the E–W direction.

NAD27: North American Datum of 1927, which is based on the Clarke 1866 spheroid and has its center at Meades Ranch, Kansas.

NAD83: North American Datum of 1983, which is based on the WGS84, or the GRS80, spheroid, and is measured from the center of the spheroid.

Parallels: Lines of latitude on the geographic coordinate system for the N–S direction.

PLSS: See *Public Land Survey System.*

Principal scale: Same as the scale of the reference globe.

Projection: The process of transforming the spatial relationship of map features on the Earth's surface to a flat map.

Public Land Survey System (PLSS): A land partitioning system used in the United States.

Reference globe: A reduced model of the Earth from which map projections are constructed. Also called a *nominal* or *generating globe.*

Reprojection: Projection of spatial data from one coordinate system to another.

Scale factor: Ratio of the local scale to the scale of the reference globe. The scale factor is 1.0 along a standard line.

SPC: See *State Plane Coordinate system.*

Spheroid: A model that approximates the Earth. Also called *ellipsoid.*

Standard line: Line of tangency between the projection surface and the reference globe. A standard line has no projection distortion and has the same scale as that of the reference globe.

Standard meridian: A standard line that follows a meridian or a line of longitude.

Standard parallel: A standard line that follows a parallel or a line of latitude.

State Plane Coordinate (SPC) system: A coordinate system developed in the 1930s to permanently record original land survey monument locations in the United States. Most states have more than one zone based on the SPC27 or SPC83 system.

Transverse Mercator projection: A common map projection, which is the basis for the UTM grid system and the SPC system for many states.

UPS grid system: Universal Polar Stereographic grid system, which divides the polar area into a series of 100,000-meter squares, similar to the UTM grid system.

UTM coordinate system: Universal Transverse Mercator coordinate system, which divides the Earth's surface between 84°N and 80°S into 60 zones, with each zone covering 6° of longitude.

WGS84 spheroid: A satellite-determined spheroid for the World Geodetic System 1984.

***x*-shift:** A value applied to *x*-coordinate readings to reduce the number of digits.

***y*-shift:** A value applied to *y*-coordinate readings to reduce the number of digits.

APPLICATIONS: MAP PROJECTIONS AND COORDINATE SYSTEMS

This applications section consists of four tasks. Task 1 shows you how to project a shapefile from the geographic grid (longitude and latitude) to a custom coordinate system. In Task 2, you will also project a shapefile from the geographic grid to a custom coordinate system but use the projection data defined in Task 1. In Task 3, you will create a shapefile from a text file containing point locations in longitude and latitude values and project the shapefile onto a predefined coordinate system. Task 4 asks you to reproject a shapefile from one coordinate system to another. All four tasks use the ArcToolbox application of ArcGIS. The projection tools in ArcToolbox serve two main functions: to define coordinate systems, and to project (or reproject) from one coordinate system to another.

Using ArcView, you can define shapefiles, geodatabase, coverages, grids, and TINs, and project (or reproject) shapefiles and geodatabase. But you cannot project coverages or grids.

Task 1: Project a Shapefile from the Geographic Grid to a Plane Coordinate System

What you need: *idll.shp*, a shapefile measured in longitude and latitude values and in decimal degrees. *Idll.shp* contains an outline map of Idaho.

ArcToolbox has three options for defining a coordinate system: selecting a predefined coordinate system, importing a coordinate system from an existing data set, or creating a new (custom) coordinate system. A projection file is provided for a predefined coordinate system. A projection file can be named and saved for a new coordinate system so that the same file can be used to define or project other data sets. For Task 1, you will select a predefined coordinate system for *idll.shp* and then project the shapefile onto the Idaho Tranverse Mercator coordinate system (IDTM). IDTM is not one of the predefined systems. IDTM has the following parameter values:

> Projection Transverse Mercator
> Datum NAD27 (based on the Clarke 1866
>   ellipsoid)
> Units meters
> Parameters
>   scale factor: 0.9996
>   central meridian: -114.0
>   reference latitude: 42.0
>   false easting: 500000
>   false northing: 100000

1. Start ArcCatalog, and make connection to the Chapter 2 database. Launch ArcToolbox. Click the plus sign next to Data Management Tools and then the plus sign next to Projection. Double-click Define Projection Wizard (shapefiles, geodatabase). You will use this wizard to define the coordinate system for *idll.shp*.

2. Use the browse button or the drag-and-drop method to add *idll.shp* as the input data set in the first panel. Click the Select Coordinate System button in the second panel. Click the Select button in the Spatial Reference Properties dialog. Double-click Geographic Coordinate Systems, double-click North America, and double-click North American Datum 1927.prj. Click OK to dismiss the Spatial Reference Properties dialog. Click Next and then Finish in the Define Projection Wizard dialog to complete defining the coordinate system for *idll.shp*.

3. Go back to the Projections menu in ArcToolbox. This time you will define the

coordinate system for the output to be called *idtm.shp*. Double-click Project Wizard (shapefiles, geodatabase). Use the browse button or the drag-and-drop method to add *idll.shp* as the data to project in the first panel. Specify *idtm.shp* as the name for the output and the Chapter 2 folder to store the output in the second panel. The third panel asks you to select the coordinate system to assign to *idtm.shp*. Click the Select Coordinate System button. Click New and then Projected in the Spatial Reference Properties dialog. In the New Projected Coordinate System dialog, first enter idtm as the Name. Then you need to provide projection information in the Projection frame and for the Geographic Coordinate System (i.e., datum and spheroid). In the Projection frame, select Tranverse_Mercator from the Name dropdown list. Enter the following parameter values: 500000 for False_Easting, 100000 for False_Northing, -114 for Central_Meridian, 0.9996 for Scale_Factor, and 42 for Latitude_Of_Origin. Make sure that the Linear Unit is Meter. Click Select for the Geographic Coordinate System. Double-click North America, and North American Datum 1927.prj. Click OK to dismiss the New Projected Coordinate System dialog. Click Save As in the Spatial Reference Properties dialog, and enter *idtm.prj* as the file name. Click OK to dismiss the Spatial Reference Properties dialog. Click Next in the Project Wizard dialog. Click Next after reviewing the Coordinate extents for the output data set. Click Finish. You have completed the projection of *idll.shp* to *idtm.shp*.

Task 2: Import a Coordinate System

What you need: *stationsll.shp*, a shapefile measured in longitude and latitude values and in decimal degrees. *Stationsll.shp* contains snow courses in Idaho.

In Task 2, you will complete the projection of *stationsll.shp* by importing the projection information on *idll.shp* and *idtm.shp* from Task 1.

1. Go back to the Projections menu in ArcToolbox. Double-click Define Projection Wizard (shapefiles, geodatabase). Use the browse button or the drag-and-drop method to add *stationsll.shp* as the input data set. Click Select Coordinate System in the next panel. Click Import in the Spatial Reference Properties dialog. Double-click *idll.shp* to add. Click OK to dismiss the Spatial Reference Properties dialog. Click Next and then Finish in the Define Projection Wizard dialog. You have completed defining the coordinate system for *stationsll.shp*.

2. Double-click Project Wizard (shapefiles, geodatabase) in ArcToolbox. Add *stationsll.shp* as the data to project. Specify *stationstm.shp* as the name for the output and the Chapter 2 folder to store the output in the next panel. Click the Select Coordinate System button in the Project Wizard dialog. Click Import in the Spatial Reference Properties dialog. Double-click *idtm.shp* to add. Click OK to dismiss the Spatial Reference Properties dialog. Click Next in the Project Wizard dialog. Click Next after reviewing the Coordinate extents for the output dataset. Click Finish. *Stationstm.shp* is now projected onto the same (IDTM) coordinate system as *idtm.shp*.

Task 3: Project a Shapefile by Using a Predefined Coordinate System

What you need: *snow.txt*, a text file containing the longitude and latitude values of 40 snow courses in Idaho.

For Task 3, you will first create a point shapefile from *snow.txt*. Then you will project the shapefile, which is still measured in longitude and latitude values, onto a predefined (UTM) coordinate system.

1. Launch ArcMap. Add *snow.txt* to the data frame Layers. Click the Tool menu and select Add XY Data. In the next dialog, make sure that *snow.txt* is the input table, longitude is the x field, latitude is the y field, and click OK. *Snow.txt Events* is added to Layers. You must convert *snow.txt Events* to a shapefile before you can project the data set. Right-click *snow.txt Events*, point to Data, and select Export Data. In the Export Data dialog, enter *snow.shp* as the output shapefile and click OK. Add *snow.shp* to Layers and make sure that the shapefile has the same features as *snow.txt Events*. You must exit ArcMap before the next step of projecting *snow.shp*.

2. Launch ArcToolbox if necessary. Double-click Define Projection Wizard (shapefiles, geodatabase). In the first panel, select *snow.shp* for the data to assign a coordinate system. Click the Select Coordinate System button in the second panel. Click Select in the Spatial Reference Properties dialog. Double-click Geographic Coordinate Systems, North America, and North American Datum 1927.prj. Click OK to dismiss the Spatial Reference Properties dialog. Click Next and then Finish in the Define Projection Wizard dialog. You have completed defining the coordinate system for *snow.shp*.

3. Double-click Project Wizard (shapefiles, geodatabase). In the first panel, select *snow.shp* for the data to project. In the second panel, specify *snowutm.shp* for the output feature class to be saved in the Chapter 2 database. Click the Select Coordinate System button in the next panel. Click Select in the Spatial Reference Properties dialog. Double-click Projected Coordinate Systems, Utm, Nad 1927, and NAD 1927 UTM Zone 11N.prj. Click OK to dismiss the Spatial Reference Properties dialog. Click Next, Next, and Finish to dismiss the Project Wizard dialog. You have completed projecting *snow.shp* onto the UTM coordinate system.

Task 4: Convert from One Coordinate System to Another

What you need: *idtm.shp* from Task 1 and *snowutm.shp* from Task 3.

Task 4 asks you to convert *idtm.shp* from the Idaho Transverse Mercator coordinate system to the UTM coordinate system so that it can be used as a reference map for *snowutm.shp* from Task 3. Because both shapefiles already have projection files that define their coordinate systems, the re-projection is relatively easy.

1. Go back to the Projections menu in ArcToolbox. Double-click Project Wizard (shapefiles, geodatabase). In the first panel, select *idtm.shp* to project. In the second panel, specify *idutm.shp* as the output

shapefile to be stored in the Chapter 2 database. Click the Select Coordinate System button in the third panel. Click Select in the Spatial Reference Properties dialog. Double-click Projected Coordinate Systems, Utm, Nad 1927, and NAD 1927 UTM Zone 11N.prj. Click OK to dismiss the Spatial Reference Properties dialog. Click Next, Next, and Finish to dismiss the Projection Wizard dialog. You have completed the re-projection of *idtm.shp*.

2. You can verify that *idutm.shp* register spatially with *snowutm.shp* in ArcMap. Launch ArcMap, and add *idutm.shp* and *snowutm.shp* to Layers. The majority of points in *snowutm.shp* should appear inside Idaho.

REFERENCES

Burkard, R. K. 1984. *Geodesy for the Layman*. Washington, DC: Defense Mapping Agency. Available at **http://www.ngs. noaa.gov/PUBS_LIB/ Geodesy4Layman/TR80003A. HTM#ZZ0/.**

Dent, B. D. 1999. *Cartography: Thematic Map Design,* 5th ed. Dubuque, IA: Wm C. Brown.

Kumar, M. 1993. World Geodetic System 1984: A Reference Frame for Global Mapping, Charting and Geodetic Applications. *Surveying and Land Information Systems* 53: 53–56.

Maling, D. H. 1992. *Coordinate Systems and Map Projections,* 2d ed. Oxford: Pergamon Press.

Moffitt, F. H., and J.D. Bossler. 1998. *Surveying,* 10th ed. Menlo Park, CA: Addison-Wesley.

Muehrcke, P. C., J. O. Muehrcke, and A. J. Kimerling. 2001. *Map Use: Reading, Analysis, and Interpretation*. Madison, WI: JP Publishers.

National Imagery and Mapping Agency. 2000. *Department of Defense World Geodetic System 1984: Its Definition and Relationships with Local Geodetic Systems*, 3rd ed. NIMA TR8350.2. Amendment 1, January 3, 2000. Available at **http://164.214.2.59/GandG/ tr8350_2.html/.**

Robinson, A. H., J. L. Morrison, P. C. Muehrcke, A. J. Kimerling, and S. C. Guptill. 1995.

Elements of Cartography, 6th ed. New York: Wiley.

Snay, R. A., and T. Soler. 2000. Modern Terrestrial Reference Systems. Part 2: The Evolution of NAD 83. *Professional Surveyor*, February 2000. Available at **http://www. profsurv.com/psarchive.htm/.**

Snyder, J. P. 1987. *Map Projections—a Working Manual*. Washington, DC: U.S. Geological Survey Professional Paper 1395.

Snyder, J. P. 1993. *Flattening the Earth: Two Thousand Years of Map Projections*. Chicago: University of Chicago Press.

3

THE VECTOR DATA MODEL

3.1 INTRODUCTION

Looking at a paper map, we can see what map features are like and how they are spatially related to one another. For example, the reference map in

Figure 3.1 shows that Idaho borders Montana, Wyoming, Utah, Nevada, Oregon, Washington, and Canada, and contains several Indian reservations. The map communicates to us through its symbols and text. We can easily see the map features and their spatial relationships, but how can the computer see these features and relationships? That is the basic question we ask in this chapter about the vector data model.

The **vector data model** uses points and their x-, y-coordinates to represent spatial features. Vector-based features are treated as discrete geometric objects over the space. The process of de-

veloping the vector data model consists of several steps. First, spatial features are represented as simple geometric objects of points, lines, and areas. Second, for some GIS applications, the spatial relationships between features must be expressed explicitly. Third, a logical structure of data files must be in place so that the computer can efficiently process data for spatial features and their spatial relationships. Fourth, land surface data, overlapping spatial features, and road networks are better represented as composites of simple geometric objects.

This chapter is divided into the following six sections. Section 3.2 covers the representation of

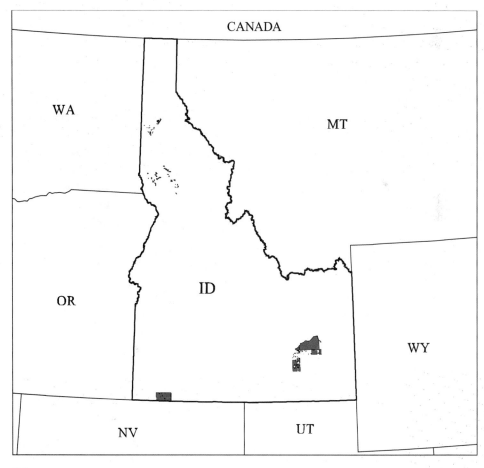

Figure 3.1
A reference map showing Idaho and land parcels held in trust by the United States for Native Americans.

vector data as geometric objects. Section 3.3 describes the data structure for computer processing. Section 3.4 discusses nontopological vector data. Section 3.5 covers spatial features that are better represented as composites of points, lines, and areas. Section 3.6 introduces the object-oriented data model. Section 3.7 discusses location data accuracy, accuracy standards, and precision.

3.2 VECTOR DATA REPRESENTATION

3.2.1 Geometric Objects

The vector data model uses x-, y-coordinates and the simple geometric objects of point, line, and area to represent spatial features. Dimensionality and property distinguish the three types of geometric objects.

- A **point** has 0 dimension and has only the property of location. A point may represent a well, a benchmark, or a gravel pit.
- A **line** is one-dimensional and has the property of length. A line may represent a road, a stream, or an administrative boundary.
- An **area** is two-dimensional and has the properties of area and perimeter or boundary. An area may represent a timber stand, a water body, or a sinkhole.

Point, line, and area are generic terms. In the GIS literature, a point may also be called a node, vertex, or 0-cell; a line, an edge, link, chain, or 1-cell; and an area, a polygon, face, zone, or 2-cell (Laurini and Thompson 1992).

The basic units of the vector data model are points and their coordinates. A line feature is made of points (Figure 3.2). Between two end points a line consists of a series of points marking the shape of the line, which may be a smooth curve or a connection of straight-line segments. Smooth curves are typically fitted by mathematical equations such as splines. Straight-line segments may represent human-made features or approximations of curves. Line features may intersect or join with other lines and may form a network.

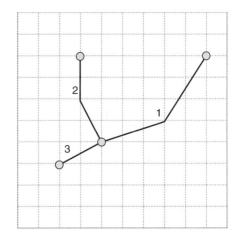

○ End point

Figure 3.2
Line objects.

An area or polygon feature is defined by lines, which are in turn defined by points (Figure 3.3). The boundary of an area feature separates the interior area from the exterior area. Area features may be isolated or connected. An isolated area feature typically has a point serving as both the beginning and end point of its boundary. Area features may form holes within other areas, such as the Indian reservations surrounded by the state of Idaho. Area features may overlap one another and create overlapped areas. For example, the burned areas from previous forest fires may overlap each other. The simple data model treats holes and overlapped areas as separate objects.

The representation of vector data using points, lines, and areas is not always straightforward because it depends on map scale and, occasionally, criteria established by government mapping agencies (Robinson et al. 1995). **Map scale** is the ratio of the map distance to the corresponding ground distance. On a 1:24,000 scale map, for example, a map distance of 1 centimeter would represent 24,000 centimeters, or 240 meters. A 1:24,000 scale map shows a smaller area but contains more

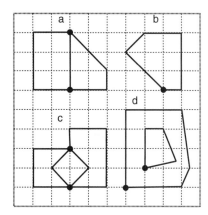

Figure 3.3
Area objects: a—contiguous areas; b—isolated area; c—three areas formed by two overlapped objects; d—a hole within an area.

details than a 1:1,000,000 scale map. This is why a city on a 1:1,000,000 scale map may appear as a point, but the same city may appear as an area on a 1:24,000 scale map.

A stream may be shown as a single line near its headwaters but as an area along its lower reaches. In this case, the width of the stream determines how it should be represented on a map. The U.S. Geological Survey (USGS) uses single lines to represent streams less than 40 feet wide on 1:24,000 scale topographic maps and double lines for larger streams. Therefore, a stream may appear as a line or an area depending on its width and the criterion used by the government agency.

3.2.2 Topology

The conceptual representation of spatial features as points, lines, and areas is the first step in building the vector data model. For some GIS applications, the next step is to turn to topology to express explicitly the spatial relationships between features. **Topology** is the study of those properties of geometric objects that remain invariant under certain transformations such as bending or stretching (Massey 1967). For example, a rubber band can be

stretched and bent without losing its intrinsic property of being a closed circuit, as long as the transformation is within its elastic limits.

Topology is often explained through **graph theory,** a subfield of mathematics that uses diagrams or graphs to study the arrangements of geometric objects and the relationships between objects (Wilson and Watkins 1990). Important to the vector data model are digraphs (directed graphs), which include points and directed lines (also called arcs). Adjacency and incidence are two relationships that can be established between the point and line objects in digraphs (Box 3.1).

A good example of topology-based, or topological, data is the TIGER (Topologically Integrated Geographic Encoding and Referencing) database from the U.S. Bureau of the Census (Broome and Meixler 1990). In the TIGER database, points are called 0-cells, lines 1-cells, and areas 2-cells (Figure 3.5). Each 1-cell in a TIGER file is a directed line, meaning that the line is directed from a starting point toward an ending point with an explicit left and right side. Each 2-cell and 0-cell has knowledge of the 1-cells associated with it. The USGS digital line graph (DLG) products, which include such features as roads, streams, boundaries, and contours, also use topology to define the spatial relationship between features. Commercial GIS vendors such as ESRI and Intergraph have proprietary topological data structures.

ESRI defines the standard, topological vector data format used in ArcInfo Workstation as **coverage** and groups coverages by point, line, and polygon. A coverage supports three basic topological relationships (Environmental Systems Research Institute, Inc. 1998):

- **Connectivity:** Arcs connect to each other at nodes.
- **Area definition:** An area is defined by a series of connected arcs.
- **Contiguity:** Arcs have directions and left and right polygons.

Other than the use of terms, the above three topological relationships are similar to the topological relationships in the TIGER database.

Box 3.1   Adjacency and Incidence

If a line joins two points, the points are said to be adjacent and incident with the line, and the adjacency and incidence relationships can be expressed explicitly in matrices. Figure 3.4 shows an adjacency matrix and an incidence matrix for a digraph. The row and column numbers of the adjacency matrix correspond to the node numbers, and the numbers within the matrix refer to the number of arcs joining the corresponding nodes in the digraph. For example, 1 in (11,12) means one arc joint from node 11 to node 12, and 0 in (12,11) means no arc joint from node 12 to node 11. The direction of the arc determines if 1 or 0 should be assigned.

The row numbers of the incidence matrix correspond to the node numbers in Figure 3.4, and the column numbers correspond to the arc numbers. The number 1 in the matrix means an arc is incident from a node, -1 means an arc is incident to a node, and 0 means an arc is not incident from or to a node. Take the example of arc 1. It is incident from node 13, incident to node 11, and not incident to all the other nodes. Thus the matrices express the adjacency and incidence relationships mathematically.

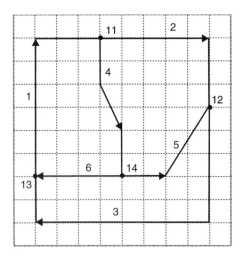

Adjacency matrix

	11	12	13	14
11	0	1	0	1
12	0	0	1	0
13	1	0	0	0
14	0	1	1	0

Incidence matrix

	1	2	3	4	5	6
11	-1	1	0	1	0	0
12	0	-1	1	0	-1	0
13	1	0	-1	0	0	-1
14	0	0	0	-1	1	1

Figure 3.4
The adjacency matrix and incidence matrix for a digraph.

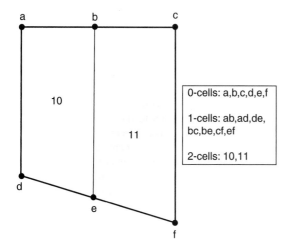

Figure 3.5
Topology in the TIGER database involves 0-cells or points, 1-cells or lines, and 2-cells or areas.

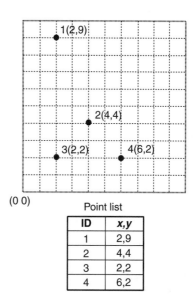

Point list

ID	x,y
1	2,9
2	4,4
3	2,2
4	6,2

Figure 3.6
The data structure of a point data model.

Topology is useful for detection of errors on digital maps. By using the topological relationships, one can detect lines that do not meet correctly, a line that is digitized more than once, or polygons that are not closed properly in a digital map. These kinds of errors must be corrected to avoid incomplete features and to ensure data integrity. For example, a shortest path analysis requires roads to meet correctly. If a gap exists on a supposedly continuous road, the analysis will take a circuitous route to avoid the gap.

Topology is also important for some types of GIS analysis. One example is traffic volume analysis, in which traffic volume measurements must follow lines with the same traffic direction. Another example is deer habitat analysis, which often involves edges between habitat types, especially edges between old growth and clear-cuts (Chang et al. 1995). Because edges are coded with left and right polygons in a topologically structured data model, specific habitat types along edges can be easily tabulated and analyzed.

Other applications may be possible if more sophisticated topological relationships can be built into a data set. For example, one might build topology between a county map and a census tract map so that counties and census tracts can share common (coincident) boundaries. As discussed later in the chapter, a new development is to implement more topological relationships in GIS data.

3.3 TOPOLOGICAL DATA STRUCTURE

So far we have examined the concepts involving the topology-based vector data model. The application of the concepts falls within the domain of data structure. Point features are simple: they can be coded with their identification numbers (IDs) and pairs of x- and y-coordinates (Figure 3.6). Topology does not apply to points because points are separate from one another.

Figure 3.7 shows the data structure of a line feature. An **arc** is a line segment, which is connected to two end points called **nodes.** The starting point is the from-node and the ending point the to-node. The arc-node list sorts out the arc–node relationship. For example, arc 2 has 12 as the from-node and 13 as the to-node. The arc-coordinate list shows the x-, y-coordinates that make up each arc.

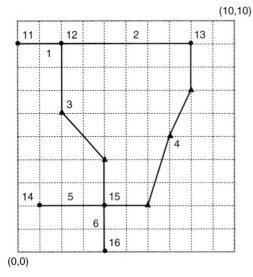

Arc-node list

Arc#	F-node	T-node
1	11	12
2	12	13
3	12	15
4	13	15
5	15	14
6	15	16

Arc-coordinate list

Arc#	x,y Coordinates
1	(0,9) (2,9)
2	(2,9) (8,9)
3	(2,9) (2,6) (4,4) (4,2)
4	(8,9) (8,7) (7,5) (6,2) (4,2)
5	(4,2) (1,2)
6	(4,2) (4,0)

Figure 3.7
The data structure of a line data model.

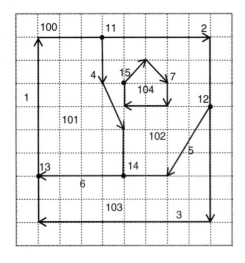

Left/right list

Arc#	L-poly	R-poly
1	100	101
2	100	102
3	100	103
4	102	101
5	103	102
6	103	101
7	102	104

Polygon/arc list

Polygon #	Arc#
101	1,4,6
102	4,2,5,0,7
103	6,5,3
104	7

Figure 3.8
The data structure of an area data model.

For example, arc 3 consists of three line segments connected at the points of (2, 6) and (4, 4).

Figure 3.8 shows the data structure of an area or polygon feature. The polygon/arc list shows the relationship between polygons and arcs. For exam-ple, arcs 1, 4, and 6 connect to define polygon 101. Polygon 104 differs from the other polygons in be-ing surrounded by polygon 102. To show that poly-gon 104 is a hole within polygon 102, the arc list for polygon 102 contains a zero to separate the

After two decades, topology has again become an issue. GIS users may wonder about the importance of topology. As described in this chapter, topology is useful for data editing and some types of spatial analysis. The topological capabilities of GIS software can differ extensively. ArcInfo Workstation uses the basic topological relationships of connectivity, area definition, and contiguity, which are sufficient for its operational needs. Sophisticated GIS operations may require more topological relationships to be present between map features.

The decision on topology therefore depends on the GIS project. For some projects, topological functions are not necessary; for others they are a must. For example, a producer of GIS data will find it absolutely necessary to use topology for error checking and for ensuring that lines meet correctly and polygons are

closed properly. Likewise, a GIS analyst working with transportation and utility networks will want to use topological data for data analysis.

Ordnance Survey (OS) is perhaps the first major GIS data producer to offer both topological and nontopological data to end-users. OS MasterMap is a new framework for the referencing of geographic information in Great Britain (**http://www.ordsvy.gov.uk/**). MasterMap has two types of polygon data: independent and topological polygon data. Independent polygon data duplicate the coordinate geometry shared between polygons. In contrast, topological polygon data include the coordinate geometry shared between polygons only once and reference each polygon by a set of line features. The reference of a polygon by line features is similar to the polygon/arc list discussed in the text.

---

external and internal boundaries. Polygon 104 is an isolated polygon consisting of only one arc (7). A node (15) is placed along the arc to be the beginning and end node. Polygon 100, which is outside the map area, is the external or universe polygon.

The left/right list in Figure 3.8 shows the relationship between arcs and their left and right polygons. For example, arc 1 is a directed line from node 13 to node 11 and has polygon 100 as the polygon on the left and polygon 101 as the polygon on the right. Finally, each polygon is assigned a label point to link the polygon to its attribute data.

The topology-based data structure facilitates the organization of data files and reduces data redundancy. The shared or common boundary between two polygons is listed once, not twice, in the arc-coordinate list. Moreover, because the shared boundary defines both polygons, updating the polygons is made easier. For example, if arc 4 in Figure 3.8 is changed to a straight line between two nodes, only the coordinate list for arc 4 needs to be changed. The way that the shared boundary is handled separates topological data from nontopological data.

3.4 NONTOPOLOGICAL VECTOR DATA

GIS developers introduced topology two decades ago to separate GIS from CAD (computer-aided design). AutoCAD by Autodesk was, and still is, the leading CAD package. A data format used by AutoCAD for transfer of data files is called DXF (drawing exchange format). DXF maintains data in separate layers and allows the user to draw each layer using different line symbols, colors, and text. But DXF files do not support topology.

The issue of topology or no topology has again returned to GIS (Box 3.2). The main advantage of using nontopological vector data is that they display more rapidly on the computer monitor than topology-based data (Throbald 2001). In recent years, nontopological data format has become one of the standard, nonproprietary data formats. Commercial GIS packages such as ArcGIS, MapInfo, and GeoMedia all have adopted nontopological data formats.

Shapefile is the standard nontopological data format used in ESRI products. Although the shapefile treats a point as a pair of x-, y-coordinates, a line

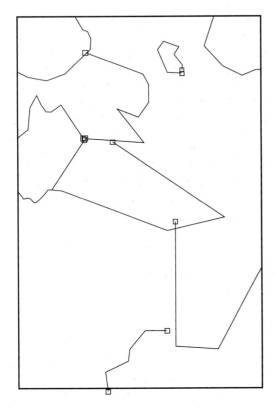

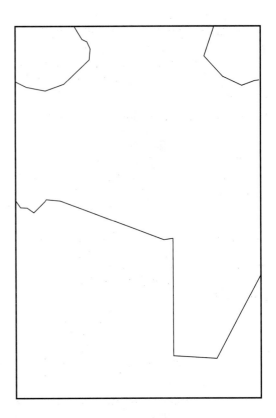

Figure 3.9a
A polygon coverage with topological errors. Each small square symbol represents an error caused by lines that do not meet correctly.

Figure 3.9b
A shapefile converted from the coverage shown in Figure 3.9a.

as a series of points, and a polygon as a series of lines, no files describe the spatial relationship among geometric objects. Shapefile polygons actually have duplicate arcs for the shared boundaries and can overlap one another. Rather than having multiple files as for an ArcInfo coverage, the geometry of a shapefile is stored in two basic files: the .shp file stores the feature geometry, and the .shx file maintains the index of the feature geometry.

Shapefiles can be converted to coverages, and vice versa. The conversion from a shapefile to a coverage requires the building of topological relationships and the removal of duplicate arcs. The conversion from a coverage to a shapefile is simpler. But if a coverage has topological errors—such as lines not joined perfectly—the errors can

lead to problems of missing features in the shapefile. Figure 3.9a shows a coverage that has errors in line joining. After converting to a shapefile, all lines that have errors disappear in the shapefile (Figure 3.9b). Figure 3.9 illustrates the importance of topology in maintaining the integrity of vector data.

3.5 HIGHER-LEVEL OBJECTS

To complete the discussion of the vector data model, this section describes the higher-level objects of TIN, regions, and dynamic segmentation. The creation and applications of TIN, regions, and dynamic segmentation are covered in Chapters 12, 15, and 16, respectively.

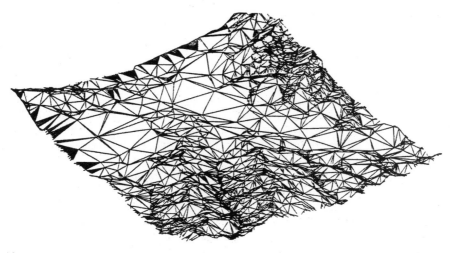

Figure 3.10
TIN in a perspective view.

3.5.1 TIN

Triangulated irregular network (TIN) is a vector data structure for terrain mapping and analysis. A TIN approximates the surface with a set of nonoverlapping triangles (Figure 3.10). Each triangle in the TIN assumes a constant gradient. These triangles are constructed using **Delaunay triangulation,** an iterative process of connecting points with their two nearest neighbors to form triangles as equiangular as possible. Different computing algorithms have been proposed for Delaunay triangulation, and these algorithms are continually being improved (Watson and Philip 1984; Tsai 1993).

A TIN consists of two basic data elements: elevation points with x, y, and z values, and edges (lines) that connect these points to form triangles. The x, y values represent the location of a point, and the z value represents the elevation at the point. The slope and aspect of each triangle are computed from the x, y, and z values at the three points that make up the triangle. These elevation points in a TIN are actually sample points selected to represent the surface. Flat areas of the surface have a small number of sample points and large triangles. Areas with high variability in elevation need a large number of sample points and small triangles.

The TIN data structure includes the triangle number, the number of each adjacent triangle, and data files showing the lists of points, edges, as well as the x, y, and z values of each elevation point.

3.5.2 Regions

Built on simple lines and areas, the **regions data model** consists of region layers and regions (Figure 3.11). A region layer is made of regions of the same attribute. The regions data model has two important characteristics. First, region layers may overlap or cover the same area. For example, a region layer representing areas burned in a 1917 forest fire may overlap with another region layer representing areas burned in a 1930 fire. When different region layers cover the same area, they form a hierarchical region structure with one layer nested within another layer.

Second, a region may have disconnected or disjoint components. The state of Hawaii may therefore be a region, including several islands. (In contrast, the simple data model requires each island to be a separate polygon.) This characteristic also applies to the void or empty area of a region. Therefore, private land parcels scattered within a national forest can be grouped as one void area.

Box 3.3 A Regions-Based Fire Coverage

The regions-based fire coverage prepared by the Idaho Panhandle National Forest includes 31 region subclasses: Pre1886, C1889, C1894, C1900, C1905, F1908–09, F1910, F1911–13, F1914–15, F1917, F1918, F1919, F1920–21, F1922–23, F1924, F1925, F1926, F1927–28, F1929, F1930, F1931, F1932–33, F1934, F1935–39, F1940–49, F1950–59, F1960–69, All_Fires, Doubles, Triples, and Quad_burns. Forest fires in 1905 and prior to 1905 are less certain and thus carry letter C for circa. Doubles, triples, and quad_burns show areas burned twice, three times, and four times, respectively. Each region subclass has its attribute table, which includes acres of burned areas. This regions-based fire coverage can be displayed and queried in ArcMap.

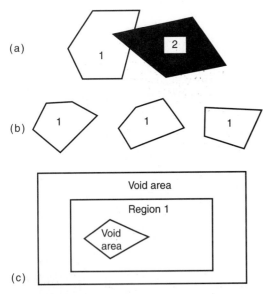

Figure 3.11
Properties of the regions data model: a—overlapped regions; b—a region with three components or rings; c—void area made of an area within the region and the external area.

ArcInfo Workstation treats region layers as subclasses in a polygon coverage. Each region subclass has its own attributes. In this way a series of region layers can be based on the same polygon coverage, which is called an **integrated coverage.** The Idaho Panhandle National Forest has built an integrated fire coverage with fire records spanning from the late 1900s to the 1960s (Box 3.3). Each fire year or period is handled as a region subclass.

The regions data structure consists of two basic elements: a file on the region–arc relationship and another on the region–polygon relationship. Figure 3.12 shows the file structure for an example with four polygons, five arcs, and two regions. The region–polygon list relates the regions to the polygons. Region 101 consists of polygons 11 and 12. Region 102 has two components: one includes polygons 12 and 13, and the other consists of polygon 14. Region 101 overlaps region 102 in polygon 12. The region–arc list links the regions to the arcs. Region 101 has only one ring, which connects arcs 1 and 2. Region 102 has two rings: one connects arcs 3 and 4, and the other consists of arc 5.

3.5.3 Dynamic Segmentation

The dynamic segmentation model combines a line coverage and a linear measurement system such as the milepost system to form higher-level objects. ArcInfo Workstation uses the basic elements of sections, routes, and events for the model. **Sections** refer directly to arcs of a line coverage and positions along arcs. Because arcs of a line coverage are made of a series of x-, y-coordinates based on a coordinate system, sections are also measured in coordinates. **Routes** are a collection of sections that represent linear phenomena such as highways, bike paths, or streams. Like regions, routes are

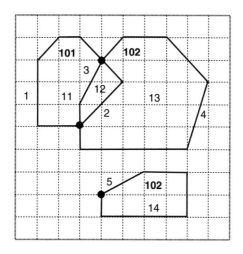

Region-polygon list

Region #	Polygon #
101	11
101	12
102	12
102	13
102	14

Region-arc list

Region #	Ring #	Arc #
101	1	1
101	1	2
102	1	3
102	1	4
102	2	5

Figure 3.12
The file structure for the regions data model.

stored as subclasses in a line coverage. Attribute data to be associated with routes are called **events** in dynamic segmentation. Events such as pavement conditions, accidents, and speed limits are measured in a linear system such as the milepost. But as long as events are accompanied by measures of their locations, the dynamic segmentation model can relate events to route systems.

Figure 3.13 shows how a route, Route 109, is coded for a route system called BIKEPATH in a line coverage called ROADS. The coverage ROADS contains a number of topologically structured arcs. Route 109 in a thick shaded line is a collection of three sections, numbered 1, 2, and 3 with BIKEPATH # (machine ID) and BIKEPATH-ID (user ID).

The section table relates the sections to the arcs in ROADS. Section 1 covers the entire length of arc 7; therefore, the from-position (F-POS) is 0 percent and the to-position (T-POS) is 100 percent. The from-measure (F-MEAS) of section 1 is 0 because it is the beginning point of Route 109. The to-measure (T-MEAS) of section 1 is 40 units (meters or feet, depending on the measurement unit of ROADS), which is measured from the line coverage. Similar to section 1, section 2 also covers the entire length of arc 8. Its from-measure and to-measure continue

from section 1. Section 3 covers 80 percent of arc 9, thus its to-position is coded 80. The to-measure of section 3 is computed by adding 80 percent of the length of arc 9 to its from-measure.

The route table in Figure 3.13 shows the route has a machine ID of 1 and a user ID of 109. The route table is linked to the section table using the ID value of BIKEPATH # and Route link #, which is 1 in the example. Attribute data for the route system, which are usually prepared as event tables, can be directly added to the route table. Events can be point events such as signs along a bike path, continuous events such as gradients along a bike path, or linear events such as stretches of poor visibility. A point event is measured by its location, a continuous event by its to measure, and a linear event by its from and to measures (Figure 3.13). Event tables include the route ID, location measures, and attributes.

The hierarchical structure of arcs, sections, and routes in the dynamic segmentation data model has several advantages over the simple arc–node data model. First, the hierarchical structure allows one-to-many relationships. For example, routes and sections can efficiently represent different bus routes using the same downtown streets. Second, dynamic segmentation combines coordinates with linear

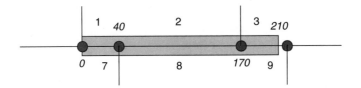

Section table

Route Link #	Arc Link #	F-MEAS	T-MEAS	F-POS	T-POS	BIKEPATH #	BIKEPATH-ID
1	7	0	40	0	100	1	1
1	8	40	170	0	100	2	2
1	9	170	210	0	80	3	3

Route table

BIKEPATH #	BIKEPATH-ID
1	109

Point event table

BIKEPATH-ID	LOCATION	ATTRIBUTE
109	40	Stop sign

Linear event table

BIKEPATH-ID	FROM	TO	ATTRIBUTE
109	100	120	Steep

Figure 3.13
The dynamic segmentation data model. Route link # in the section table and BIKEPATH # in the route table link the two tables. BIKEPATH-ID links the route table to the event tables.

measures in the section table. The from-measure and to-measure function as linear measures, like mileposts, in a linear reference system. But because the from-measure and to-measure are based on a line coverage, they are also georeferenced, meaning that they are referenced to a coordinate system. This georeferencing makes it easier to register route systems with other georeferenced coverages in the database. Third, dynamic segmentation offers a much more efficient means for working with segmented data, such as pavement conditions along

highways. Using the arc–node data model would require adding and removing nodes to store segmented data.

3.6 OBJECT-ORIENTED DATA MODEL

Our discussion so far about vector data has focused on the georelational data model. The **georelational data model** separates spatial data from attribute data, and stores spatial data in graphic files and attribute data in tables. The two types of

Box 3.4 Object-Oriented User Interface and Programming

An object-oriented user interface uses icons, dialog boxes, and glyphs (i.e., graphical objects) in place of command lines. This kind of user interface allows the user to communicate with a GIS package by point and click and, at the same time, restricts user choice of operations to those defined for the selected data set, thus avoiding operational errors. This is why users often find menu selections differ depending on the input maps.

Visual Basic (VB) and C++ are examples of the object-oriented computer language that have been adopted by GIS vendors for software development. Some GIS companies have also developed object-oriented macro languages such as Avenue for Arc-

View 3.x and MapX for MapInfo. These macro languages allow users to modify the user interface, develop specialized functionality, and extend the analytical capabilities of the GIS package.

ArcGIS Desktop uses the Component Object Model (COM) technology (Ungerer and Goodchild 2002). Therefore, ArcGIS users can customize the desktop applications through Visual Basic for Applications (VBA), which is embedded within ArcCatalog and ArcMap. VBA programmers can write codes in VB to work with ArcObjects, the development platform for ArcGIS Desktop, and to create customized commands, menus, and tools (Zeiler 2001).

data are linked through the feature IDs. The arc–node list, the polygon–arc list, and the region–polygon list are examples of graphic files.

Object-oriented technology has impacted many fields for the past two decades. GIS is no exception. GIS users are familiar with the change of user interface and programming language (Box 3.4). This section focuses on another application of object-oriented technology: the object-oriented data model. The **object-oriented data model** uses objects to organize spatial data. Unlike a geometric object of a point, line, or area, an **object** is defined here as something that has a set of properties and can perform operations upon requests on a computer screen. By this definition, almost everything one uses in a GIS is an object. For example, a land-use map is an object, which has properties such as its coordinate system and feature type and can respond to requests such as zoom in, zoom out, and query.

3.6.1 Structural Aspects of Objects

An important issue in applying the object-oriented data model to GIS is the structural aspects of objects (Worboys et al. 1990; Worboys 1995). The

principles that can be used to group objects include association, aggregation, generalization, instantiation, and specialization (Larman 1997; Zeiler 1999).

Association describes the relationships between objects of two types. If owner and land parcel represent two types of objects, the relationships between them can follow the rules that (1) an owner can own one or more parcels, and (2) a parcel can have one or more owners. Aggregation is an asymmetric association in a whole-part relationship. For example, census block groups are connected to form a census tract and census tracts are connected to form a county. Generalization uses the commonality among objects to group objects of similar types into a higher-order type. For example, parcel, zoning, and census tract maps may be grouped into a higher-order class called boundary. The grouping of objects forms a hierarchical structure, which organizes objects into classes and classes into superclasses and subclasses. Instantiation means that an object of a class can be created from an object of another class. For example, a high-density residential area object may be created from a residential area object. Specialization differentiates objects of a given class by a set of rules.

For example, roads may be separated by average daily traffic volume.

3.6.2 Behavioral Aspects of Objects

Another issue in applying the object-oriented data model to GIS is the behavioral aspects of objects (Egenhofer and Frank 1992). Inheritance is the basic principle in explaining the behaviors of objects: subclasses inherit properties and operations from a superclass, and objects inherit properties and operations from a subclass. Suppose that residential area is a superclass and low-density area and high-density area are the subclasses. All properties of the class residential area are inherited by its two subclasses. Through inheritance, properties need only be defined once in the class hierarchy. Additional properties (i.e., not inherited) may be defined for a particular subclass to separate it from other subclasses. For example, the lot size may be added as a property to separate low-density and high-density areas.

Encapsulation and polymorphism are two other principles closely related to the behavioral aspects of objects. Encapsulation refers to the mechanism to hide the properties and operations of an object so that the object can only be accessed through predefined interfaces. For example, to get the area extent of a polygon defined by xmin, ymin, xmax, and ymax, one will use the interface called envelope in a Visual Basic (VB) code in ArcGIS Desktop. Polymorphism means that the same operation can produce different effects, depending on the kind of object, to which the operation is being applied. For example, the AddLayer method in ArcGIS Desktop can be sent to a coverage or a shapefile but the result differs depending on the data source. A coverage data source will add a coverage and a shapefile data source will add a shapefile.

3.6.3 The Geodatabase Model

The geodatabase model, offered alongside the georelational data model in ArcGIS, is a new entry in applying object-oriented technology to the GIS data model. Companies specializing in utilities and

communications have previously adopted the technology. They include SmallWorld (now GE-SmallWorld, **http://www.gepower.com/network solutions/**) and Laser-Scan (**http://www.laser-scan.com/**).

The geodatabase model uses the geometries of point, polyline, and polygon to represent vector-based spatial features (Zeiler 1999). A point feature may be a simple feature with a point or a multipoint feature with a set of points. A polyline feature is a set of line segments, which may or may not be connected. A polygon feature may be made of one or many rings. A ring is a set of connected, closed, nonintersecting line segments. A polygon feature with many rings would be the same in concept as a region with disjoint components in the regions data model.

The geodatabase model organizes spatial features into feature classes and feature datasets. A feature class stores spatial data of the same geometry type. A feature dataset stores feature classes that share a common coordinate system. A feature class is like a shapefile. With one or more layers, a feature dataset is similar to a coverage in data structure. But a feature dataset can contain maps of different themes, so long as they are based on the same coordinate system, whereas a coverage contains different parts of a single map such as arcs, nodes, and tics. One can build a geodatabase from imported coverages (including coverages with subclasses), shapefiles, and tables (Box 3.5).

The geodatabase model stores the geometries of spatial features in a field. Therefore, instead of storing spatial data and attribute data in a split data system, the geodatabase model stores both data in a single database. One advantage of this data structure is that it eliminates the complexity of coordinating disparate data stores, thus reducing the processing overhead.

Another major advantage of the geodatabase model is that objects in the database can have properties, methods, and relationships. ArcGIS, for example, provides four general validation rules for the grouping of objects: attribute domains, relationship rules, connectivity rules, and custom rules (Zeiler 1999). Attribute domains group objects into

Box 3.5   Coverage, Shapefile, and Geodatabase

T he following table compares the terms used to describe geometries of the geodatabase, coverage, and shapefile.

Geodatabase	Shapefile	Coverage
Point	Point	Point
Multipoints	Multipoints	NA
Polyline	Polyline	Arc
Polygon	Polygon	Polygon
Composite polygon	Composite polygon	Region
Polyline with measure	Polyline with measure	Route
Feature class	Shapefile	Point, arc, polygon, tic
Feature dataset	NA	Coverage

Box 3.6   Topology for the Geodatabase

E SRI Inc. plans to deliver 25 topology rules with ArcGIS 8.3. The coverage model, as discussed in the text, only includes continuity, area definition, and contiguity. These topology rules may be applied to features in a feature class or to features between two or more feature classes in the same feature dataset. The latter case means that the user can apply topology rules to two or more maps such as county, census tract, and block group maps to make sure that their boundaries are shared (coincident).

classes, or subtypes, by a valid range of values or a valid set of values for an attribute. Relationship rules connect objects that are associated. Connectivity rules let users build geometric networks such as streams, roads, and water and electric utilities. And custom rules allow users to create custom features for advanced applications. To paraphrase a promotional ad, the object-oriented data model turns "dumb" lines into "smart" objects.

Two new developments with the geodatabase model are worth watching. The first is topology: the upcoming ArcGIS 8.3 will add topology to the geodatabase model. Unlike the built-in topology for the coverage model, the geodatabase model will allow the user to select the topological relationships needed and implement them as rules. For example, one rule may state that all roads must be connected and another may state that no gaps are allowed between soil polygons. Other differences include the number of topology rules available and the target of topology rules (Box 3.6).

The second is the development of geodatabase models for specific industries and applications. Real-world objects all have different properties and behaviors. It is therefore impossible to apply, for example, the behavior and properties of transportation-related objects to forestry-related objects. In other words, one cannot expect to have a "universal" geodatabase model. A solution is to develop geodatabase models for specific fields. ESRI Inc. has set up a website that updates the progress in the development of models for forestry, transportation, hydrology, land parcel, environmental facilities, and other fields (**http://www. esri.com/software/arcgisdatamodels/**).

3.7 LOCATION DATA ACCURACY

Because the vector data model uses points and their *x*-, *y*-coordinates to represent discrete spatial features, an important issue for vector data is accuracy. For example, one may ask about the accuracy of a land parcel's boundary on a map. **Location data accuracy** defines how close the apparent location of a map feature is to its true ground location. Map scale is an indicator of map accuracy on paper maps. The accuracy of a map feature is less reliable on a 1:100,000 scale map than on a 1:24,000 scale map. Map scale also influences the level of detail on a map. As the map scale becomes smaller, the amount of map detail decreases, and the degree of line generalization increases (Monmonier 1996). For example, a meandering stream on a large-scale map becomes less sinuous on a small-scale map.

Although paper maps are still the most common source for spatial data entry, new data entry methods using global positioning systems (GPS) and remote sensing imagery can bypass maps and the practice of various methods of map generalization. The accuracy of spatial data collected by GPS or satellite images is directly related to the resolution of the measuring instrument. The spatial resolution of satellite images can range from 1 meter to 1 kilometer. Similarly, the spatial resolution of GPS point data can range from several millimeters to 100 meters. Map scale does not apply to GPS and imagery data.

It is also difficult to apply map scale to digital spatial data because they can be easily manipulated and output to any scale. Increasingly, GIS users must turn to accuracy standards and reports to determine the quality of spatial data.

3.7.1 Spatial Data Accuracy Standards

In the United States, the development of spatial data accuracy standards has gone through three phases. Revised and adopted in 1947, the U.S. National Map Accuracy Standard (NMAS) sets the accuracy standard for published maps such as topographic maps from the USGS (U.S. Bureau of the Budget 1947). The standards for horizontal accuracy require that no more than 10 percent of the well-defined map points tested shall be more than 1/30 inch at scales larger than 1:20,000, and 1/50 inch at scales of 1:20,000 or smaller. This means that the threshold value is 40 feet on the ground for 1:24,000 scale maps and about 167 feet on the ground for 1:100,000 scale maps.

In 1990 the American Society for Photogrammetry and Remote Sensing (ASPRS) published accuracy standards for large-scale maps (American Society for Photogrammetry and Remote Sensing 1990). The ASPRS defines the horizontal accuracy in terms of the root mean square (RMS) error, a measure derived by comparing map coordinate values and coordinate values from an independent source of higher accuracy for identical points (Box 3.7). Examples of a higher-accuracy data source include digital or hard-copy map data, GPS, or survey data. The ASPRS standards stipulate that the threshold RMS error for class 1 maps is, for example, 16.7 feet for 1:20,000 scale maps and 2 feet for 1:2,400 scale maps.

In 1998 the Federal Geographic Data Committee (FGDC), a committee representing 17 federal agencies, established the National Standard for Spatial Data Accuracy (NSSDA) to replace the NMAS. The NSSDA follows the ASPRS accuracy standards but extends to map scales smaller than 20,000 (Federal Geographic Data Committee 1998). A major difference between the NSSDA and NMAS or the ASPRS accuracy standards is that the NSSDA omits threshold accuracy values that spatial data, including paper maps and digital data, must achieve. Instead, agencies are encouraged to establish accuracy thresholds for their products and to report the NSSDA statistic, a statistic based on the RMS error (Box 3.7). Given the new accuracy standards, GIS users must ask for the product spec-

Box 3.7 | **National Standard for Spatial Data Accuracy (NSSDA) Statistic**

To use the ASPRS standards or the NSSDA statistic, one must first compute the root mean square (RMS) error, which is defined by

$$RMS = \sqrt{\Sigma[(x_{data,i} - x_{check,i})^2 + (y_{data,i} - y_{check,i})^2]/n}$$

where $x_{data,i}$ and $y_{data,i}$ are the coordinates of the ith check point in the data set; $x_{check,i}$ and $y_{check,i}$ are the coordinates of the ith check point in the independent source of higher accuracy; n is the number of check points tested; and i is an integer ranging from 1 to n.

The NSSDA suggests that a minimum of 20 check points shall be tested. After the RMS is computed, it is multiplied by 1.7308, which represents the standard error of the mean at the 95 percent confidence level. The product is the NSSDA statistic. A handbook on how to use NSSDA to measure and report geographic data quality has been published by the Land Management Information Center at Minnesota Planning (**http://www.lmic.state.mn.us/**).

ifications and determine if the level of data accuracy is acceptable for their applications.

3.7.2 Location Accuracy and Topological Accuracy

Topological accuracy defines how well the topological relationships between spatial features are maintained. Map scale and the data entry process largely determine the location accuracy of spatial features. Topological accuracy, on the other hand, depends on data entry, the capability of a GIS package to detect errors, and the ability of a GIS data producer to remove errors.

3.7.3 Location Data Accuracy and Precision

Location data accuracy measures how close the recorded location of a spatial feature is to its ground location, whereas **precision** measures how exactly the location is recorded. The number of significant digits used in data recording expresses the precision of a recorded location. For example, distances may be measured with decimal digits or rounded off to the nearest meter or foot. The precision of location data stored in the computer is dictated by the hardware's word length (single or double precision) and by whether the data are stored as integers or floating points. Coverage coordinates in ArcInfo Workstation, for example, are stored as either single-precision real numbers with 6 to 7 significant digits or double precision with 13 to 14 significant digits.

KEY CONCEPTS AND TERMS

Arc: A line segment with two end points.

Area: A spatial feature that is represented by a series of lines and has the geometric properties of size and perimeter. Also called *polygon*, *face*, or *zone*.

Area definition: An area is defined by a series of connected arcs.

Connectivity: A topological relationship that stipulates that arcs connect to each other at nodes.

Contiguity: Arcs have directions and left and right polygons.

Coverage: The standard, topological vector data format used in ESRI products.

Delaunay triangulation: An iterative process of connecting points with their two nearest neighbors to form triangles as equiangular as possible in a triangulated irregular network (TIN).

Dynamic segmentation: A data model that is built upon arcs of a line coverage and allows the use of coordinates with linear measures.

Event: A basic element of the dynamic segmentation model that associates attribute data such as pavement conditions, accidents, and speed limits with routes.

Georelational data model: A GIS data model that stores spatial data and attribute data in two separate but related file systems.

Graph theory: A subfield of mathematics that uses diagrams or graphs to study the arrangements of objects and the relationships between objects.

Integrated coverage: A polygon coverage that contains the geometries of the built-in region layers.

Line: A spatial feature that is represented by a series of points and has the geometric properties of location and length. Also called *arc, edge, link,* or *chain.*

Location data accuracy: A measure of how close the apparent location of a map feature is to its true ground location.

Map scale: A ratio of map distance to ground distance.

Node: The beginning or end point of a line.

Object: An entity such as a land parcel that has a set of properties and can perform operations upon requests.

Object-oriented data model: A data model that uses objects to organize spatial data.

Point: A spatial feature that is represented by a pair of coordinates and has only the geometric property of location. Also called *node* or *vertex.*

Precision: A measure of how exactly the location data are recorded.

Regions data model: A data model that is built upon lines and polygons and allows disjoint components and overlapped areas.

Route: A basic element of the dynamic segmentation model that uses a collection of sections to represent a linear phenomenon such as a highway, bike path, or stream.

Section: A basic element of the dynamic segmentation model that refers directly to arcs and positions along arcs.

Shapefile: The standard, nontopological vector data format used in ESRI products.

Topological accuracy: An assessment of how well the topological relationships between spatial features are maintained.

Topology: A subfield of mathematics that studies properties of geometric objects, which remain invariant under certain transformations such as bending or stretching.

Triangulated irregular network (TIN):   A vector data model that approximates the terrain with a set of nonoverlapping triangles.

Vector data model: A data model that uses points and their *x*-, *y*-coordinates to construct spatial features.

APPLICATIONS: THE VECTOR DATA MODEL

This applications section consists of three tasks. In Task 1, you will import an ArcInfo interchange file to a coverage, convert the coverage to a geodata-base feature class, and convert the geodatabase feature class to a shapefile. Unlike ArcInfo, ArcView does not have a tool for converting a coverage directly to a shapefile. Part of Task 1 is to examine the data structure of a coverage, a geodatabase feature class, and a shapefile. Task 2 lets you view the data models of TIN, regions, and dynamic segmentation in ArcCatalog. In Task 3, you will use the data models of TIN, regions, and dynamic segmentation in ArcInfo Workstation.

Task 1: Examine the Data File Structure of Coverage, Geodatabase Feature Class, and Shapefile

What you need: *land.e00*, an ArcInfo interchange file.

In Task 1, you will use the interchange file *land.e00* to create a coverage, a geodatabase feature class, and a shapefile. You can then examine the data file structure of these three data types.

1. Start ArcCatalog, and make connection to the Chapter 3 database. Launch ArcToolbox. Double-click on Conversion Tools, Import to Coverage, and ArcView Import from Interchange File.

2. Use the browse button to add *land.e00* as the input file. Save the Output data set as *land_arc* in the Chapter 3 database. Click OK.

3. *Land_arc* now appears in the Catalog tree. (If not, select Refresh from the View menu in ArcCatalog.) Click the plus sign next to *land_arc*. *Land_arc* contains four feature classes: *arc*, *label*, *polygon*, and *tic*. You can preview each class. *Arc* shows lines that make up *land_arc*. *Label* shows label points for each polygon in *land_arc*. *Polygon* shows polygons in *land_arc*. And *tic* shows the tics or control points in *land_arc*.

4. Right-click *land_arc* in the Catalog tree and select Properties. The Coverage Properties dialog shows the tabs of General, Projection, Tics and Extent, and Tolerances. Right-click *polygon* and select Properties. The Coverage Feature Class Properties dialog shows the tabs of General, Items, and Relationships.

5. Data files associated with *land_arc* reside in two folders in the Chapter 3 database:

Land_arc and INFO. You can use the Windows Explorer to view these files. The Land_arc folder contains files such as arc.adf, bnd.adf, pat.adf, tic.adf, and so on. The INFO folder, which is shared by other coverages in the same database, contains files such as arc0000.dat, arc0000.nit, and so on. All these files are binary files and cannot be read.

6. This step is to create a personal geodatabase. Right-click the Chapter 3 database in the Catalog tree, point to New, and select Personal Geodatabase. *New Personal Geodatabase.mdb* should appear in the Catalog tree. (If the extension .mdb does not appear, do the following: select Options from the Tools menu, click the General tab, and uncheck the box next to Hide file extensions at the bottom of the Options dialog.) Change the name of the new personal geodatabase to *land.mdb*.

7. Now you can add a new feature class to *land.mdb*. Right-click *land.mdb* in the Catalog tree, point to Import, and select Coverage to Geodatabase Wizard. In the first panel, enter *land_arc* by using the browse button or the drag-and-drop method and make sure that the box for polygon (the feature class to convert) is checked. In the second panel, opt to create a stand-alone feature class and click Next. In the third panel, accept default parameters and click Next. Click Finish to dismiss the Coverage to Geodatabase Wizard dialog.

8. Click the plus sign next to *land.mdb*. *Land_arc_polygon* is the polygon feature class added to the geodatabase. (If *land_arc_polygon* does not appear, select Refresh from the View menu.) Click *land_arc_polygon*, and you can preview the geography and table of the feature class.

9. Right-click *land.mdb* in the Catalog tree and select Properties. The Database Properties dialog shows Domains. A domain can be used to establish valid values or ranges of values for an attribute to minimize data entry

errors. Right-click *land_arc_polygon* and select Properties. The Feature Class Properties shows the tabs of General, Fields, Indexes, Subtypes, and Relationships. Subtypes allow a feature class to be subdivided into different classifications, each with a special attribute. Relationships can relate objects in a subtype to objects in another subtype.

10. You can use the Windows Explorer to find *land.mdb* in the Chapter 3 database. *Land.mdb* is a Microsoft Access database. Double-click *land.mdb* to see the contents of the database. Among the tables is *land_arc_polygon*. You can view the table by double-clicking on it. The table is the same as the table you have previewed in the previous step.

11. This step is to convert *land_arc_polygon* to a shapefile. Right-click *land_arc_polygon*, point to Export, and select Geodatabase to Shapefile. In the next dialog, make sure that *land_arc_polygon* is the feature class to convert and enter *land.shp* as the output shapefile. Click OK. *Land.shp* should be added to the Catalog tree. If not, select Refresh from the View menu.

12. Right-click *land.shp* in the Catalog tree and select Properties. The Shapefile Properties dialog shows the tabs of General, Fields, and Index.

13. The shapefile *land* is associated with several other files including *land.dbf* and *land.shx*. You can use the Windows Explorer to view these files in the Chapter 3 database. *Land.dbf* is a dBASE file containing the attributes of the shapefile, and *land.shx* is an index file.

Task 2: View TIN, Regions, and Dynamic Segmentation in ArcCatalog

What you need: *emidatin*, a TIN prepared from a digital elevation model; *fire*, a polygon coverage with region subclasses; and *highway*, a network coverage based on the dynamic segmentation model.

Task 2 lets you view the data models of TIN, regions, and dynamic segmentation and their data structure in ArcCatalog.

1. Make sure that ArcCatalog is connected to the Chapter 3 database. Click *emidatin* in the Catalog tree, and preview the TIN. You can use the Zoom In tool to view a small area of *emidatin* and the triangles that make up the area.

2. Click the plus sign next to *fire*. Seven feature classes appear. *Arc*, *label*, *polygon*, and *tic* are separate features of the polygon coverage. *Region.fire1*, *region.fire12*, and *region.fire123* are separate region layers built on the polygon coverage. You can preview each feature class.

3. Click the plus sign next to *highway*. Five feature classes appear. *Route.fastroute* is a route built upon the arcs of *highway*. You can preview both *arc* and *route.fastroute*. The table of *route.fastroute* lists only the ID value and does not contain attribute data.

Task 3: Use TIN, Regions, and Dynamic Segmentation in ArcInfo Workstation

What you need: *emidatin*, a TIN prepared from a digital elevation model; *fire*, a polygon coverage with region subclasses; and *highway*, a network coverage.

Task 3 lets you work with the data models of TIN, regions, and dynamic segmentation using ArcInfo Workstation. You will examine the data structure of these data models. You will also create a route system on a network.

1. TIN

Like a coverage, a TIN is stored as a directory in ArcInfo Workstation. Change the workspace to *emidatin* and list data files in the directory. You should see the edge list (tedg.adf), node list (tnod.adf), node/xy list (tnxy.adf), and node/z list (tnz.adf). Change the workspace to the level above *emidatin*. Go to Arcplot and type the following commands:

Arcplot: display 9999
Arcplot: mapextent emidatin
Arcplot: tin emidatin

The TIN, *emidatin*, is displayed as a network of triangles. Triangles near the border are elongated rather than compact like those inside the border. This is caused by a lack of data points near the border. Unlike a coverage, a TIN does not have INFO files associated with it. Therefore you cannot further examine the TIN data model.

2. Regions

The coverage *fire* contains areas that have been burned once, and some twice or even three times. The regions data model is ideal for working with overlapped polygons, such as overlapped polygons of past forest fires. Change the workspace to *fire* and list data files in the directory. Three region subclasses reside in the directory. *Fire1* consists of areas burned at least once, *fire12* consists of areas burned at least twice, and *fire123* consists of areas burned three times. Each region subclass has a set of data files including the region–arc list (pal) and the region–polygon list (rxp). Change the workspace to the level above *fire*.

To display the three region subclasses, go to Arcplot and type the following commands:

Arcplot: mapextent fire
Arcplot: arcs fire /*shows polygon boundaries of the coverage *fire*
Arcplot: regionshades fire fire1 3 /*shade areas burned at least once in green
Arcplot: regionshades fire fire12 4 /*shade areas burned at least twice in blue
Arcplot: regionshades fire fire123 2 /*shade areas burned three times in red

The three subclasses form nested regions on the computer monitor: *fire123* is nested within *fire12*, and *fire12* is nested within *fire1*. Exit ArcPlot.

Next, go to Tables and type the following commands:

Tables: select fire.patfire1
Tables: list

The list shows those areas that were burned at least once and the year of the fire.

Tables: select fire.patfire12
Tables: list

The list shows those areas that were burned at least twice and the years (under fire1 and fire2) of the fires.

Tables: select fire.patfire123
Tables: list

Finally, the display shows areas that were burned three times and the years of those fires.

3. Dynamic Segmentation

For this part of Task 3, you will define a route system on a network first and then examine the data structure for the route system. The coverage *highway* shows the highways in Idaho. To define a route system, go to ArcPlot and view the transportation network:

Arcplot: mapextent highway
Arcplot: arcs highway /* plot the network
Arcplot: mapextent * /* zoom in a portion of the network
Arcplot: clear
Arcplot: arcs highway
Arcplot: markercolor 2 /* set the node color as red
Arcplot: nodes highway

Now you can define a route system on *highway* by typing the following commands:

Arcplot: netcover highway fastroute /* called the route system on *highway* as *fastroute*
Arcplot: path * /* path is the command to define a route system interactively
Enter point
Enter point
9 to exit
/* Enter two nodes as the beginning and end nodes of the route system. Enter 9 to exit.
Arcplot: routelines highway fastroute 3 /* draw *fastroute* in green
Arcplot: sectionlines highway fastroute /* draw sections of *fastroute* in different colors

The display shows the relationships between arcs, sections, and the route system: Sections are based on arcs in *highway*, and *fastroute* is a collection of sections.

To view the data structure of the route system, exit ArcPlot and change the workspace to *highway* at the Arc prompt. List data files in the *highway* directory. You should see the route (rat) and section (sec) data files for *fastroute*. Change the workspace to the level above *highway*. Go to Tables and type the following commands:

Tables: select highway.secfastroute
Tables: items

You should see items that were discussed under the dynamic segmentation data model in the chapter.

Tables: list

The first record in the listing should have the F-MEAS value of 0 because this is the beginning point of *fastroute*. The T-MEAS value in the first record represents the length of the first section or arc in real-world measurement units (feet, in this case). The second record is for the second section in *fastroute*. Its T-MEAS value should be the sum of the lengths of the first two sections. This continues until the last record whose T-MEAS value should be the total length of the route system. The F-POS value of each record should be either 0 or 100, depending on the direction of the arc. But in either case each section should cover 100% of the arc because the route system was defined between two nodes on a network.

Tables: select highway.ratfastroute
Tables: items

You should see the items of FASTROUTE# and FASTROUTE-ID. Use the LIST command to see the item values. The DROPFEATURES command in Arc can remove the route system and its sections.

REFERENCES

American Society for Photogrammetry and Remote Sensing. 1990. ASPRS Accuracy Standards for Large-Scale Maps. *Photogrammetric Engineering and Remote Sensing* 56: 1068–70.

Broome, F. R., and D. B. Meixler. 1990. The TIGER Data Base Structure. *Cartography and Geographic Information Systems* 17: 39–47.

Chang, K., D. L. Verbyla, and J. J. Yeo. 1995. Spatial Analysis of Habitat Selection by Sitka Black-Tailed Deer in Southeast Alaska. *Environmental Management* 19: 579–89.

Egenhofer, M. F., and A. Frank. 1992. Object-Oriented Modeling for GIS. *Journal of the Urban and Regional Information System Association* 4: 3–19.

Environmental Systems Research Institute, Inc. 1998. *Understanding GIS: The ARC/INFO Method.* Redlands, CA: ESRI Press.

Federal Geographic Data Committee. 1998. Part 3: *National Standard for Spatial Data Accuracy, Geospatial Positioning Accuracy Standards*, FGDC-STD-007.3-1998. Washington, DC: Federal Geographic Data Committee. Also available at **http://www. fgdc.gov/standards/documents/ standards/accuracy/chapter3. pdf/.**

Larman, C. 1997. *Applying UML and Patterns: An Introduction to Object-Oriented Analysis and Design.* Upper Saddle River, NJ: Prentice Hall PTR.

Laurini, R., and D. Thompson. 1992. *Fundamentals of Spatial Information Systems.* London: Academic Press.

Massey, W. S. 1967. *Algebraic Topology: An Introduction.* New York: Harcourt, Brace & World.

Monmonier, M. 1996. *How to Lie with Maps,* 2d ed. Chicago: Chicago University Press.

Robinson, A. H., J. L. Morrison, P. C. Muehrcke, A. J. Kimerling, and S. C. Guptill. 1995. *Elements of Cartography,* 6th ed. New York: Wiley.

Theobald, D.M. 2001. Topology Revisited: Representing Spatial Relations. *International Journal of Geographical Information Science* 15: 689–705.

Tsai, V. J. D. 1993. Delaunay Triangulations in TIN Creation: An Overview and a Linear Time Algorithm. *International Journal of Geographical Information Systems* 7: 501–24.

Ungerer, M.J., and M.F. Goodchild. 2002. Integrating Spatial Data Analysis and GIS: A New Implementation Using the Component Object Model (COM). *International Journal of Geographical Information Science* 16: 41–53.

U.S. Bureau of the Budget. 1947. *United States National Map Accuracy Standards*. Washington, DC: U.S. Bureau of the Budget.

Watson, D. F., and G. M. Philip. 1984. Systematic Triangulations. *Computer Vision, Graphics, and Image Processing* 26: 217–23.

Wilson, R. J., and J. J. Watkins. 1990. *Graphs: An Introductory Approach*. New York: Wiley.

Worboys, M. F. 1995. *GIS: A Computing Perspective*. London: Taylor & Francis.

Worboys, M. F., H. M. Hearnshaw, and D. J. Maguire. 1990. Object-Oriented Data Modelling for Spatial Databases. *International Journal of Geographical Information Systems* 4: 369–83.

Zeiler, M. 1999. *Modeling Our World: The ESRI Guide to Geodatabase Design*. Redlands, CA: ESRI Press.

Zeiler, M., ed. 2001. *Exploring ArcObjects*. Redlands, CA: ESRI Press.

CHAPTER

4

VECTOR DATA INPUT

4.1 INTRODUCTION

The most expensive part of a GIS project is database construction. Converting from paper maps to digital maps used to be the first step in constructing a database. But in recent years this situation has changed as digital data clearinghouses have become commonplace on the Internet. Now the strategy for a GIS user is to look at what exists in the public domain before deciding to create new data. Private companies have also entered the GIS market. Some of them produce new GIS data for their customers; others produce value-added GIS data from public data.

The proliferation of available GIS data has made it easier to organize a GIS project but has not reduced the importance of data input in GIS. GIS data must still be produced to be put on the Internet or to be sold to customers. GIS users may not find the digital data they want and may have to create their own data. New GIS data can be created from satellite images, GPS (global positioning system) data, street addresses, or paper maps. GIS users can also choose manual digitizing or scanning for converting paper maps to digital maps.

A newly digitized map has the same measurement unit as the source map used in digitizing or scanning. To make the digitized map useful in a GIS project, it must be converted to real-world coordinates through geometric transformation. Using a set of control points in geometric transformation of a digitized map makes the process somewhat uncertain. The root mean square (RMS) error helps determine the quality of transformation.

This chapter is presented in the following five sections. Section 4.2 discusses existing GIS data on the Internet, including examples from different levels of government and private companies. Sections 4.3 and 4.4 cover metadata and the data exchange methods, topics important to the use of existing data. Section 4.5 provides an overview of creating new GIS data from satellite images, GPS data, survey data, street addresses, text files with x-, y-coordinates, and paper maps, and the methods of manual digitizing and scanning. Section 4.6 concludes the chapter with geometric transformation and the interpretation of the RMS error.

4.2 EXISTING GIS DATA

To find existing GIS data for a project is often a matter of knowledge, experience, and luck. Government agencies at different levels have set up websites for sharing public data and for directing users to the source of the desired information (Onsrud and Rushton 1995). The Internet is also a medium for finding existing data from nonprofit organizations and private companies. But searching for GIS data on the Internet can be difficult. A keyword search with GIS will probably result in thousands of matches, but most hits are irrelevant to the user. Internet addresses may be changed or discontinued. Data on the Internet may be in a format that is incompatible with the GIS package used for a project, or to be usable for a project, the data may need extensive processing such as clipping the study area from a large data set.

Most GIS data on the Internet are data that many organizations regularly use for GIS activities. These are called **framework data**, which typically include seven basic layers: geodetic control (accurate positional framework for surveying and mapping), orthoimagery (rectified imagery such as ortho photos), elevation, transportation, hydrography, governmental units, and cadastral information (**http://www.fgdc.gov/framework/frame work.html/**).

4.2.1 Public Data

4.2.1.1 Federal Geographic Data Committee
Public data are often free and downloadable from the Internet. All levels of government let GIS users access their public data through clearinghouses in the United States. To GIS users looking for data, the website maintained by the **Federal Geographic Data Committee (FGDC) (http://www. fgdc.gov/)** is a good start. Consisting of 17 federal agencies, the FGDC coordinates the development of the National Spatial Data Infrastructure (NSDI), which is aimed at the sharing of geospatial data throughout all levels of government, the private and nonprofit sectors, and the academic community. The FGDC website provides a link to the

Geospatial Data Clearinghouse, a collection of more than 250 spatial data nodes in the United States and overseas. Also available is a listing of websites that provide mostly free U.S. geospatial and attribute data.

4.2.1.2 U.S. Geological Survey

Through its national mapping program, the U.S. Geological Survey (USGS) is the major provider of GIS data in the United States. Its website (**http://mapping.usgs.gov/**) offers pathways to USGS national mapping and remotely sensed data and to thematic data clearinghouses on national biological information, geologic information, and water resources information. One can download from the USGS website **digital line graphs (DLGs)** and land-use/land-cover data.

DLGs are digital representations of point, line, and area features from the USGS quadrangle maps at the scales of 1:24,000, 1:100,000, and 1:2,000,000. DLGs include such data categories as hypsography (i.e., contour lines and spot elevations), hydrography, boundaries, transportation, and the U.S. Public Land Survey System. DLGs contain attribute data and are topologically structured.

The USGS compiled land-use/land-cover (LULC) data in the 1970s and early 1980s from aerial photographs and earlier land-use maps and field surveys. Level-1 LULC data have nine categories: urban or built-up land, agricultural land, rangeland, forested land, water, wetland, barren land, tundra, and perennial snow and ice. A modified Anderson level II scheme subdivides each level-1 category (Anderson et al. 1976). For example, level-2 streams and canals, lakes, reservoirs, and bays and estuaries are the subdivisions of level-1 water. LULC data use a minimum-mapping unit of 10 acres or 4 hectares. These data are available at scales of 1:250,000 and 1:100,000 and sorted by state.

4.2.1.3 U.S. Census Bureau

The U.S. Census Bureau offers the TIGER/Line files, which are extracts of geographic/cartographic information from its **TIGER (Topologically Integrated Geographic Encoding and Referencing)** database. The TIGER/Line files contain legal and statistical area boundaries such as counties, census tracts, and block groups, which can be linked to the census data, as well as roads, railroads, streams, water bodies, power lines, and pipelines (Sperling 1995). TIGER/Line attributes include the address range on each side of a street segment that can be used for address matching.

Several versions of the TIGER/Line files are available for download at the Census Bureau's website (**http://www.census.gov/**). The latest version corresponds to Census 2000 and covers all counties and county equivalents for each state.

4.2.1.4 Statewide Public Data: An Example

Most states in the United States have designated websites for statewide GIS data (Appendix C). The Montana State Library, for example, maintains a website on GIS data in Montana (**http://www.nris.state.mt.us/**). This website contains both statewide and regional data. Statewide data include such categories as people/places, land/water, administration, map sheets/grids, and environmental monitoring. These data are available in ArcInfo export files and shapefiles for downloading.

4.2.1.5 Regional Public Data: An Example

The Greater Yellowstone Area Data Clearinghouse (GYADC) (**http://www.sdvc.uwyo.edu/gya/**) is a FGDC data node sponsored by a group of federal agencies, state agencies, universities, and nonprofit organizations. This data clearinghouse is unique in that it focuses on basic framework data for Yellowstone and Grand Teton national parks.

4.2.1.6 Metropolitan Public Data: An Example

Sponsored by 19 local governments in the San Diego region, the San Diego Association of Governments (SANDAG) (**http://www.sandag.cog.ca.us/**) is an example of a metropolitan data clearinghouse. Data that can be downloaded free from SANDAG's website include administrative boundaries, base map features, district boundaries, land cover and activity centers, transportation, and sensitive lands/natural resources.

4.2.2 GIS Data from Private Companies

Many GIS companies are engaged in software development, technical service, consulting, and data production. Some also provide free data or samples of commercial data on the Internet or can direct GIS users to suitable sources. MapInfo offers sample data at (**http://www.mapinfo.com/free/index.cfm/.** ESRI Inc. offers the Geography Network (**http://www.geographynetwork.com/**), a clearinghouse with data provided by organizations worldwide that can be accessed directly from ArcMap.

Some companies provide specialized GIS data for their customers. GDT (geographic data technology) offers street, address, postal, and census databases at its website (**http://www.geographic.com/**). Tele Atlas North America (**http://www.etak.com/**) offers road and address databases for urban centers and rural areas. In contrast, online GIS data stores tend to carry a variety of digital spatial data. Examples of GIS data stores include GIS Data Depot (**http://data.geocomm.com/**), Map-Mart (**http://www.mapmart.com/**), and LAND INFO International (**http://www.landinfo.com/**).

4.3 METADATA

Metadata provide information about spatial data (Guptill 1999). They are particularly important to GIS users who want to use public data for their projects. First, metadata let GIS users know if the data meet their specific needs for area coverage, data quality, and data currency. Second, metadata show GIS users how to transfer, process, and interpret spatial data. Third, metadata include the contact information for GIS users who need more information.

The FGDC has developed the content standards for metadata. These standards have been adopted by federal agencies in developing their public data. FGDC metadata standards describe a data set based on the following:

- Identification information—basic information about the data set, including title, geographic data covered, and currency.

- Data quality information—information about the quality of the data set, including positional and attribute accuracy, completeness, consistency, sources of information, and methods used to produce the data.
- Spatial data organization information—information about the data representation in the data set such as method for data representation (e.g., raster or vector) and number of spatial objects.
- Spatial reference information—description of the reference frame for and means of encoding coordinates in the data set such as the parameters for map projections or coordinate systems, horizontal and vertical datums, and the coordinate system resolution.
- Entity and attribute information—information about the content of the data set, such as the entity types and their attributes and the domains from which attribute values may be assigned.
- Distribution information—information about obtaining the data set.
- Metadata reference information—information on the currency of the metadata information and the responsible party.

The FGDC website (**http://www.fgdc.gov/metadata/metadata.html/**) contains detailed information about the above metadata standards. Appendix D shows a complete example of metadata that describe a map downloaded from a statewide clearinghouse. Each of the entries in Appendix D must be collected from the appropriate information resources and entered.

FGDC standards do not specify how metadata should be organized in a computer system or in a data transfer, or how metadata should be transmitted to the user. To assist the entry of metadata, many metadata tools have been developed over the past several years (**http://www.fgdc.gov/metadata/metatool.html/**). These tools are typically designed for different operating systems. Some tools are free and some are designed for specific GIS packages. ESRI Inc. offers a metadata creation tool in ArcCatalog. Intergraph offers the Spatial Metadata Management System (SMMS) for GeoMedia.

4.4 CONVERSION OF EXISTING DATA

Data conversion, also called data transfer, converts data obtained from government or private sources to a format compatible with a GIS package. Because a large variety of GIS packages and data formats are in use, conversion of existing data can be difficult at times. The choice of a conversion method basically depends upon the specificity of the data format. Proprietary data formats require special translators for data conversion, whereas neutral or public formats require a GIS package to have translators that can work with the formats (Robinson et al. 1995).

4.4.1 Direct Translation

Direct translation uses a translator or algorithm in a GIS package to directly convert spatial data from one format to another (Figure 4.1). Direct translation used to be the only method for data conversion before the development of data standards and open GIS. GIS users still prefer direct translation because it is easier to use than other methods. The ArcToolbox application of ArcGIS, for example, can translate ArcInfo's interchange files, MGE and Microstation's DGN files, Auto-CAD's DXF and DWG files, and MapInfo files into shapefiles. Likewise, GeoMedia can access and integrate data from ArcGIS, AutoCAD, MapInfo, MGE, and Microstation.

4.4.2 Neutral Format

A **neutral format** is a public or de facto format for data exchange. DLG is a familiar neutral format to GIS users. The USGS provides DLG files, and other government agencies such as the Natural Resources Conservation Service (NRCS) distribute digital soil maps in the DLG format. This is because most GIS packages have translators to work with the DLG format (Figure 4.2).

Figure 4.1
The IGDSARC command in ArcInfo Workstation converts a DGN file prepared by an Intergraph user to an ArcInfo coverage.

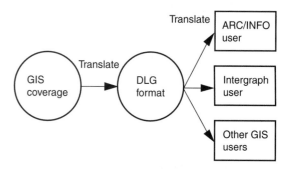

Figure 4.2
To accommodate users of different GIS packages, a government agency can translate public data into a neutral format such as the DLG format. Using the translator in the GIS package, the user can convert the neutral data file into the format used in the GIS.

The **Spatial Data Transfer Standard (SDTS)** is a neutral format approved by the Federal Information Processing Standards (FIPS) Program in 1992 **(http://mcmcweb.er.usgs.gov/sdts/).** SDTS is now mandatory for federal agencies, and will probably become the standard format for public data from state and local agencies as well. GIS vendors have already developed translators for SDTS data.

In practice, SDTS uses "profiles" to transfer spatial data. Each profile is targeted at a particular type of spatial data. Currently there are five SDTS profiles:

- The Topological Vector Profile (TVP) covers DLG, TIGER, and other topology-based vector data.

- The Raster Profile and Extensions (RPE) accommodate digital orthophotos, digital elevation models, and other raster data.
- The Transportation Network Profile (TNP) covers vector data with network topology.
- The Point Profile supports geodetic control point data.
- The Computer Aided Design and Drafting Profile (CADD) supports vector-based CADD data, with or without topology.

The **vector product format (VPF)** is a standard format, structure, and organization for large geographic databases that are based on a georelational data model. The National Imagery and Mapping Agency (NIMA) uses VPF for digital vector products developed at a variety of scales (**http://www.nima.mil/**). NIMA's vector products for drainage system, transportation, political boundaries, and populated places are also part of the global database that is being developed by the International Steering Committee for Global Mapping (ISCGM) (**http://www.iscgm.org/**).

Although a neutral format is typically used for public data from government agencies, it can also be found with "industry standards" in the private sector. A good example is the DXF (drawing interchange file) format of AutoCAD. Another example is the ASCII format. Many GIS packages can import ASCII files, which have point data with x-, y-coordinates, into digital maps.

4.5 CREATING NEW DATA

This section covers a variety of data sources that can be used to create new digital spatial data.

4.5.1 Remotely Sensed Data

Remotely sensed data such as digital orthophoto quads (DOQs) and satellite images are data acquired by a sensor from a distance. Although they are raster data, DOQs and satellite images are useful for vector data input. DOQs are digitized aerial photographs that have been differentially rectified or corrected to remove image displacements by

camera tilt and terrain relief. Black-and-white DOQs have a 1-meter ground resolution (i.e., each pixel in the image measures 1-by-1 meter on the ground) and have pixel values representing 256 gray levels. Because DOQs combine the image characteristics of a photograph with the geometric qualities of a map, they can be effectively used as a background for digitizing or updating of existing digital maps (Figure 4.3). GIS users can process satellite images and extract data for a variety of maps in vector format such as land use and land cover, hydrography, water quality, and areas of eroded soils.

4.5.2 Global Positioning System (GPS) Data

Using satellites in space as reference points, a GPS receiver can determine its precise position on the Earth's surface (Moffitt and Bossler 1998). **GPS data** include the horizontal location based on the geographic grid or a coordinate system and, if chosen, the elevation of the point location (Box 4.1). When taking GPS positions along a line such as a road, one can determine the line feature by the collection of positional readings. By extension, a polygon can be determined by a series of lines measured by GPS. This is why GPS has become a useful tool for spatial data input (Kennedy 1996).

The method by which a GPS receiver determines its position on the Earth's surface is similar to triangulation in surveying. The GPS receiver measures its distance (range) from a satellite using the travel time of radio signals it receives from the satellite. With three satellites simultaneously available, the receiver can determine its position in space (x, y, z), relative to the center of mass of the Earth. But to correct timing errors, a fourth satellite is required to get precise positioning (Figure 4.4). The receiver's position in space can then be converted to latitude, longitude, and height based on a reference spheroid.

The U.S. military maintains a constellation of 24 NAVSTAR (Navigation Satellite Timing and Ranging) satellites in space, and each satellite follows a precise orbit. This constellation gives GPS

Figure 4.3
A digital orthophoto (DOQ) can be used as the background for digitizing or updating of existing maps.

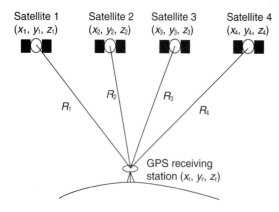

Figure 4.4
Use four GPS satellites to determine the coordinates of a receiving station. x_i, y_i, and z_i are coordinates relative to the center of mass of the Earth. R_i represents the distance (range) from a satellite to the receiving station.

An important aspect of using GPS for spatial data entry is the need to correct errors in GPS data. The first type of error may be described as noise, which can come from different sources. Although a satellite orbit is precisely defined, it may involve slight position or ephemeris (orbital) errors between monitoring times. Distance between a GPS receiver and a satellite is calculated by multiplying the travel time of a radio signal by the speed of light, but the speed of light is not constant as the signal passes through the outer space and the atmosphere to the Earth's surface. When it does reach the surface, the signal may bounce off obstructions before reaching the receiver, thus creating a problem called multipath interference.

The second type of error is intentional. For example, to make sure that no hostile force can get precise GPS readings, the U.S. military used to degrade their accuracy under the policy called "Selective Availability" or "SA" by introducing noise into the satellite's clock and orbital data. On May 2, 2000, SA was switched off. Without the effect of SA, the horizontal accuracy of GPS readings is improved by almost a factor of 10.

users between five and eight satellites visible from any point on the Earth's surface. A good practice is to pick satellites that are widely separated to minimize errors in GPS readings.

Box 4.1   **An Example of GPS Data**

The following printout is an example of GPS data. The header information shows that the datum used is NAD27 (North American Datum 1927) and the coordinate system is UTM (Universal Transverse Mercator). The GPS data include seven point locations. The record for each point location includes the UTM zone number (i.e., 11), Easting (*x*-coordinate), and Northing (*y*-coordinate). The GPS data do not include the height or Alt value.

```
H  R DATUM
M  G NAD27 CONUS
H  Coordinate System
U  UTM UPS
```

H	IDNT	Zone	Easting	Northing	Alt	Description
W	001	11T	0498884	5174889	-9999	09-SEP-98
W	002	11T	0498093	5187334	-9999	09-SEP-98
W	003	11T	0509786	5209401	-9999	09-SEP-98
W	004	11T	0505955	5222740	-9999	09-SEP-98
W	005	11T	0504529	5228746	-9999	09-SEP-98
W	006	11T	0505287	5230364	-9999	09-SEP-98
W	007	11T	0501167	5252492	-9999	09-SEP-98

With the aid of a reference or base station, **differential correction** can significantly reduce noise errors. A reference station is located at a point that has been accurately surveyed. Using its known position, the reference receiver can calculate what the travel time of the GPS signals should be. The difference between the predicted and actual travel times thus becomes an error correction factor.

The reference receiver computes error correction factors for all visible satellites. These correction factors are then available to GPS receivers within about 300 miles (500 kilometers) of the reference station. GIS applications usually do not need real-time transmission of error correction factors. Differential correction can be made later as long as records are kept of measured positions and the time each position is measured. Reference stations are operated by private companies and by public agencies such as the U.S. Coast Guard. GIS users considering use of differential GPS data should find out their closest reference stations.

Equally important as correcting errors in GPS data is the type of GPS receiver. Most GIS users use code-based receivers (Figure 4.5). With differential correction, code-based GPS readings can easily achieve the accuracy of 3 to 5 meters, and some newer receivers are even capable of submeter accuracy. Carrier phase receivers and dual-frequency receivers are mainly used in surveying and geodetic control. They are capable of subcentimeter differential accuracy (Lange and Gilbert 1999).

4.5.3 Survey Data

Survey data consist of distances, angles, and curves. For example, a legal description of a land parcel may include "north 45°30'00″east 500.5 feet" to describe a course (line segment) of the parcel boundary. North 45°30'00″east is the angle measured 45 degrees 30 minutes 00 seconds from the north in the clockwise direction, and 500.5 feet is the distance. The parcel (polygon) boundary is

Figure 4.5
A portable GPS receiver.
(Courtesy of Trimble.)

comprised of courses, each with an angle and distance. In some cases, a polygon may contain curves rather than just straight-line segments. **Coordinate geometry (COGO),** a study of geometry by means of algebra, provides the methods for creating digital spatial data of points, lines, and polygons from survey data.

4.5.4 Street Addresses

One can create point features from street addresses in a process called **geocoding.** Address geocoding requires two sets of data: individual street addresses, and a database that consists of a street map and the range of address numbers along each side of each street segment. The TIGER/Line files or commercial products can be used to prepare such a database. Address geocoding first locates the street segment that contains an address. It then interpolates where the address falls within the address range. For example, if the address is 620 and the address range is from 600 to 640 in the database, address geocoding will locate 620 at the midpoint along the even-number side of the street segment. The accuracy of the input data is crucial to the success in geocoding addresses. For exam-

ple, outdated TIGER/Line files will not have the necessary address numbers or even streets to locate new street addresses.

4.5.5 Text Files with *x-*, *y*-Coordinates

Digital spatial data can be created from a text file that contains *x-*, *y*-coordinates. The *x-*, *y*-coordinates can be geographic or projected. Each pair of *x-*, *y*-coordinates is used to create a point. Therefore, one can create a digital map from a file that records the locations of weather stations, epicenters, or a hurricane track.

4.5.6 Digitizing Using a Digitizing Table

Digitizing is the process of converting data from analog to digital format. Manual digitizing uses a digitizing table (Figure 4.6). A **digitizing table** has a built-in electronic mesh, which can sense the position of the cursor. To transmit the *x-*, *y*-coordinates of a point to the connected computer, the operator simply clicks on a button on the cursor after lining up the cursor's cross hair with the point. Large-size digitizing tables typically have an absolute accuracy of 0.001 inch (0.003 centimeter).

Many GIS packages have a built-in digitizing module for manual digitizing. A topology-based GIS package has functionalities to ensure the topology of the digitized map. ArcInfo Workstation, for example, offers several commands to help users in digitizing. ARCSNAP allows the user to specify a distance within which an arc will be snapped to an existing arc (Figure 4.7). NODESNAP works the same way as ARCSNAP except it snaps a node to an existing node (Figure 4.8). INTERSECTARCS calculates arc intersections and adds nodes at intersections (Figure 4.9).

Digitizing usually begins with a set of control points, which are later used for converting the digitized map to real-world coordinates. Digitizing point features is simple: each point is clicked once to record its location. Digitizing line or area features can follow either point mode or stream mode. The operator selects points to digitize in point

Figure 4.6
A large digitizing table and a cursor with a 16-button keypad.
(Courtesy of GTCO Calcomp, Inc.)

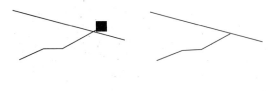

—— Arcsnap tolerance

Figure 4.7
The arcsnap tolerance allows the end of an arc to be
snapped to an existing arc. The left diagram shows the
overextension of a digitized arc. Because the
overextension is smaller than the arcsnap tolerance, the
end of the digitized arc is snapped to an existing arc, as
shown in the right diagram.

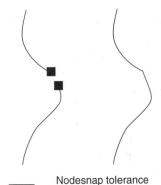

—— Nodesnap tolerance

Figure 4.8
The nodesnap tolerance allows the snapping of nodes.
The left diagram shows the digitized arc not reaching
its intended end point. Because the gap between the
nodes are smaller than the nodesnap tolerance, the end
node of the arc is snapped to the other node.

mode. In stream mode, lines are digitized at a preset time or distance interval. For example, lines
can be automatically digitized at a 0.01-inch interval. Most GIS users prefer point mode because
point mode creates a smaller data file than stream
mode and is more efficient in digitizing simple line
features with straight-line segments.

Digitizing line or polygon features can also
follow either discrete or continuous mode. In discrete mode the operator observes the arc–node

topological relationship. Points are digitized as
nodes if they are where lines meet or intersect. In
continuous mode, also called spaghetti digitizing,
the operator digitizes long, continuous lines and
lets the GIS package build the arc–node relationship in processing.

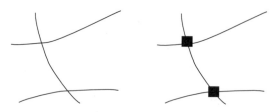

Figure 4.9
The INTERSECTARCS command calculates arc intersections and adds nodes at the intersections. The left diagram shows the arcs were digitized without adding nodes at the intersections. The right diagram shows the creation of nodes with the INTERSECTARCS command.

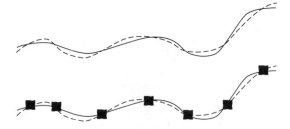

Figure 4.10
Duplicate lines create problems for editing. The top diagram shows the same arc was digitized twice. After the topology of the coverage is built, nodes are created at the intersections of the duplicate lines and a series of small polygons are formed between the lines in the bottom diagram. The extra arcs must be removed in editing.

Because the vector data model treats a polygon as a series of lines, digitizing polygon features is the same as digitizing line features except that each polygon requires a label (feature ID) in addition to its boundary. The label is needed to link the polygon with its attribute data.

Although digitizing itself is mostly manual, the quality of digitizing can be improved with planning and checking. An integrated approach is useful in digitizing different map layers of a GIS database that share common boundaries. For example, soils, vegetation types, and land-use types may share some common boundaries in a study area. Digitizing these boundaries only once and using them on each of the map layers not only saves time in digitizing but also ensures the matching of the layers.

A rule of thumb in digitizing line or polygon features is to digitize each line once and only once to avoid duplicate lines. Duplicate lines are seldom on top of one another because of the high accuracy of a digitizing table. In fact, duplicate lines form a series of polygons between them and are difficult to detect and correct in editing (Figure 4.10). One way to reduce the number of duplicate lines is to put a transparent sheet on top of the source map and to mark off each line on the transparent sheet after the line is digitized. This method can also reduce the number of missing lines.

4.5.7 Scanning

Scanning is a digitizing method that converts an analog map into a scanned file, which is then converted back to vector format through tracing (Verbyla and Chang 1997). A scanner (Figure 4.11) converts an analog map into a scanned image file in raster format. The map to be scanned is typically a black-and-white map: black lines represent map features, and white areas represent the background. The map may be a paper map or a Mylar map, and inked or penciled. Scanning converts the map into a binary scanned file in raster format; each pixel has a value of either 1 (map feature) or 0 (background). Map features are shown as raster lines, a series of connected pixels on the scanned file (Figure 4.12). The pixel size depends on the scanning resolution, which is often set at 300 dots per inch (dpi) or 400 dpi for digitizing. A raster line representing a thin inked line on the source map may have a width of 5 to 7 pixels (Figure 4.13).

Scribe sheets and color maps can also be scanned for digitizing. Because a scribed sheet is essentially a negative of a black-and-white map, the values of 1 and 0 need to be reversed for the scanned file. Expensive scanners can recognize colors. The digital raster graph (DRG) is a raster

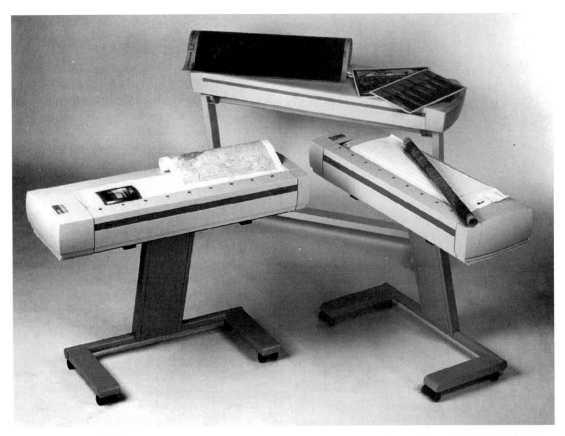

Figure 4.11
Large format drum scanners.
(Courtesy of GTCO Calcomp, Inc.)

image scanned from the USGS quadrangle map. The DRG has 13 different colors, each representing one type of map feature on the quadrangle.

To complete the digitizing process, a scanned file must be vectorized. **Vectorization** turns raster lines into vector lines in a process called tracing. Commercial GIS packages such as ArcInfo Workstation offer raster-to-vector algorithms for tracing. Tracing involves three basic elements: line thinning, line extraction, and topological reconstruction (Clarke 1995). Lines in the vector data model have length but no width. Lines in a scanned file (raster lines), however, usually occupy several pixels in width. Raster lines must be thinned, ideally to a 1-cell width, for vectorization. Line extraction is the process of determining where individual lines begin and end. Finally, topology is built between lines extracted from the raster model. Results of raster-to-vector conversion often show steplike features along diagonal lines. A line smoothing operation can eliminate those artifacts from raster data.

Tracing can be semiautomatic or manual. In semiautomatic mode, the user selects a starting point on the image map and lets the computer trace all the connecting raster lines (Figure 4.14). In manual mode, the user determines the raster line to be traced and the direction of tracing.

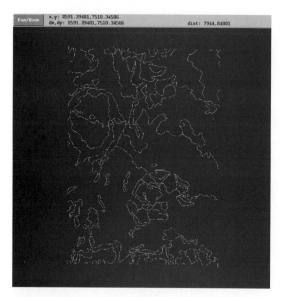

Figure 4.12
A binary scanned file: the lines are soil lines, and the black areas are the background.

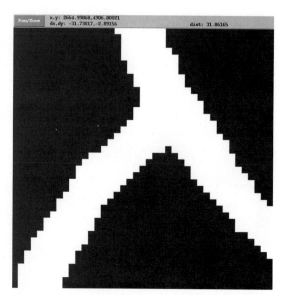

Figure 4.13
A raster line in a scanned file has a width of several pixels.

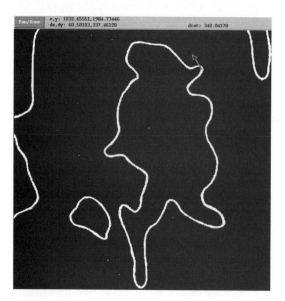

Figure 4.14
Semiautomatic tracing starts at a point (shown with an arrow) and traces all lines connected to the point.

Large data producers prefer scanning to manual digitizing for several reasons. First, scanning

uses the machine and computer algorithm to do most of the work, thus avoiding human error caused by fatigue or carelessness. Second, tracing has both the scanned image and vector lines on the same screen, making tracing more flexible than manual digitizing. With tracing, the operator can zoom in or out and can move around the image map with ease. Manual digitizing requires the operator's attention on both the digitizing table and the computer monitor, and the operator can easily get tired. Third, scanning has been reported to be more cost effective than manual digitizing. The cost of scanning by a service company has dropped significantly in recent years. Scanning a black-and-white map the size of a 1:24,000 scale USGS quadrangle costs less than $10.

The tracing algorithm for raster-to-vector conversion plays a key role in digitizing by scanning. No single tracing algorithm will work satisfactorily with different types of maps under different conditions. The algorithm must decide how to trace a vector line within a raster line that encompasses five to seven pixels. The width of a raster line actually doubles or triples when lines meet or

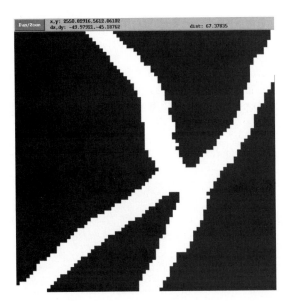

Figure 4.15
The width of a raster line doubles or triples when lines meet or intersect.

intersect (Figure 4.15). The algorithm must also decide where to continue when a raster line is broken or when two raster lines are close together. Small mistakes are common in tracing. ArcInfo Workstation users can adjust parameter values used in tracing. For example, the line-straightening properties, which apply to lines at intersections or corners, may be adjusted depending on if the map contains human-made features (i.e., more straight lines) or natural features (i.e., more smooth, continuous lines). ArcInfo Workstation users can also specify the maximum raster line width to be traced and the maximum distance between raster line segments to be jumped.

4.5.8 On-Screen Digitizing

On-screen digitizing is manual digitizing on the computer monitor using a data source such as a DOQ as the background. On-screen digitizing is an efficient method for editing or updating an existing map such as adding new trails or roads that are not on an existing map but are on a new DOQ. Likewise, the method can be used for editing a

vegetation map based on new information from a new DOQ that shows recent clear-cuts or burned areas.

4.5.9 Importance of Source Maps

Despite the increased availability of high-resolution remotely sensed data and GPS data, maps are still the predominant source for creating new GIS data. Digitizing, either manual digitizing or scanning, converts an analog map to its digital format. The accuracy of the digital map is therefore directly related to the accuracy of the source map. The digital map can be only as good or as accurate as its source map.

A variety of factors can affect the accuracy of the source map. Maps such as USGS quadrangle maps are secondary data sources because these maps have gone through the cartographic processes of compilation, generalization, and symbolization. Each of these processes can affect the accuracy of the mapped data. For example, if the compilation of the source map contains errors, the errors will be passed on to the digital map.

Paper maps generally are not good source maps because they tend to shrink and expand with changes in temperature and humidity. In even worse scenarios, GIS users may use copies of paper maps or mosaics of paper map copies for digitizing. Such source maps will not yield good results. Their plastic backing makes Mylar maps much more stable than paper maps for digitizing.

The quality of line work on the source map will determine not only the accuracy of the digital map but also the GIS user's time and effort in digitizing and editing. The line work should be thin, continuous, and uniform, as expected from inking or scribing. One should never use felt-tip markers to prepare the line work. Penciled source maps may be adequate for manual digitizing but are not recommended for scanning. Scanned files are binary data files, separating only the map feature from the background. Because the contrast between penciled lines and the background (i.e., surface of paper or Mylar) is not as sharp as inked lines, one may have to adjust the scanning parameters to increase the contrast. But the adjustment often results in the scanning of erased lines and smudges and messes up the scanned file. For words or symbols on the source map to give supplemental information, they can be drawn in orange color, which will not be scanned.

4.6 GEOMETRIC TRANSFORMATION

A newly digitized map is measured in the same measurement unit as the source map used for digitizing or scanning. Before the digitized map can be used in a GIS project, it must be converted to real-world coordinates such as UTM coordinates. **Geometric transformation** is the process of converting a digital map from one coordinate system to another by using a set of control points and transformation equations.

Figure 4.16 illustrates a typical scenario of geometric transformation. The process begins with digitizing a set of control points, which are points with known longitude and latitude values or real-world coordinates. For example, a USGS 1:24,000 scale quadrangle map has 16 points with known longitude and latitude values: 12 points along the border, and four additional points within the quadrangle. These 16 points divide the quadrangle into 2.5 minutes in longitude and latitude. The next step is to convert the longitude and latitude readings of the control points to real-world coordinates. This step can be saved if the real-world coordinates of the control points are already known. For example, a state agency in charge of GIS services may have already compiled a master tic (control point) file that lists the UTM coordinates of quadrangle corner points in a state. A map without map features is then copied from the digitized map, and its control points are updated to their real-world coordinates. Finally, the digitized map is converted into real-world coordinates through the use of a transformation method and the control points.

Different methods are available for geometric transformation (Taylor 1977). Each method is distinguished by the geometric properties it can pre-

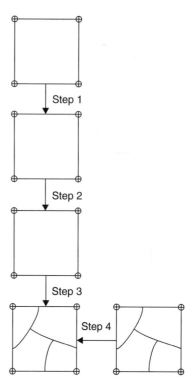

Figure 4.16
A typical scenario of geometric transformation. Step 1 updates the tics (control point) from digitizer units to longitude and latitude values. Step 2 projects the tics to real-world coordinates. Step 3 creates a coverage with tics only. Step 4 converts the digitized map features to real-world coordinates and places them in the output coverage.

serve and by the operations or changes it allows on an object such as a rectangle corresponding to a USGS quadrangle (Figure 4.17).

- Equiarea (congruence) Transformation: The method allows rotation of the rectangle and preserves its shape and size.
- Similarity Transformation: The method allows rotation of the rectangle and preserves its shape but not size.
- **Affine Transformation:** The method allows angular distortion but preserves the parallelism of lines.

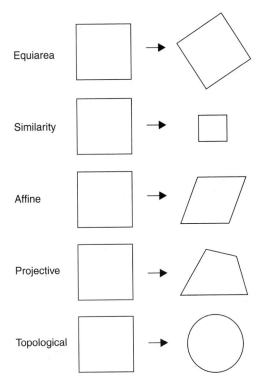

Figure 4.17
Different types of geometric transformation between coordinate systems.

- Projective Transformation: The method allows angular and length distortion, thus allowing the rectangle to be transformed into an irregular quadrilateral.
- Topological Transformation: The method preserves the topological properties of an object but not the shape, thus allowing the rectangle to be transformed into a circle.

4.6.1 Affine Transformation

Although GIS packages such as ArcInfo Workstation and Intergraph's MGE offer similarity, affine, and projective transformation methods, most GIS users use the affine transformation on digital maps and satellite images and the projective transforma-

tion on aerial photographs with displacement. The following discussion thus focuses on the operations allowed in affine transformations.

While preserving line parallelism, the affine transformation allows rotation, translation, skew, and differential scaling on the rectangular object (Pettofrezzo 1978). Rotation rotates its x and y axes from the origin. Translation shifts its origin to a new location. Skew changes its shape to a parallelogram with a slanted direction. And differential scaling changes the scale by expanding or reducing in the x and/or y direction (Figure 4.18).

Because the geometric transformation of a map is based on the control points, the affine transformation and its operations are first applied to the control points. In other words, the control points are transformed from their locations in the digitized map (the input or estimated control points) to their locations in real-world coordinates (the output or actual control points). Mathematically, the affine transformation is expressed as a pair of linear equations:

(4.1)

$$x' = Ax + By + C$$

(4.2)

$$y' = Dx + Ey + F$$

where x and y are in digitizer units, x' and y' are in real-world coordinates, and A, B, C, D, E, and F are the transformation coefficients. The six coefficients can be estimated by using the digitized location and the real-world coordinates of the control points. A minimum of three control points is sufficient for the estimation. But often four or more control points can reduce problems in measurement errors. With four or more control points, one can use a least-squares solution, similar to that of a regression analysis, to estimate the transformation coefficients (Box 4.2). After the coefficients are estimated, one can then apply the transformation equations to the digitized map and convert the coordinates of map features from digitizer units to real-world coordinates.

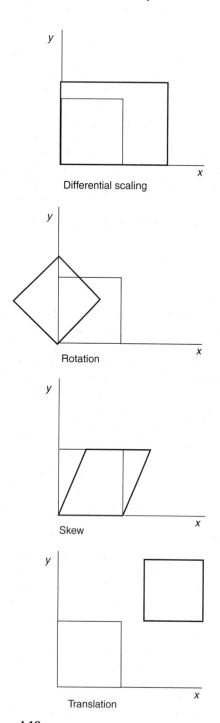

Figure 4.18
Differential scaling, rotation, skew, and translation in the affine transformation.

4.6.2 Geometric Interpretation of Transformation Coefficients

The geometric properties of the affine transformation can be interpreted from the transformation coefficients. The coefficient C represents the translation in the x direction, and F the translation in the y direction. The coefficients of A, B, D, and E are related to rotation, skew, and scaling in the following equations:

(4.3)

$$A = Sx \cos(t)$$

(4.4)

$$B = Sy\,[k \cos(t) - \sin(t)]$$

(4.5)

$$D = Sx \sin(t)$$

(4.6)

$$E = Sy\,[k \sin(t) + \cos(t)]$$

where Sx is the change of scale in x, Sy is the change of scale in y, t is the rotation angle, and k is the shear factor. The skew angle can be derived from arctan (k). The equations for A and D can be used to solve for t and, in turn, Sx. The equations for B and E can be used to solve for k and, in turn, Sy and the skew angle. Box 4.3 shows the output from an affine transformation including the transformation coefficients and the geometric properties of transformation.

4.6.3 Root Mean Square (RMS) Error

Control points play a crucial role in geometric transformation. The quality of the control points with respect to their estimated and actual locations determines the accuracy of geometric transformation and the positional accuracy of digitized map features (Bolstad et al. 1990). One measure of the goodness of the control points is the **root mean square (RMS)** error, which measures the deviation between the actual location of the control points, as projected from their longitude and

Box 4.2   Estimation of Transformation Coefficients

The example used here and later in the chapter is a third-quadrangle soil map (one-third of a USGS 1:24,000 scale quadrangle map), which has been scanned at 300 dpi. The map has four control points marked at the corners: Tic 1 at the NW corner, Tic 2 at the NE corner, Tic 3 at the SE corner, and Tic 4 at the SW corner. X and Y denote the control points' real-world (output) coordinates in meters based on the UTM coordinate system, and x and y denote the control points' digitized (input) locations. The measurement unit of the digitized locations is 1/300 of an inch, corresponding to the scanning resolution. The following table shows the input and output coordinates of the control points:

Tic-ID	x	y	X	Y
1	465.403	2733.558	518843.844	5255910.5
2	5102.342	2744.195	528265.750	5255948.5
3	5108.498	465.302	528288.063	5251318.0
4	468.303	455.048	518858.719	5251280.0

One method for deriving the transformation coefficients is to run two multiple regression analyses: the first regresses X on x and y, and the second regresses Y on x and y. The results show

$$A = 2.032, B = -0.004, C = 517909.198,$$
$$D = 0.004, E = 2.032, F = 5250353.802$$

The other method is to estimate the transformation coefficients using the following equation in matrix form:

$$\begin{bmatrix} C & F \\ A & D \\ B & E \end{bmatrix} = \begin{bmatrix} n & \Sigma x & \Sigma y \\ \Sigma x & \Sigma x^2 & \Sigma xy \\ \Sigma y & \Sigma xy & \Sigma y^2 \end{bmatrix}^{-1} \cdot \begin{bmatrix} \Sigma X & \Sigma Y \\ \Sigma xX & \Sigma xY \\ \Sigma yX & \Sigma yY \end{bmatrix}$$

where n is the number of control points and all other notations are the same as previously defined. The transformation coefficients derived from the equation are the same as from the regression analysis.

latitude readings into the output map, and the estimated location, as digitized on the input map. Typically, RMS errors are listed for each control point, and for the average from all control points. The equations for computing the RMS errors are as follows:

(4.7)

$$\text{RMS for a tic} = \sqrt{(x_{act} - x_{est})^2 + (y_{act} - y_{est})^2}$$

(4.8)

$$\text{Average RMS} = \sqrt{\left[\sum_{i=1}^{n} (x_{act,i} - x_{est,i})^2 + \sum_{i=1}^{n} (y_{act,i} - y_{est,i})^2\right]\Big/ n}$$

where n is the number of control points. Box 4.4 shows an example of the average RMS and x error

and y error for each control point from an affine transformation. To ensure the accuracy of geometric transformation, the RMS error should be within a tolerance value. The tolerance value, which is defined by the data producer, should vary depending on the accuracy and the map scale of the source map. If the RMS error exceeds the specified tolerance, either the control points need to be redigitized or a new set of control points has to be entered. If the RMS is within the acceptable range, then the assumption is that this same level of accuracy can also be applied to map features, which are usually bounded within the control points.

The above assumption relating RMS to the location accuracy of digitized map features can be wrong if gross errors are made in digitizing the control points or in inputting the longitude and latitude readings of the control points. As an example,

Box 4.3   Output from an Affine Transformation

Using the data from Box 4.2, the following shows the portion of the output from an affine transformation that is related to the transformation coefficients and the geometric transformation:

Scale (X, Y) = (2.032, 2.032);
skew (degrees) = (−0.014)

Rotation (degrees) = (0.102);
translation = (517909.198, 5250353.802)

Affine   X = Ax + By + C
          Y = Dx + Ey + F

A = 2.032; B = −0.004; C = 517909.198;
D = 0.004; E = 2.032; F = 5250353.802

The positive rotation angle means a rotation counterclockwise from the x-axis, and the negative skew angle means a shift clockwise from the y-axis. Both angles are very small, meaning that the change from the original rectangle to a parallelogram through the affine transformation is very slight.

Box 4.4   RMS from an Affine Transformation

The following shows a RMS report using the data from Box 4.3.

RMS error (input, output) = (0.138, 0.280)

Tic-ID	Input x Output x	Input y Output y	x error	y error
1	465.403	2733.558		
	518843.844	5255910.5	−0.205	−0.192
2	5102.342	2744.195		
	528265.750	5255948.5	0.205	0.192
3	5108.498	465.302		
	528288.063	5251318.0	−0.205	−0.192
4	468.303	455.048		
	518858.719	5251280.0	0.205	0.192

The output shows that the average deviation between the input and output locations of the control points is 0.280 meter based on the UTM coordinate system, or 0.00046 inch (0.138 divided by 300) based on the digitizer unit. This RMS error is well within the acceptable range. The individual x and y errors suggest that the error is slightly lower in the y direction than the x direction and the average RMS error is equally distributed among the four control points.

we can shift the locations of control points 2 and 3 (the two control points to the east) on the previous third quadrangle by increasing their x values by a constant. The RMS error would remain about the same because the object formed by the four control points retains the shape of a parallelogram. But the soil lines would deviate from their locations on the source map. The same problem occurs, if we in-

Figure 4.19
Inaccurate location of soil lines because of input or estimated tic location errors. The thin lines represent correct soil lines and the thick lines incorrect soil lines. In this case, the x values of the upper two tics were increased by 0.2″ while the x values of the lower two tics were decreased by 0.2″ on a third quadrangle (15.4″ × 7.6″).

crease the x values of control points 1 and 2 (the upper two control points) by a constant and decrease the x values of control points 3 and 4 (the lower two control points) by a constant (Figure 4.19). In fact, the RMS error would be well within the tolerance value as long as the object formed by the shifted control points remains as a parallelogram.

Longitude and latitude readings printed on paper maps are sometimes erroneous. This would lead to acceptable RMS errors but significant location errors of digitized map features. Suppose the latitude readings of control points 1 and 2 (the upper two control points) are off by 10″ (e.g., 47°27′20″ instead of 47°27′30″). The RMS error from transformation would be acceptable but the soil lines would deviate from their locations on the source map (Figure 4.20). The same problem occurs if the longitude readings of control points 2 and 3 (the two control points to the east) are off by 30″ (e.g., 2116°37′00″ instead of 2116°37′30″). Again, this happens because the affine transformation works with parallelograms. Although we tend to take for granted the accuracy of published maps,

erroneous longitude and latitude readings are quite common, especially with inset maps (maps that are smaller than the regular size) and oversized maps (maps that are larger than the regular size).

GIS users typically use the four corner points of the source map as control points. This practice makes sense because the exact readings of longitude and latitude are usually shown at those points. Moreover, using the corner points as control points helps the process of joining the map with its adjacent maps. The practice of using four corner control points, however, does not preclude the use of more control points if additional points with known locations are available. As mentioned earlier, a least-squares solution is used in an affine transformation when more than three points are used as control points. Therefore, the use of more control points means a better coverage of the entire map in transformation. In other situations, control points that are closer to the map features of interest should be used instead of the corner points. This ensures the location accuracy of these map features.

Figure 4.20
Incorrect location of soil lines because of output or actual tic location errors. The thin lines represent correct soil lines and the thick lines incorrect soil lines. In this case, the latitude readings of the upper two tics were off by 10″ (e.g., 47°27′20″ instead of 47°27′30″) on a third quadrangle.

KEY CONCEPTS AND TERMS

Affine transformation: A geometric transformation method commonly used in GIS, which allows rotation, translation, skew, and differential scaling on a rectangular object, while preserving line parallelism.

Coordinate geometry (COGO):   A branch of geometry that provides the methods for creating digital spatial data of points, lines, and polygons from survey data.

Data conversion: Conversion of spatial data to the format compatible with the GIS package in use.

Differential correction: A method that uses data from a base station to correct noise errors in GPS data.

Digital line graphs (DLGs):   Digital representations of point, line, and area features from USGS quadrangles including contour lines, spot elevations, hydrography, boundaries, transportation, and the U.S. Public Land Survey System.

Digitizing: The process of converting data from analog to digital format.

Digitizing table: A table with a built-in electronic mesh that can sense the position of the cursor and can transmit its x-, y-coordinates to the connected computer.

Direct translation: Use of a translator or algorithm in a GIS package to directly convert spatial data from one format to another.

Federal Geographic Data Committee (FGDC): A U.S. multiagency committee that coordinates the development of standards for spatial data.

Framework data: Data that many organizations regularly use for GIS activities.

Geocoding: The process of creating points from street addresses.

Geometric transformation: The process of converting a map or an image from one coordinate system to another by using a set of control points and transformation equations.

Global positioning system (GPS) data:
Longitude, latitude, and elevation data for point locations made available through a navigational satellite system and a receiver.

Metadata: Data that provide information about spatial data.

Neutral format: A public format such as SDTS that can be used for data exchange.

On-screen digitizing: Manual digitizing on the computer monitor by using a data source such as a DOQ as the background.

Remotely sensed data: Data such as digital orthophotos and satellite images that are acquired by a sensor from a distance.

Root mean square (RMS) error: In geometric transformation, the RMS measures the deviation between the actual location and the estimated location of the tics.

Scanning: A digitizing method that converts an analog map into a scanned file in raster format, which can then be converted back to vector format through tracing.

Spatial Data Transfer Standard (SDTS):
Public data formats for transferring spatial data such as DLGs, DEMs, and digital orthophotos from the USGS.

TIGER (Topologically Integrated Geographic Encoding and Referencing): A database prepared by the U.S. Bureau of the Census that contains legal and statistical area boundaries, which can be linked to the census data.

Vector product format (VPF): A standard format, structure, and organization for large geographic databases that are based on a georelational data model.

Vectorization: The process of converting raster lines into vector lines through tracing.

APPLICATIONS: VECTOR DATA INPUT

This applications section covers methods for vector data input. Task 1 uses existing data on the Internet. Task 2 lets you digitize several polygons directly on the computer screen. Task 3 shows you how you can create a shapefile from a table that contains x-, y-coordinates. Task 4 shows you how you can convert street addresses to point features and save them to a shapefile. The last two tasks require use of ArcInfo Workstation. Task 5 is an exercise on raster-to-vector conversion using a scanned file. Task 6 is an exercise on geometric transformation.

Task 1: Download a Digital Map from the Internet

Like many other states, Idaho has a website that lets GIS users download digital maps in either ArcInfo export (e00) format or shapefile format. Maintained by the Idaho Department of Water Resources, the website also provides metadata. Task 1 shows you how to import an export file and convert the imported coverage to a shapefile.

1. Go to **http://www.idwr.state.id.us/gisdata/,** and click GIS Data. Then click Statewide Data.

2. The Statewide Data table lists coverages that can be downloaded and the choices of Preview gif, e00, Shapefile, and Metadata. Click on Metadata for County Boundaries. The metadata gives you information about the coverage. Right-click on the dot for e00 and select "Save Link As" (or "Save Target As" using the Internet Explorer) from the menu. Name the file *idcounty.e00* and provide the path to save the file.

3. The first part of this exercise is to import *idcounty.e00*. Start ArcCatalog and make connection to the Chapter 4 database. Launch ArcToolbox. Double-click on Conversion Tools and then Import to Coverage. Double-click on ArcView Import from Interchange File. In the next dialog, use the browse buttons to select *idcounty.e00* for the Input file and to enter *idcounty* for the Output dataset to be stored in the Chapter 4 database. Click OK.

4. *Idcounty* should appear in the Catalog tree; if not, select Refresh from the View menu. Click the plus sign next to *idcounty* and click on the polygon feature class to preview the county boundary map.

5. You can use the same procedure to download the shapefile version of *idcounty* from the same website.

Task 2: Digitize On-screen in ArcMap

What you need: *landuse.shp*, a background map for digitizing. *Landuse.shp* is measured in digitizer units (inches). Therefore, you can ignore the message about missing spatial reference information from ArcMap.

On-screen digitizing is technically similar to manual digitizing. The differences are (1) you use the mouse rather than the digitizer's cursor for digitizing, (2) the resolution of the computer monitor is coarser than a digitizer, and (3) you need a coverage, a shapefile, or an image (e.g., a digital orthophoto) as the background for digitizing. This task lets you digitize several polygons off *land use.shp* and make a new shapefile.

1. Make sure that ArcCatalog is still connected to the Chapter 4 folder. Right-click the Chapter 4 folder, point to New, and select Shapefile. In the Create New Shapefile dialog, enter *trial* for the Name and select Polygon from the Feature Type dropdown list. Click OK. *Trial.shp* is added to the Catalog tree.

2. Launch ArcMap. Right-click Layers in the Table of Contents, select Properties from Layers' context menu, click the General tab, and enter Task 2 for the Name. Add *landuse.shp* and *trial.shp* to Task 2. Before digitizing, you want to show the background map of *landuse* in red with labels and *trial* in black for differentiation. Select Properties from the context menu of *landuse*. Click the Symbology tab. Click Symbol, and change it to a Hollow symbol with the Outline Color in red. Click the Labels tab. Check the box to Label Features in this layer and select LANDUSE_ID from the Label Field dropdown list. Click OK to dismiss the Layer Properties dialog. Click the symbol of *trial* in the Table of Contents. Select the Hollow symbol and the Outline Color of black. Click OK to dismiss the Symbol Selector dialog.

3. This step is to set up the snapping tolerance for digitizing. Click the Tools menu and make sure that Editor Toolbar is checked. The alternative is to click the Editor Toolbar button. Select Start Editing from the Editor dropdown list. Make sure that the Task is to Create New Feature, and the Target is *trial*. Select Options from the Editor dropdown list. Under the General tab, enter 0.05 and select map units from the dropdown list for the Snapping tolerance. Click OK. Click the Editor dropdown arrow and select Snapping. Check the box for End for the Layer of *trial*.

4. You are ready to digitize. Zoom in the area around polygon 73. Click the Create New Feature tool (the Sketch tool is renamed to correlate to the task). Digitize a start point of polygon 73 by left-clicking the mouse. Continue to digitize other vertices of polygon 73, and double-click the start point to complete the polygon. The completed polygon 73 appears in cyan with an x inside it. Digitize polygon 74 in the same way as polygon 73.

5. Polygon 75 has a shared border with 74. So that you do not have to digitize the same border twice, click the Task dropdown arrow and select Auto Complete Polygon. Start by digitizing one of the end points of the shared

border. Continue digitizing other vertices of polygon 75. Double-click the other end point of the shared border to complete polygon 75. You can digitize the end points inside polygon 74 because the overshoots will be trimmed automatically.

6. After you have digitized polygons 72–76, select Stop Editing from the Editor dropdown list and save the edits.

Task 3: Add X Y Data in ArcMap

What you need: *events.txt*, a text file containing GPS readings.

In Task 3, you will use ArcMap to create a new shapefile from *events.txt*, a text file that contains *x*-, *y*-coordinates of a series of points collected by GPS readings.

1. Select Data Frame from the Insert menu in ArcMap. Rename the new data frame Task 3, and add *events.txt*. (*Events.txt* is shown as a data set in a geodatabase.) Select Add XY Data from the Tools menu. Make sure that *events.txt* is the table to be added as a layer. Use the dropdown lists to select EASTING for the X Field and NORTHING for the Y Field. Click OK. *Events.txt Events* is added to the Table of Contents.

2. *Events.txt Events* can be saved as a shapefile. Right-click *events.txt Events*, point to Data, and select Export Data. Opt to export all features and save the output as *events.shp* in the Chapter 4 database.

3. You can also convert *events.txt* into a shapefile directly in ArcCatalog. Right-click *events.txt* in the Catalog tree, point to Create Feature Class, and select From XY Table. The Create Feature Class From XY Table dialog allows you to specify the X Field, the Y Field, and the output shapefile.

Task 4: Geocode Street Addresses

What you need: *address.txt*, a text file containing street addresses; *latah_streets.shp*, a street shape-file for Latah County, Idaho, converted from the 1995 TIGER files.

In Task 4, you will learn how to create point features from street addresses. Address geocoding requires an address table and a reference data set. The address table contains a list of street addresses to be located. The reference data set has the address information for locating street addresses. For this task, you will use *latah_streets.shp* for the reference data set. The process of geocoding addresses involves four steps. First, view the input data to geocoding addresses in ArcMap. Second, create a geocoding service in ArcCatalog. Third, match addresses in the address table with reference data attributes in ArcMap. Fourth, rematch unmatched addresses from the second step.

1. Select Data Frame from the Insert menu in ArcMap. Rename the new data frame Task 4, and add *latah_streets.shp* and *address.txt*. Right-click *latah_streets.shp* and open its attribute table. Because *latah_streets.shp* is derived from the TIGER files, it has all the attributes from the original data. The following attributes are important for geocoding addresses in this task:

PREFIX is a direction that precedes a street name.
STREETNAME is the name of a street.
TYPE is the street name type, which may be St, Rd, Ln, and so on.
L_ADD_FROM and L_ADD_TO are the beginning and ending address numbers on the left side of a street segment.
R_ADD_FROM and R_ADD_TO are the beginning and ending address numbers on the right side of a street segment.
ZIP_LEFT and ZIP_RIGHT are the ZIP Codes for the left and right sides of a street segment.

Right-click *address.txt*, and open the table. The table contains three fields: Name, Address, and Zip. Close the tables.

2. Start ArcCatalog if necessary. Click the folder symbol next to Geocoding Services.

Double-click Create New Geocoding Service. Select US Streets with Zone (File) in the next dialog. The US Streets with Zone style is closest to the geocoding style of *latah_streets.shp*, and File means that the reference data set is a shapefile and not a geodatabase. Click OK to dismiss the dialog.

3. Start with the left side of the next dialog. Highlight the default name and type Task 4. Then use the browse button to specify *latah_streets.shp* for the reference data. Move to the right side of the dialog. In the Input Address Fields frame, first match Street with Address and delete others. To delete Addr, highlight it and click the Delete button. Delete Street. Next match Zone with Zip. Use the same procedure to delete Zipcode, City, and Zone. Keep the default values for spelling sensitivity, minimum candidate score, and minimum match score in the Matching Options frame. Click OK to dismiss the dialog.

4. The new geocoding service you have just created shows up in the Catalog tree with Task 4 prefixed by the user name on your computer.

5. Activate ArcMap. Click the Tools menu, point to Geocoding, and select Geocoding Addresses. Click Add in the next dialog, and use the browse button to add the geocoding service of xxx.Task 4. Click OK.

6. In the next dialog, make sure that *address.txt* is the address table and the address input fields are correct. Provide the path and save the output shapefile as *geocoding_result.shp* in the Chapter 4 folder.

7. The Review/Rematch Addresses dialog shows that 5 out of 6 addresses in *address.txt* are matched with score 80 - 100 and 1 is unmatched. Dismiss the dialog by clicking the Done button. *Geocoding_result*, with a prefix of Geocoding Result, is added to the Table of Contents in ArcMap. Zoom in to see the five point features created from the five matched street addresses. Right-click *geocoding_result* and open its attribute table. The table has the field Status showing whether a record is matched (M) or not (U). It also has the field Score showing the match score. In this case, the five addresses that are matched all have the perfect score and the sole unmatched address has the score of 0.

8. The last part of the task is to examine the unmatched address and rematch it if possible. Click the Tools menu, point to Geocoding, point to Review/Rematch Addresses, and select Geocoding Result: Geocoding Result. Click Yes to start editing.

9. Click Match Interactively in the Review/Rematch Addresses dialog. This opens a dialog with the unmatched address of 308 W 6TH ST highlighted. Click the Modify button to open the Edit Standardization form. Highlight 6TH in the StreetName field and enter Sixth. Two match candidates appear at the bottom of the Review/Rematch Addresses dialog. The first one is the match with a perfect score. Highlight it, click Match, and close the dialog. Apparently, the street is spelled Sixth in *latah_streets.shp* rather than 6TH. Now all six addresses are geocoded successfully, and the result is saved in *geocoding_result.shp*.

Task 5: Use Scanning for Spatial Data Input in ArcInfo Workstation

What you need: *hoytmtn_gd*, an ArcInfo grid converted from a scanned file.

As explained in the text, scanning for digitizing involves two steps: scanning and raster-to-vector conversion. Task 5 performs the second step. A scanned file is an image file, commonly in TIFF (tagged image file format) or RLC (run length compressed) format. Working with a grid, which uses ESRI's proprietary format for raster data, is easier than working with an image file in ArcInfo Workstation. A scanned file of a soil map has been converted to a grid called *hoytmtn_gd* for Task 5.

In the following, you will convert *hoytmtn_gd* to an ArcInfo coverage.

1. Before going to ArcTools, you need to create an empty coverage *hoytmtntrace* that will eventually be filled with arcs from the tracing of *hoytmtn_gd*:

    Arc: create hoytmtntrace hoytmtn_gd

    By specifying *hoytmtn_gd* as an argument in the CREATE command ensures that *hoytmtntrace* has the same map extent (BND) as *hoytmtn_gd*.

2. Start ArcTools:

    Arc: arctools

3. Select Edit Tools in the ArcTools menu and click OK.

4. Select File in the Edit Tools menu, and select Grid Open in the File pulldown.

5. Choose *hoytmtn_gd* in the Select an Edit Grid menu. Click OK. This action results in the display of *hoytmtn_gd* in the ArcEdit window, and the opening of the Grid Editing menu. The grid *hoytmtn_gd* contains text on data source. What you do in Steps 6–9 is to remove the text using the grid editing tools in ArcScan.

6. Click the Sketch button in the Grid Editing menu to open the Sketch menu.

7. Type 0 as the Fill value in the Sketch menu.

Note: Hoytmtn_gd is a bi-level, or black-and-white, grid file. The fill value of 0 is the same as the background.

8. Go to the ArcEdit window and use Extent in the Pan/Zoom pulldown to find the data source. Use Fullview to get the full view of *hoytmtn_gd*.

9. After you see the data source clearly in the ArcEdit window, click on the fill tool (the Solid Square icon) in the Sketch menu. Press the mouse and drag a box around the data source. The data source disappears as you release the mouse.

10. Select File in the Edit Tools menu, and select Coverage Open from the File pulldown.

11. In the Select an Edit Coverage menu, click on *hoytmtntrace* in the Coverages list. Click the Create New Feature button to open the Create Feature menu.

12. In the Create Feature menu, select Arc in the Feature Class list. Click Apply. This action adds Arc to the Available Features list. Highlight Arc in the Available Features list. Click OK to dismiss the menu.

13. Zoom into *hoytmtn_gd* so that you can see lines in the ArcEdit window.

14. Click on the Trace button in the Edit Arcs & Nodes menu. Press the cursor at the start of a line along the border of *hoytmtn_gd* that is well connected with other lines. A green line appears at the starting point; it is called the tracer.

Note: As in manual digitizing, you do not want to trace the same line twice. That could happen if you select a starting point in the middle of an arc.

15. Press 8 on the keyboard for semi-automatic tracing. Lines that have been traced turned into green. The green lines then turn into yellow after all the connected lines to the starting point have been traced and the built-in algorithms for line straightening and generalization have been applied to the traced lines.

16. Lines that remain untraced in *hoytmtn_gd* are those of island polygons, polygons along the border, and tic marks. Click the Trace button in the Edit Arcs & Nodes menu, press the cursor along an untraced line, and press 8 on the keyboard.

17. Repeat the above step until all lines are traced. Each time you press the cursor at a point, the tracer appears; you can change the tracer direction by middle-clicking on the mouse. Left-click on the mouse switches to the option of manual tracing. Manual tracing allows you to trace one line segment at a time.

Note: You need to change the tracer direction often in manual tracing because you trace one arc at a time.

18. Select File in the Edit Tools menu, and select Coverage Save in the File pulldown. This saves *hoytmtntrace* with its traced lines. Exit Arctools.

Task 6: Use TRANSFORM on a Newly Digitized Map in ArcInfo Workstation

What you need: a digitized soils map called *hoytmtn*, and a text file called *tic.geo*.

The digitized map *hoytmtn* is measured in inches. It has four control points that correspond to the four corner points, with the following longitude and latitude values in degrees-minutes-seconds (DMS):

Tic-id	Longitude	Latitude
1	−116 00 00	47 15 00
2	−115 52 30	47 15 00
3	−115 52 30	47 07 30
4	−116 00 00	47 07 30

The text file *tic.geo* contains the longitude and latitude values of the four tics. (Open *tic.geo* to see how the file is prepared.) The first step is to use the PROJECT command with the file option to convert the longitude and latitude values in *tic.geo* into UTM coordinates and save the new coordinates in *tic.utm*.

```
Arc: project file tic.geo tic.utm
Project: input
Project: projection geographic
Project: units dms
Project: parameters
Project: output
Project: projection utm
Project: units meters
Project: zone 11
Project: parameters
Project: end
```

The choice of a real-world coordinate system is made through the PROJECT command. PROJECT is a dialog command, meaning that after you enter the command, you must provide projection parameters such as the UTM zone by using PROJECT's subcommands. The dialog, as shown in the preceding paragraphs, is organized into two sections. The first section describes the input data, and the second the output data. Each section ends with the subcommand of parameters to see if any parameters are required for the specified projection. In this case, no parameters are required for the geographic grid (i.e., longitude and latitude values) or the UTM coordinate system.

The output file, *tic.utm*, has the UTM coordinates of the four tics as follows:

Tic-id	x	y
1	575672.2771	5233212.6163
2	585131.2232	5233341.4371
3	585331.3327	5219450.4360
4	575850.1480	5219321.5730

The next step is to create an empty coverage, *hoytmtn2*, by copying the tics and the BND from *hoytmtn*:

```
Arc: create hoytmtn2 hoytmtn
```

The tics of *hoytmtn2* are measured in inches, similar to those of *hoytmtn*. But with the output from the PROJECT command you can now use the UPDATE command in Tables to update the tics of *hoytmtn2* to their UTM coordinates. The final step is to use TRANSFORM to convert the rest of the coverage into UTM coordinates by using the tics as control points:

```
Arc: transform hoytmtn hoytmtn2
```

REFERENCES

Anderson, J. R., et al. 1976. A Land Use and Land Cover Classification System for Use with Remote Sensor Data. *U. S. Geological Survey Professional Paper 964*. Washington, DC: U.S. Government Printing Office.

Bolstad, P. V., P. Gessler, and T. M. Lillesand. 1990. Positional Uncertainty in Manually Digitized Map Data. *International Journal of Geographical Information Systems* 4: 399–412.

Clarke, K. C. 1995. *Analytical and Computer Cartography,* 2d ed. Englewood Cliffs, NJ: Prentice Hall.

Guptill, S. C. 1999. Metadata and Data Catalogues. In P. A. Longley, M. F. Goodchild, D. J. MaGuire, and D. W. Rhind, eds., *Geographical Information*

Systems, 2d ed., pp. 677–92. New York: Wiley.

Kennedy, M. 1996. *The Global Positioning System and GIS.* Ann Arbor, MI: Ann Arbor Press.

Lange, A. F., and C. Gilbert. 1999. Using GPS for GIS Data Capture. In P. A. Longley, M. F. Goodchild, D. J. MaGuire, and D. W. Rhind, eds., *Geographical Information Systems,* 2d ed., pp. 467–79. New York: Wiley.

Moffitt, F. H., and J. D. Bossler. 1998. *Surveying,* 10th ed. Menlo Park, CA: Addison-Wesley.

Onsrud, H. J., and G. Rushton, eds. 1995. *Sharing Geographic Information.* New Brunswick, NJ: Center for Urban Policy Research.

Pettofrezzo, A. J. 1978. *Matrices and Transformations.* New York: Dover Publications.

Robinson, A. H., J. L. Morrison, P. C. Muehrcke, A. J. Kimerling, and S. C. Guptill. 1995. *Elements of Cartography,* 6th ed. New York: Wiley.

Sperling, J. 1995. Development and Maintenance of the TIGER Database: Experiences in Spatial Data Sharing at the U.S. Bureau of the Census. In H. J. Onsrud and G. Rushton, eds., *Sharing Geographic Information,* pp. 377–96. New Brunswick, NJ: Center for Urban Policy Research.

Taylor, P. J. 1977. *Quantitative Methods in Geography: An Introduction to Spatial Analysis.* Boston: Houghton Mifflin.

Verbyla, D. L., and K. Chang. 1997. *Processing Digital Images in GIS.* Santa Fe, NM: OnWord Press.

CHAPTER

5

SPATIAL DATA EDITING

5.1 INTRODUCTION

Spatial data editing refers to the process of adding, deleting, and modifying features in digital maps. A major part of spatial data editing is to remove digitizing errors. Newly digitized maps, no matter how carefully prepared, always have some errors. Digital maps from any source may contain errors from initial digitizing or from outdated data sources. Therefore, roads, land parcels, forest inventory, and other data require regular revision and updating. The updating process is basically the same as correcting errors on a newly digitized map.

There are two types of digitizing errors. Location errors such as missing polygons or distorted lines relate to inaccuracies of map features, whereas topological errors such as dangling arcs and unclosed polygons relate to logical inconsistencies among map features. To correct location errors, one often has to reshape individual arcs and digitize new arcs. To correct topological errors, one must be knowledgeable about the topological relationships and use a topology-based GIS package to help make corrections. Correcting errors may extend beyond individual maps. When a study area covers two or more source maps, one must match features across the map border.

Spatial data editing is more than removing digitizing errors. Modification of map features may take the form of simplification and smoothing. New features may be created using existing map features. And features from different maps may be integrated to share common boundaries. The use of both topological and nontopological data in GIS has significantly expanded the scope of spatial data editing.

This chapter is organized into the following six sections. Section 5.2 discusses location and topological errors. Section 5.3 considers topological editing and nontopological editing. Section 5.4 provides an overview of topological editing and the global and local methods for correcting topological errors. Section 5.5 covers edgematching, an editing operation to make sure that lines from adjacent maps meet perfectly along the border. Section 5.6 discusses editing operations for nontopological digital maps. Section 5.7 concludes the chapter with a discussion of other feature manipulations.

5.2 DIGITIZING ERRORS

5.2.1 Location Errors

Location errors relate to the location of map features on a digitized map. Assuming that the source map is correct, the ultimate goal of digitizing is to duplicate the source map in digital format. To determine how well the goal has been achieved, one can make a **check plot** of the digitized map at the same scale as the source map, superimpose the plot on the source map, and see how well they match.

Typically a producer of digital maps decides on the tolerance of location error. The U.S. Federal Geographic Data Committee (FGDC) does not specify threshold accuracy values that spatial data must achieve but encourages agencies to establish accuracy thresholds for their products and to report accuracy statistics (Chapter 3). The Natural Resources Conservation Service (NRCS), for example, stipulates that each digitized soil boundary in the Soil Survey Geographic (SSURGO) database shall be within 0.01inch (0.254 millimeter) line width of the source map. At the scale of 1:24,000, this tolerance represents 20 feet (6 to 7 meters) on the ground.

Several scenarios may explain the discrepancies between digitized lines and lines on the source map. The first is human errors in manual digitizing. Human error is not difficult to understand: when a source map has hundreds of polygons and thousands of lines, one can easily miss some lines or connect the wrong points.

The second scenario consists of errors in scanning and tracing. As described in Chapter 4, a tracing algorithm usually has problems when raster lines meet or intersect, are too close together, are too wide, or are too thin and broken. Digitizing errors from tracing include collapsed lines, misshapen lines, and extra lines (Figure 5.1).

The third scenario consists of errors in converting the digitized map to real-world coordinates (Chapter 4). To make a check plot at the same scale as the source map, one must convert the digitized map to real-world coordinates by using a set of control points. With erroneous control points, this conversion can cause discrepancies between digitized lines and source lines. Unlike seemingly random errors from the first two scenarios, discrepancies from geometric transformation often exhibit regular patterns. To correct this type of location errors, one must redigitize control points and rerun geometric transformation.

Location errors also include duplicate lines from manual digitizing or tracing. Each arc on a line or polygon map is supposed to be digitized only once. But a complex map may have lines that were digitized twice or even more times because of human errors. Duplicate lines also occur frequently in tracing because semiautomatic tracing

follows continuous lines even if some of the lines have already been traced. Duplicate lines from either manual digitizing or tracing are not going to be on top of one another because of the high resolution of a digitizing table in manual digitizing and the multiple-pixel raster line in tracing. If the topology of the map is built, these duplicate lines intersect one another to form a series of tiny polygons or slivers, and each polygon has a missing label (Figure 5.2). Duplicate lines are sometimes difficult to detect without zooming in.

5.2.2 Topological Errors

Topological errors violate the topological relationships used in a GIS package. A common error occurs when arcs that are supposed to meet at a point (node) do not meet perfectly. This type of error becomes an **undershoot** if a gap exists between the arcs, and an **overshoot** if one arc is overextended (Figure 5.3). The result of both cases is a **dangling arc,** which has the same polygon on its left and right sides, and a **dangling node** at the end of the arc. In general, it is easier to identify and remove an overshoot with its dangling node than an undershoot. This explains a common practice in digitizing that deliberately creates small overshoots rather than falls short of the intersection of two arcs. Dangling nodes also occur when a polygon is not closed perfectly (Figure 5.4).

Dangling nodes are acceptable in special cases. Manual digitizing using a topology-based GIS package requires that the starting and end points of an arc be nodes. Therefore, small tributaries on a stream map and dead-end streets on a

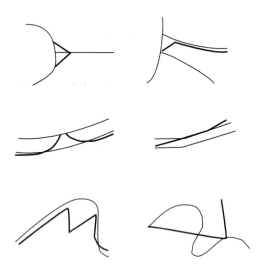

Figure 5.1
Common types of digitizing errors from tracing. The thin lines are lines on the source map, and the thick lines are lines from tracing.

Figure 5.2
Digitizing errors from duplicate lines include slivers and missing labels for the sliver polygons. Slivers are exaggerated for the purpose of illustration.

street map have dangling nodes attached to the arcs. These are acceptable dangling nodes.

A **pseudo node** appears along a continuous arc and divides the arc unnecessarily into separate arcs (Figure 5.5). Some pseudo nodes are, however, acceptable or even necessary. For example, to show different segments of a road having different attributes such as speed limit, pseudo nodes may be placed at points where attribute values change. Another example applies to isolated polygons not connected to other polygons. An arbitrary starting point (and end point) must be chosen to digitize the polygon. That point then becomes a pseudo node.

The arc direction may become a source of topological errors. For example, in hydrologic applications, all streams must follow the downstream direction and the starting point (the from-node) of a stream must be at a higher elevation than the end point (the to-node). Another example is a street coverage. The direction of an arc or street is not an issue for two-way streets but is an issue for one-way streets (Figure 5.6).

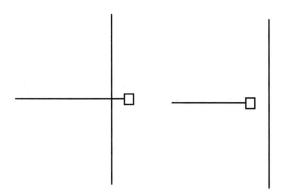

Figure 5.3
Digitizing errors of overshoot (left) and undershoot (right).

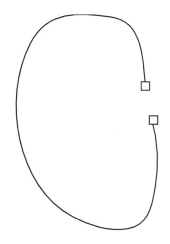

Figure 5.4
Digitizing errors of an unclosed polygon.

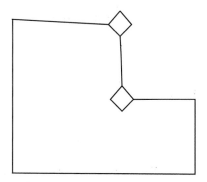

Figure 5.5
Pseudo nodes, shown by the diamond symbol, are nodes that are not located at line intersections.

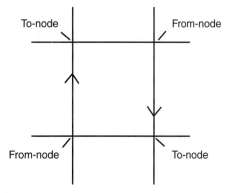

Figure 5.6
The from-node and to-node of an arc determine the arc's direction.

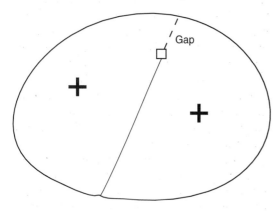

Figure 5.7
Digitizing error of multiple labels because of unclosed polygons.

A label links a polygon to its attribute data. A polygon therefore should have one, and only one, label. Unclosed polygons are often the cause for multiple label errors (Figure 5.7). When a gap exists between polygons, the two polygons are treated as a single polygon topologically. Consequently, the two labels, which have already been assigned to each polygon, become a case of multiple labels. The multiple label error disappears when the gap is filled.

5.3 TOPOLOGICAL AND NONTOPOLOGICAL EDITING

The nature of spatial data editing varies among GIS users: it depends on whether they use topological or nontopological GIS data, whether they are GIS data producers or users, and what GIS package they use. A topology-based GIS package can detect and display topological errors, and has commands to remove them. Examples of topology-based GIS packages include ArcInfo Workstation, AutoCAD Map, MGE, ILWIS, and SPANS (Box 5.1). They have similar capabilities in making sure that lines meet correctly, polygons are closed, and polygons are properly labeled. Some GIS packages are not topology-based, but

have the capability of performing topological editing. For example, IDRISI has a separate software product for defining the spatial relationship between map features in vector format.

A nontopological GIS package cannot detect topological errors or build topology, although it can be used for modifying map features and creating new features from existing features. Examples of nontopological GIS packages include ArcGIS Desktop and MapInfo. The ArcMap application of ArcGIS is designed for data display and editing, but its current version (8.2) does not have the functionality of displaying topological errors.

5.4 TOPOLOGICAL EDITING

Topological editing ensures that map features follow the topological relationships that are either built into a data model or specified by the user. Assuming that a map has just been digitized, this section describes the process of using a GIS package such as ArcInfo Workstation to edit the map. ArcInfo Workstation uses three topological relationships of continuity, area definition, and contiguity (Chapter 3). Therefore, editing operations discussed in this section are directly related to those relationships. Use of the object-oriented data model in GIS, such as the geodatabase model in ArcGIS, will change topological editing operations, although the basic concepts remain the same (Box 5.2). (The upcoming ArcGIS 8.3 will add topology to the geodatabase model.)

5.4.1 An Overview

The editing process begins with construction of the topology of the map to be edited. This step ensures that the computer can recognize individual nodes, arcs, and polygons on the map. Also, the tolerances used for building the topology between map features can remove some of the digitizing errors.

The next step is to display what types of digitizing errors exist on the map. A topology-based GIS package can detect topological errors and denote these errors with special symbols. Location

Box 5.1 GIS Packages for Topological Editing

ArcInfo Workstation, AutoCAD Map, MGE, and SPANS are all capable of performing topological editing. They are similar in editing operations. This chapter references only ArcInfo Workstation commands, but the following list highlights the editing capabilities of the other three packages:

- AutoCAD Map: build topology, repair undershoots, snap clustered nodes, remove duplicate lines, simplify linear objects, and correct other errors.

- MGE (Base Mapper): build topology, remove redundant linear data segments, fix undershoots and overshoots, and create an intersection at a line crossing.
- SPANS: identify and repair knots, overshoots, and dangles; build topology for line and area data; and use edgematching to join adjacent line data.

Box 5.2 Topological Editing for the Object-Oriented Data Model

A new development with the geodatabase model is the addition of topology in the upcoming ArcGIS 8.3. The topological relationships will be implemented as rules. The user can select topology rules and apply them to features within a feature class or to features between two or more feature classes. Rules applied to features within a feature class will be similar to the editing operations covered in this chapter such as polygons must not overlap, lines must not have dangles, and polygons must not have gaps. Rules applied to features between two or more feature classes such as coincident boundaries between soil and land-use maps will be new. Use of topology rules means less manual work and a faster editing process.

Laser-Scan, a company in Cambridge, England, launched an extension to Oracle9i called Radius Topology in June 2002 (**http://www.radius.laser-scan.com/**). Like topology for the geodatabase model, Radius Topology also implements the topological relationships as rules between features. But the rules are stored as tables in the database. When these tables are activated, topology rules automatically apply to spatial features in the enabled classes. An earlier version of Radius Topology was used to create the Ordnance Survey's MasterMap (**http://www.ordsvy.gov.uk/**).

errors such as missing lines or polygons require manual checking by comparing a check plot with its source map.

The third step is to remove the digitizing errors. Digitizing errors may be removed individually or as a group. To remove a specific type of digitizing error usually requires a specific action.

After removing digitizing errors, one must rebuild the map's topology to accommodate changes made during editing. Only rarely can one remove all digitizing errors in a single editing session. Typically, one must repeat the editing process before the map is free of digitizing errors.

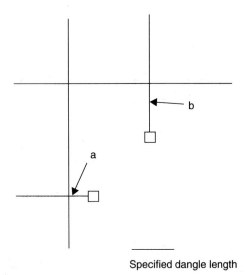

Specified dangle length

Figure 5.8
The dangle length specified by the CLEAN command can remove an overshoot if the overextension is smaller than the specified length. In this diagram, the overshoot a is removed and the overshoot b remains.

5.4.2 Correcting Digitizing Errors

5.4.2.1 Global Method

Topology-based GIS packages typically include commands that will apply some specified tolerances to an entire map to remove digitizing errors. One example of these global methods is the CLEAN command in ArcInfo Workstation. CLEAN uses two tolerances on an input coverage: dangle length and fuzzy tolerance. The **dangle length** specifies the minimum length for dangling arcs on the output coverage. A dangling arc is removed if it is shorter than the specified dangle length. Therefore, the dangle length is useful in removing minor overshoot errors (Figure 5.8). The **fuzzy tolerance** specifies the minimum distance between points (vertices) and arcs on the output coverage. The tolerance applies to points that make up an arc as well as points along two nearby arcs. The latter case can be useful in removing duplicate lines within the specified tolerance (Figure 5.9). CLEAN also automatically inserts a node at a line

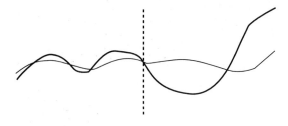

—— Specified fuzzy tolerance

Figure 5.9
The fuzzy tolerance specified by the CLEAN command can snap duplicate lines if the gap between the duplicate lines is smaller than the specified fuzzy tolerance. In this diagram, the duplicate lines to the left of the dashed line will be snapped but not those to the right.

intersection to enforce the topological arc–node relationship.

Because it applies to an entire coverage, the global method must be used cautiously (Box 5.3). If the dangle length is set too large, it will remove undershoots as well as overshoots (Figure 5.10). Instead of being removed, undershoots should be extended to fill the gap. If the fuzzy tolerance is set too large, it will snap arcs that are not duplicate arcs, thus distorting the shape of map features (Figure 5.11).

5.4.2.2 Local Method

The local method deals with selected digitizing errors from a digitized map. To facilitate local editing, GIS users can use tolerances. **Nodesnap** in ArcInfo Workstation specifies the tolerance for snapping nodes, which is as useful in editing as in digitizing. The nodesnap tolerance should not be set too large as to snap wrong nodes. Also, when a node is snapped to a new location within the nodesnap tolerance, the node moves the arc segment that is attached to it. Therefore, the positional accuracy of arcs may suffer because of the snapping of nodes. This is another reason why the nodesnap tolerance must be set carefully.

Editdistance in ArcInfo Workstation specifies a search radius for selecting features for editing

Box 5.3   Default Parameter Values in CLEAN

Unless specified, the parameters of dangle length and fuzzy tolerance are given the default values in ArcInfo Workstation. The default value of dangle length is 0. The default value of fuzzy tolerance depends on whether the coverage is measured in digitizer units (inches) or in real-world coordinates (meters or feet). It is 0.002 inch in digitizer units, or, in real-world coordinates, 1/10,000 of the width or height of the map extent, whichever is greater. The default fuzzy tolerance value therefore increases as the size of the coverage increases in area. For example, the default value is 2 meters if the larger dimension of a coverage measures 20,000 meters, and 20 meters if the dimension measures 200,000 meters. This change in the default value presents a problem to an organization, which may have to piece together a coverage from smaller coverages. A large fuzzy tolerance can easily distort map features. The solution is to use a constant fuzzy tolerance for a project by entering its value with the CLEAN command.

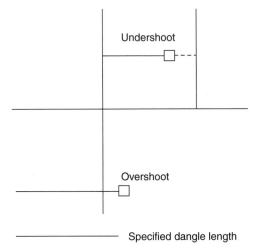

Specified dangle length

Figure 5.10
If the dangle length is set too large, it can remove overshoots, which should be removed, and undershoots, which should not be removed. In this diagram, the dangling arc at the top represents an undershoot. It will be removed using the specified dangle length.

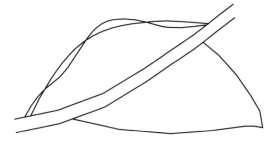

——— Specified fuzzy tolerance

Figure 5.11
If the fuzzy tolerance is set too large, it can remove duplicate lines (top) as well as map features such as a small stream channel (middle) as shown in the diagram.

(Figure 5.12). If the editdistance is too small, the computer cannot select a feature for editing. If it is too large, the computer may select a feature that is within the search radius but not the target feature.

Both nodesnap and editdistance may be set with a specific value or interactively. Because nodesnap affects the geometry of spatial data, an agency may use a threshold value in editing. If set interactively, nodesnap and editdistance must be adjusted frequently during an editing session while the user zooms in and out. Zooming in provides a clearer picture of features to be edited, whereas zooming out allows the user to move to other parts of the coverage for editing. Each zooming in or zooming out changes the map scale and the applicability of nodesnap and editdistance.

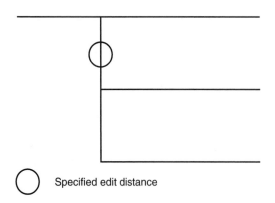

○ Specified edit distance

Figure 5.12
Editdistance should be set large enough for the computer to select the arc to be edited. But, if it is set too large, the computer may select a wrong arc.

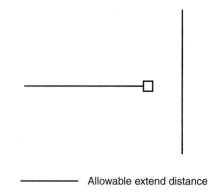

—— Allowable extend distance

Figure 5.13
The EXTEND command uses a specified allowable distance to extend a dangling arc to meet with an existing arc.

Most GIS packages allow GIS users to undo edits that are made inadvertently. ArcInfo Workstation's UNDO command, also known as OOPS, can undo the last edit made. Repeated use of UNDO can remove all edits made in a single edit session.

The following summarizes the use of local methods in removing different types of digitizing errors. Some errors cannot be easily categorized. Often they represent errors made on top of other errors. Checking the source map is the first step to take in removing that kind of error.

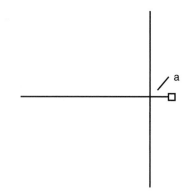

Figure 5.14
To remove the overshoot, simply select the overextension a and delete it.

- Undershoot: One can extend the dangling arc to meet with the target arc at a new node by using a command such as EXTEND in ArcInfo Workstation (Figure 5.13).
- Overshoot: One can remove an overshoot by selecting the extension, which is a separate arc after the coverage has been "cleaned," and deleting it (Figure 5.14).
- Duplicate Arcs: One solution is to carefully select extra arcs and delete them, and the other is to delete all duplicate arcs in a box area and redigitize.
- Wrong Arc Directions: One can alter the arc direction by using a command such as FLIP in ArcInfo Workstation to change the relative position of the from-node and to-node.

- Pseudo Nodes: One can remove a pseudo node by first setting the two arcs on each side of the node to have the same ID value, before using a command such as UNSPLIT in ArcInfo Workstation (Figure 5.15). Occasionally, pseudo nodes represent the end points of missing arcs, which need to be connected.
- Label Errors: One can add new labels with proper ID values for missing labels on a polygon coverage. If a polygon has multiple labels, the extra labels can be selected and

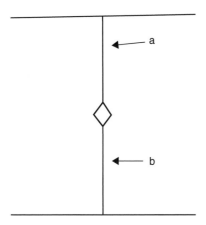

Figure 5.15
To remove the pseudo node, first select a and b, the arcs on both sides of the pseudo node. Make sure the arcs have the same ID value. Then use the UNSPLIT command to remove the pseudo node.

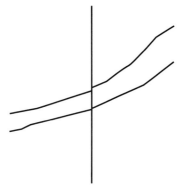

Figure 5.16
The diagram shows lines from two adjacent maps that do not meet perfectly. The mismatch is only visible after zooming in.

deleted. Multiple labels may also indicate an unclosed polygon.
- Reshaping Arcs: To reshape an arc, one can move, add, or delete points (vertices) that make up the arc by using a command such as VERTEX in ArcInfo Workstation.

5.5 EDGEMATCHING

The discussion has so far concentrated on correcting digitizing errors on a single map. Many GIS projects, however, use multiple maps. For example, a forest fire coverage for a national forest may consist of a number of maps, each representing a unit area or a USGS 1:24,000 scale quadrangle map. Each map is digitized and edited separately. After all maps are finished, they must be joined to make the final coverage. **Edgematching** is a necessary operation before joining maps to make sure that lines from two maps meet perfectly along the border (Figure 5.16).

Edgematching involves a snap coverage and an edit coverage. Nodes and arcs on the edit coverage are moved to match those on the snap coverage. A snapping tolerance can assist in snapping

nodes at ends of arcs from the two coverages. Edgematching can be performed on the entire edit coverage or one section of the coverage at a time. After edgematching, one can join the snap coverage and edit coverage together to form a single coverage. The border separating the two coverages stays on the joined coverage, but it can be removed by using a command such as DISSOLVE in ArcInfo Workstation. Task 2 in the Applications section covers the working details of edgematching.

5.6 NONTOPOLOGICAL EDITING

Nontopological editing is defined here as editing on nontopological spatial data. This definition encompasses a wide variety of operations similar to those in CAD (Computer Aided Design). Many of them involve correcting location errors such as extending lines and splitting polygons. These operations are conceptually similar to topological editing. And, similar to topological editing, one can set the snapping tolerances for snapping points and lines and use the line and polygon sketches to select features.

Some nontopological operations are, however, quite different from topological editing. For example, one can create new features using existing

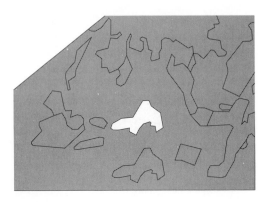

Figure 5.17
After a polygon of a shapefile is moved, a void area appears in its location.

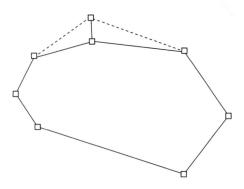

Figure 5.18
Reshape a line by moving a selected vertex.

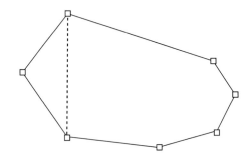

Figure 5.19
Reshape a line by deleting a selected vertex.

map features. Often these new features have disjoint, multiple parts, which are allowed in nontopological data. One can also integrate features from different maps so that they share common boundaries. The geodatabase model, for example, allows operations to be performed on two or more feature classes (maps) within a feature dataset.

5.6.1 Edit Existing Features

The following summarizes nontopological operations on existing features.

- Extend/Trim Lines: One can extend or trim a line to meet a target line.
- Delete/Move Features: One can delete or move one or more selected features, which may be points, lines, or polygons. Because each polygon in nontopological data is a unit, separate from other polygons, moving a polygon means placing the polygon on top of an existing polygon while creating a void area in its original location (Figure 5.17).
- Reshaping Features: One can alter the shape of a line by moving (Figure 5.18), deleting (Figure 5.19), or adding (Figure 5.20) points (also called vertices) on the line. The same method can be used to reshape a polygon. If

the reshaping is intended for a polygon and its connected polygons, one must use a different tool (called the shared edit tool in ArcMap) so that, when a boundary is moved, all polygons that share the same boundaries are reshaped simultaneously.
- Split Lines and Polygons: One can sketch a new line that crosses an existing line to split the line, or sketch a split line through a polygon to split the polygon (Figure 5.21).

5.6.2 Create Features from Existing Features

The following summarizes nontopological operations that create new features from existing features.

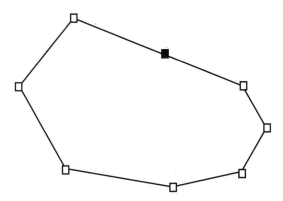

Figure 5.20
Add a vertex to a line.

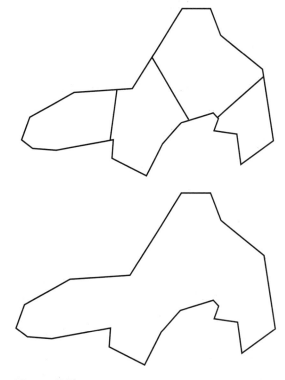

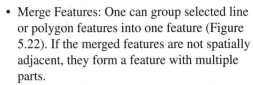

Figure 5.21
A polygon can be split into two by drawing a split line across the polygon boundary.

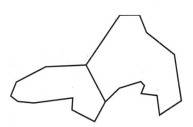

Figure 5.22
Two or more selected polygons can be merged into one.

- Merge Features: One can group selected line or polygon features into one feature (Figure 5.22). If the merged features are not spatially adjacent, they form a feature with multiple parts.
- Buffer Features: One can create a buffer around a line or polygon feature at a specified distance.
- Union Features: One can combine features from different maps into one feature. This operation differs from the merge operation because it works with different maps.
- Intersect Features: One can create a new feature from the intersection of overlapped features from different maps.

5.6.3 Integrate Features

The integrate command in ArcMap uses a user-specified **cluster tolerance** to snap all points or lines that fall within the tolerance (Figure 5.23). The cluster tolerance is functionally similar to the fuzzy tolerance. But the cluster tolerance can be applied to a feature dataset, which may contain two or more feature classes. In other words, the integrate command can snap points and lines that exist in different maps so long as these maps have the same coordinate system. For example, a boundary feature dataset may contain census tract, county, and state boundaries. If these boundaries fall within the specified cluster tolerance, they are considered to be identical and integrated. Like the fuzzy tolerance, the cluster tolerance must be used cautiously because it can alter the geometries of

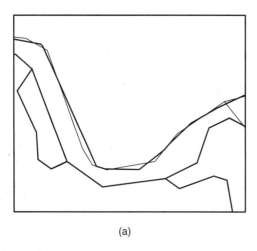
(a)

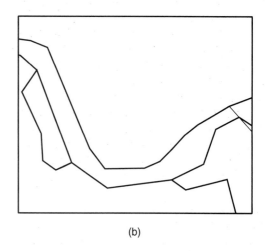
(b)

Figure 5.23
If the boundaries of two maps, such as those along the northern border in (a), fall within a specified cluster tolerance, they are considered to be identical and integrated, as shown in (b). In this case, the integrate command adjusts the boundary of the map in thinner line to be coincident with that of the map in thicker line.

map features. Task 4 in the Applications section covers the integrate command.

5.7 OTHER TYPES OF MAP FEATURE MANIPULATION

5.7.1 Line Simplification, Densification, and Smoothing

Line simplification refers to the process of simplifying or generalizing a line by removing some of its points. Line simplification is a common practice in map display (Robinson et al. 1995) and a common topic under generalization in digital cartography and GIS (McMaster and Shea 1992; Weibel and Dutton 1999). When a map digitized from the 1:100,000 scale source map is to be displayed at the scale of 1:1,000,000, lines become jumbled and fuzzy because of the reduced map space. Line simplification may also be important for some GIS analysis that requires measurements from each point that makes up a line. One example of such analysis is buffering (Chapter 10). Lines with too many points require more computer-

processing time but do not necessarily change or improve the result of analysis. In either scenario, lines can be simplified by removing some of their points. A new research direction in GIS is to store vector data at multiple scales, each of which is associated with a proper level of generalization. Idevio, a Swedish company, has recently developed a compressed, multiresolution format for vector databases (**http://www.idevio.com/).**

The **Douglas-Peucker algorithm,** perhaps the best-known algorithm for line simplification, is used in ArcInfo Workstation and other GIS packages (Douglas and Peucker 1973). The algorithm works line by line and with a specified tolerance. The line-by-line approach is important because point density may vary across a map. For example, parts of a state boundary line may follow physical features and have higher point densities, whereas others have straight-line segments. For a specific line to be simplified, the algorithm starts with a trend line connecting the end points of the line (Figure 5.24). Deviations of the intermediate points from the trend line are then calculated. In the case of Figure 5.24A, the points with the largest deviation for each curve are connected to

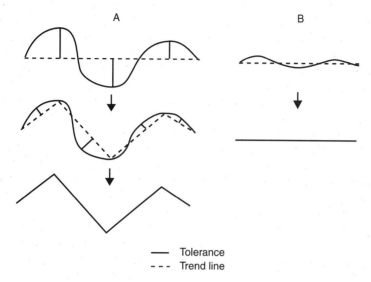

— Tolerance
- - - Trend line

Figure 5.24

The Douglas-Peucker line simplification algorithm is an iterative process, which requires use of a tolerance, trend lines, and calculation of deviations of vertices to the trend line. See text for explanation.

the end points of the original line to form new trend lines, followed by calculation of deviations of the points from the new trend lines. This process continues until no deviations exceed the tolerance, and the result is a simplified line (often with sharp angles) that connects the trend lines. In the case of Figure 5.24B, because deviations of the points are all smaller than the tolerance, the simplified line is the straight line connecting the end points. Figure 5.25 shows the result of line simplification using ArcInfo Workstation's GENERAL-IZE command.

Line densification is the process of adding new points to a line or lines in a map at a specified interval. The operation is relatively simple. Starting at the starting point of a line, the operation adds a new point at every specified interval (Figure 5.26). Existing points are retained, and the shape of the line is not altered.

Line smoothing also adds new points to lines, but the location of new points is generated by mathematical functions. Line smoothing is perhaps most important for data display. Lines derived from computer processing such as isolines on a

precipitation map are sometimes jagged and unappealing. These lines can be smoothed for data display purposes. Figure 5.27 shows the result of smoothing using ArcInfo Workstation's SPLINE command. The most common method for spline consists of cubic polynomials, which have the general form (Davis 1986):

(5.1)

$$y = b_1 + b_2x + b_3x^2 + b_4x^3$$

A cubic polynomial can pass exactly through four points. For an arc with more than four points, one must use a succession of polynomial segments and smooth out abrupt changes in curvature between segments.

5.7.2 Transferring Map Features from One Map to Another

The GIS user will find it useful to be able to transfer map features from one map to another if the maps share common features. For example, a soil map may share some polygon boundaries with a

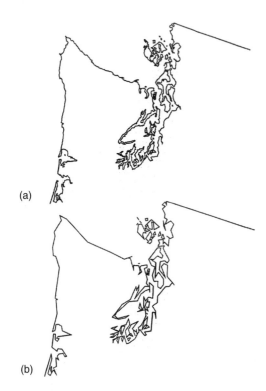

(a)

(b)

Figure 5.25
Two views of the Puget Sound area in Washington.
Created by ArcInfo Workstation's GENERALIZE
command, (b) is a generalized version of (a).

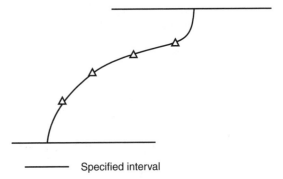

——— Specified interval

Figure 5.26
Line densification adds new vertices to an arc at a
specified interval.

Figure 5.27
The ArcInfo Workstation SPLINE command smoothes
the polygon boundary by adding new vertices
generated mathematically. The smoothed boundary is
shown in thicker line.

land-use map. After the land-use map is digitized,
it is desirable to transfer those shared boundaries
from the land-use map to the soil map. Not only
does this transfer save time in digitizing, but it also
ensures the maps match spatially for data analysis.

ArcInfo Workstation, for example, has the
commands of GET and PUT for transferring map
features. GET copies all features of the specified
type (i.e., point, line, or polygon) from one coverage to another. PUT copies only selected features
from one coverage to another. The copy and paste
operation in ArcMap performs the same function
for nontopological data by copying selected features from one map to another.

KEY CONCEPTS AND TERMS

Check plot: A plot of a digitized map for error
checking.

Cluster tolerance: A tolerance for snapping
points and lines. Functionally similar to the fuzzy
tolerance.

Dangle length: A tolerance used in ArcInfo Workstation that specifies the minimum length for dangling arcs on an output coverage.

Dangling arc: An arc that has the same polygon on both its left and right sides and a dangling node at the end of the arc.

Dangling node: A node at the end of an arc that is not connected to other arcs.

Douglas-Peucker algorithm: A computer algorithm for line simplification.

Edgematching: An operation in joining adjacent coverages.

Editdistance: A tolerance used in ArcInfo Workstation that specifies a search radius for selecting features for editing.

Fuzzy tolerance: A tolerance used in ArcInfo Workstation that specifies the minimum distance between arc vertices on an output coverage.

Line densification: The process of adding new points to a line.

Line simplification: The process of simplifying or generalizing a line by removing some of the line's points.

Line smoothing: The process of smoothing a line by adding new points, which are typically generated by a mathematical function such as spline, to the line.

Location errors: Errors related to the location of map features such as missing lines or missing polygons.

Nodesnap: A tolerance used in ArcInfo Workstation for snapping nodes.

Nontopological editing: Editing on nontopological data.

Overshoot: One type of digitizing error that results in an overextended arc.

Pseudo node: A node appearing along a continuous arc.

Spatial data editing: The process of adding, deleting, and modifying features in digital maps.

Topological editing: Editing on topological data to make sure that they follow the required topological relationships.

Topological errors: Errors related to the topology of map features such as dangling arcs and missing or multiple labels.

Undershoot: One type of digitizing error that results in a gap between arcs.

APPLICATIONS: SPATIAL DATA EDITING

This applications section covers five tasks. Task 1 covers correction of topological errors using ArcInfo Workstation. Task 2 uses ArcTools, ArcInfo Workstation's menu-driven utility program, to perform edgematching. Task 3 reviews the editing capabilities of ArcMap for shapefiles. Task 4 lets you integrate map features in two different maps by using the integrate command in ArcMap. Task 5 uses ArcMap to perform editing of coverage features while maintaining the topological relationship between features. Task 5 requires ArcEditor or ArcInfo.

Task 1: Correct Topological Errors in ArcInfo Workstation

What you need: *editmap1* (Figure 5.28), a coverage with topological errors.

The coverage *editmap1* has several types of digitizing errors: overshoot, undershoot, unclosed polygon, missing label, and multiple labels. Task 1 shows you how you can correct these digitizing errors. While working on this task, you need to zoom in to get a better look at map features for editing and zoom out to move from one part of the coverage to another part. You also need to define

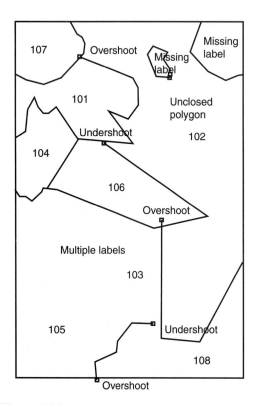

Figure 5.28
Editmap1 has topological errors.

nodesnap, editdistance, and other tolerances and redefine them as you zoom in and out.

1. Go to ArcEdit and set the environment for *editmap1*:

   Arc: arcedit
   Arcedit: display 9999
   Arcedit: mapex editmap1
   Arcedit: editcov editmap1
   Arcedit: drawenv arc node errors label
   Arcedit: nodecolor dangle 2 /* color dangling nodes in red
   Arcedit: nodecolor pseudo 3 /* color pseudo nodes in green
   Arcedit: draw

2. To remove an overshoot:

   Arcedit: editfeature arc /* specify arc as the feature to be edited

   Arcedit: select /* zoom in an overshoot and select the overextended arc
   Arcedit: delete

3. To remove an undershoot:

   Arcedit: editfeature arc
   Arcedit: select /* zoom in an undershoot and select the arc
   Arcedit: extend /* specify a distance so that the dangling arc can meet its target

4. To close an unclosed polygon:

   Arcedit: editfeature node
   Arcedit: nodesnap closest * /* specify a circle for nodes to be snapped
   Arcedit: move /* select the node to move, press 4 to move, select the node to move to

Note: An alternative to moving a node to close an unclosed polygon is to digitize an arc connecting the two dangling nodes.

5. To add a label:

   Arcedit: editfeature label
   Arcedit: add /* click the mouse within the polygon that has a missing label

Note: If the label value is not what you want, press 8 for digitizing options. This will bring up a menu, which allows you to define the new user-id or label value.

6. To remove a label:

   Arcedit: editfeature label
   Arcedit: select /* select the label to remove
   Arcedit: delete

Note: An alternative to removing a label is to move it to a polygon that has a missing label. To move a label, you will first select the label and then type "move."

7. After you have removed digitizing errors in *editmap1*, you need to save the changes before quitting ArcEdit. Then, you must rebuild the topology of *editmap1* by using CLEAN or BUILD.

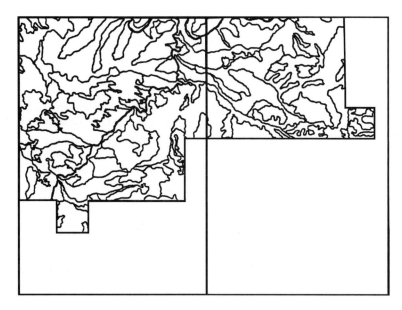

Figure 5.29
Qhoytmtn (right) and qmrblemtn (left) are two coverages to be edgematched and joined.

Task 2: Perform Edgematching in ArcInfo Workstation

What you need: *qhoytmtn* and *qmrblemtn,* two soil coverages that need to be matched and joined (Figure 5.29). Both coverages are based on the Universal Transverse Mercator (UTM) coordinate system and measured in meters.

The soil coverages have been digitized from two separate quads. They must be edgematched and joined if the study area of a project covers both quads. EDGEMATCH is a menu-driven operation in ArcEdit. MAPJOIN in Arc is a dialog command, which can join up to 50 coverages.

1. Go to ArcEdit and start the EDGEMATCH command:

   Arc: arcedit
   Arcedit: display 9999
   Arcedit: edgematch

2. In the Coverage menu, specify *qhoytmtn* as the Edit Coverage, *qmrblemtn* as the Snap Coverage, and Node as the snap coverage feature class. Click Apply. This action initiates the opening of the Edgematching menu and the display of *qhoytmtn* (in white) and *qmrblemtn* (in red) in the ArcEdit window.

3. The Edgematching menu includes several parameter settings, of which the Snap Environment is probably the most important. Click the Snap Environment button. The Snapping Distance specifies the tolerance within which the features from *qhoytmtn* and *qmrblemtn* are linked. A default value of 14.02 meters is given to the Snapping Distance, but it can be changed interactively or by typing in a new value. You can use 7 meters as the snapping distance for this task. For the node to be snapped to within the search area, you can choose the closest one.

4. Click the Add Automatically button in the Edgematching menu. Use Zoom/Pan to check each link. All features from *qhoytmtn* and *qmrblemtn* are linked correctly except for the top two lines, which need to be linked

interactively. Click the Add Interactively button in the menu. Zoom in the area of the first line to be linked so that you can see the end node of the line in each quad. Click on the end node in *qhoytmtn* and then click on its corresponding end node in *qmrblemtn*. An arrow linking the two nodes should appear. Do the same for the other line. Press 9 to exit.

5. Click the Adjust button. The arrow symbols now become square symbols. Again, examine the result of Adjust by using Zoom/Pan. If everything looks right, click the Save button. Exit Edgematch and ArcEdit.

6. Edgematching has altered the topology of the edit coverage. Therefore, you need to rebuild the topology for *qhoytmtn*:

   Arc: build *qhoytmtn* poly

7. Now the two coverages have been edgematched. The next step is to join *qhoytmtn* and *qmrblemtn* to create *qandq*:

   Arc: mapjoin qandq poly all /* enter *qhoytmtn* and *qmrblemtn* as the coverages to join

8. You can display *qandq* in either ArcPlot or ArcEdit. Although the two quads are joined, the quad boundary still remains. To remove the artificial quad boundary, you can use the DISSOLVE command in Arc:

   Arc: dissolve qandq qandq2 #all /* use all attribute items to dissolve *qandq*

9. Figure 5.30 shows *qandq2*.

Task 3: Edit a Shapefile in ArcMap

What you need: *editmap2.shp* (Figure 5.31), *editmap3.shp* (Figure 5.32).

   Task 3 covers three common edit functions in ArcMap for shapefiles: merging polygons, splitting a polygon, and reshaping the polygon boundary. You will work with *editmap2.shp* and use *editmap3.shp* as a reference, which shows how *editmap2.shp* will look like after editing.

1. Start ArcCatalog and make connection to the Chapter 5 database. Launch ArcMap. Right-click Layers in the Table of Contents, select Properties from Layers' context menu, click the General tab, and enter Task 3 for the Name. Add *editmap3.shp* and *editmap2.shp* to Task 3. To edit *editmap2* by using *editmap3* as a guide, they must be shown in different outline symbols.

2. Select Properties from the context menu of *editmap2*. Click the Symbology tab. Click the Symbol, and change the symbol to Hollow with the Outline Color in black. Click the Labels tab. Check the box to Label Features in this layer. Click the Label Field dropdown arrow and select LANDED_ID. Click OK to dismiss the Properties dialog.

3. Click the symbol of *editmap3* in the Table of Contents. Choose the Hollow symbol and the Outline Color of red. Make sure that *editmap2* is above *editmap3* in the Table of Contents.

4. Make sure that Editor Toolbar is checked. Click the Editor dropdown arrow and choose Start Editing. The first edit is to merge polygons 74 and 75. Make sure that the Target layer is *editmap2*. Click the Edit tool. Click inside polygon 75, and then click inside polygon 74 while pressing the Shift key. The two polygons are highlighted in cyan. Click the Editor dropdown arrow and choose Merge. Polygons 74 and 75 are merged into one with the label of 75. (To unselect polygon 75, click the empty space outside the map.)

5. The second edit is to cut polygon 71. Zoom in the area around polygon 71. Use the Edit tool to select polygon 71. Click the Task dropdown arrow and choose Cut Polygon Features. Click the Cut Polygon Features tool. (The ToolTip for the Sketch tool changes depending on the task selection.) To

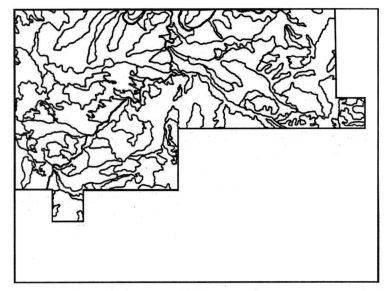

Figure 5.30
Qandq2 is a seamless coverage created from qhoytmtn and qmrblemtn.

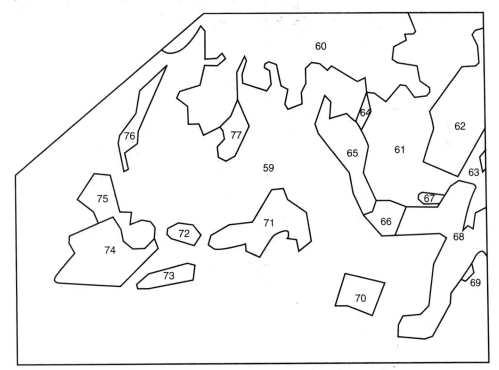

Figure 5.31
Editmap2.shp is a polygon shapefile to be edited in ArcMap for polygon merging, polygon splitting, and reshaping of the polygon boundary.

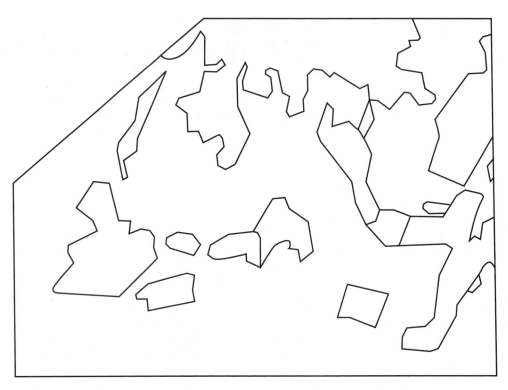

Figure 5.32
Editmap2.shp after editing.

cut a polygon, the cut line should cross over the polygon boundary. Click the left mouse button where you want the cut line to start, click each vertex that makes up the cut line, and double-click the end vertex. Polygon 71 is cut into two, each labeled 71.

6. The third edit is to reshape polygon 73 by extending its southern border in the form of a rectangle. The strategy in reshaping the polygon is to add three new vertices and to drag the vertices to form the new shape. Zoom in the area around polygon 73. Click the Shared Edit tool because polygon 73 shares its boundary with polygon 59. Double-click the boundary of polygon 73 to select it. The edit sketch of Polygon 73 with vertices and segments appears in green except for a red square, which is the last digitized vertex. ArcMap has automatically

changed the Task to Modify Feature. Move the mouse pointer to about the midpoint of the southern border, right-click the point, and select Insert Vertex. A new vertex appears. Position the mouse pointer over the new vertex (vertex 1). When the pointer changes to a four-headed arrow, drag it to where the new border is going to be (use *editmap3* as a guide), and release the mouse button. (The original border of polygon 73 remains in place as a reference. It will disappear when you click anywhere outside polygon 73.)

7. Next, add another vertex (vertex 2) along the line that connects vertex 1 and the original SE corner of polygon 73. Position the mouse pointer over vertex 2. When the pointer changes, drag it to the SE corner of the new boundary. Do the same to form the SW corner of the new boundary.

8. Select Stop Editing from the Editor dropdown list, and save the edits.

Task 4: Integrate Map Features in a Feature Dataset in ArcMap

What you need: *landuse.shp* and *soils.shp*, two shapefiles based on the same coordinate system. In this task, you will integrate map features in a geodatabase feature dataset. As explained in the text, the integrate command uses a user-specified cluster tolerance to snap all points or lines that fall within the tolerance. Because the command will alter the geometries of map features, you must use it with caution. Try to use the smallest cluster tolerance possible to integrate map features. The task takes you through three steps. First, you will create a personal geodatabase and a feature dataset. Second, convert *landuse.shp* and *soils.shp* to feature classes and store them in the feature dataset. Feature classes in the same feature dataset must be based on the same coordinate system. Third, integrate the shoreline boundaries in *landuse.shp* and *soils.shp* to be coincident. Because the shorelines were digitized separately for the shapefiles, they are not on top of one another. In this task, you will make sure that the two maps share the common shoreline.

1. Make sure that ArcCatalog is still connected to the Chapter 5 database. Right-click the Chapter 5 database in the Catalog Tree, point to New, and select Personal Geodatabase. Rename the new geodatabase *Integrate.mdb*. Right-click *Integrate.mdb*, point to New, and select Feature Dataset. In the next dialog, type *trial* in the Name box. Then click Edit to define the coordinate system for the new feature dataset. Click Import in the Spatial Reference Properties dialog. Use the browser and select *landuse.shp* (or *soils.shp*) as the dataset to add. Both *landuse.shp* and *soils.shp* are based on the UTM coordinate system. Click OK to dismiss both dialogs.

2. This step is to convert *landuse.shp* and *soils.shp* to feature classes and store them in the feature dataset *trial*. Right-click

landuse.shp, point to Export, and select Shapefile to Geodatabase. Use the browser in the next dialog to specify *Integrate.mdb* for the output geodatabase. Click the dropdown arrow and select *trial* for the feature dataset. Click OK to dismiss the dialog. Follow the same procedure to convert *soils.shp*. Double-click *Integrate.mdb* in the Catalog Tree and make sure that *landuse* and *soils* reside in the *trial* feature dataset.

3. Select Data Frame from the Insert menu in ArcMap. Rename the new data frame Task 4, and add *trial* to Task 4. *Landuse* and *soils* appear in the map. To better observe how the integrate command works, display *landuse* and *soils* in outline symbols: *landuse* in red and *soils* in black.

4. The shoreline to be integrated follows the northern border of *landuse* and *soils*. Zoom in the northern border, and you will see that *landuse* and *soils* do not share the same shoreline. Use the Measure tool to measure the discrepancy along the shoreline between *landuse* and *soils*. The discrepancy at most locations is smaller than 6 meters.

5. Select Starting Editing from the Editor menu. Then select Options from the Editor menu. Click the Topology tab in the Editing Options dialog. Leave the box checked to integrate the visible extent only. If the box is unchecked, the map feature in the entire feature dataset including soil and land-use boundaries will be integrated. That is not the purpose of this task. Enter 5 (meters) for the Cluster Tolerance. Click OK to dismiss the dialog.

6. Click Integrate in the Editor menu. The shorelines in *soils* and *landuse* are integrated if they fall within 5 meters of each other. You can use the UNDO and REDO Integrate buttons to examine the effects of the integrate command.

7. Pan to another area and follow the same procedure to integrate the shorelines of *soils* and *landuse*. When you are done, select Stop Editing from the Editor menu and save edits.

Task 5: Edit a Coverage in ArcMap

What you need: *editmap2*, a coverage with the same polygon features as *editmap2.shp* in Task 3. Task 5 covers three common edit functions in Arc-Map for polygon features in a coverage: merging polygons, splitting a polygon, and deleting a polygon. The topology between coverage features is automatically rebuilt after editing. ArcView cannot perform topological editing. Therefore, Task 5 requires ArcEditor or ArcInfo.

1. Select Data Frame from the Insert menu in ArcMap. Rename the new data frame Task 5, and add *editmap2 polygon*. Select Properties from the context menu of *editmap2 polygon*. Click the Symbology tab, and change symbology to Single symbol in black. Click the Labels tab. Check the box next to Label Features in this layer, and select LANDED_ID for the Label Field. The map now shows the polygon boundaries and the polygon label IDs. Add *editmap3.shp* to Task 5, and change its symbol to a red outline symbol. You will use *editmap3* as a guide.

2. Click the Editor Toolbar button. Select Starting Editing from the Editor dropdown menu and choose *editmap2 polygon* for editing. First, you will merge polygons 74 and 75. Click the Shared Edit tool. Use the mouse pointer to select the boundary between the two polygons. After the

boundary is highlighted, click the Delete button. The merged polygon carries the label ID of 75.

3. This step is to split polygon 71 into two. Click the Task dropdown arrow and select Auto Complete Polygon, and make sure that *editmap2 polygon* is the Target layer. Click the Sketch (Auto Complete Polygon) tool. Create a line sketch that splits polygon 71 into two, and double-click to finish the sketch. Both polygons carry the same label ID of 71.

4. This step is to delete polygon 72. Click the Shared Edit tool. Use the mouse pointer to select the boundary of the polygon, and click Delete.

5. The last edit is to change the label of one of the two split polygons from polygon 71. Select Open Attribute Table from the context menu of *editmap2 polygon*. Click the Edit tool, and click the polygon that you want to change the label ID from 71 to 78. The record corresponding to the selected polygon is now highlighted in the attribute table. Click the LANDED_ID cell of the highlighted record and enter the ID value of 78.

6. Click Stop Editing from the Editor menu, and save edits. The topology of *editmap2* is updated with the edits you have made.

REFERENCES

Davis, J. C. 1986. *Statistics and Date Analysis in Geology,* 2d ed. New York: Wiley.

Douglas, D. H., and T. K. Peucker. 1973. Algorithms for the Reduction of The Number of Points Required to Represent a Digitized Line or Its Caricature. *The Canadian Cartographer* 10: 110–22.

McMaster, R. B., and K. S. Shea. 1992. *Generalization in Digital Cartography.* Washington, DC: Association of American Geographers.

Robinson, A. H., J. L. Morrison, P. C. Muehrcke, A. J. Kimerling, and S. C. Guptill. 1995. *Elements of Cartography,* 6th ed. New York: Wiley.

Weibel, R., and G. Dutton. 1999. Generalizing Spatial Data and Dealing with Multiple Representations. In P. A. Longley, M. F. Goodchild, D. J. MaGuire, and D. W. Rhind, eds., *Geographical Information Systems,* 2d ed., pp. 125–55. New York: Wiley.

ATTRIBUTE DATA INPUT AND MANAGEMENT

6.1 INTRODUCTION

GIS involves both spatial and attribute data: spatial data relate to the geometry of map features, whereas attribute data describe the characteristics of the map features. Attribute data are stored in tables. Each row of a table represents a map feature, and each column represents a characteristic. The intersection of a column and a row shows the value of a particular characteristic for a particular map feature.

The difference between spatial and attribute data is well defined with vector-based map features. The **georelational data model**, which is still the dominant model in GIS, stores spatial data and attribute data separately and links the two using the feature ID (Figure 6.1). The two data sets must be synchronized so that the spatial and attribute data of map features can be queried, analyzed, and displayed in unison. In contrast, the new **object-oriented data model** combines both spatial and attribute data in a single database. Each map fea-

ture has a unique object ID and an attribute to store its geometry. Synchronization between data sets is not necessary, thus saving the processing overhead and ensuring the integrity of data.

Chapters 4 and 5 cover spatial data input and spatial data editing, respectively. To complete the discussion of vector data, this chapter covers attribute data input and management. Materials covered in this chapter apply to the georelational data model and the object-oriented data model because both models operate in the same relational database environment.

This chapter is divided into the following five sections. Section 6.2 provides an overview of attribute data in GIS. Section 6.3 discusses the relational database model, data normalization, and types of data relationship. Sections 6.4 and 6.5 cover attribute data input and verification, respectively. Section 6.6 discusses creation of new attribute data by data classification and computation.

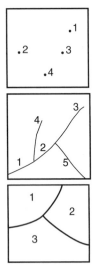

Point-ID	Field 1	Field 2	. . .
1			
2			
3			
4			

Line-ID	Field 1	Field 2	. . .
1			
2			
3			
4			
5			

Polygon-ID	Field 1	Field 2	. . .
1			
2			
3			

Figure 6.1
Attribute data on the right are linked to spatial data by the label ID of map features.

6.2 ATTRIBUTE DATA IN GIS

6.2.1 Organization of Attribute Data

Attribute data are stored in feature attribute tables. An attribute table is organized by row and column. Each row represents a map feature, which has a unique label ID or object ID. Each column describes an attribute of the map feature (Figure 6.2). Line features have the default attribute of length, and area features have the default attributes of area and perimeter. A row is also called a **record** or a *tuple,* and a column is also called a **field** or an *item.*

A feature attribute table may be the only table needed if a map has only several attributes to be associated. But this is not the case with most GIS projects. For example, a soil map unit can have over 80 estimated physical and chemical properties, interpretations, and performance data. To store all these attributes in a single table will require many repetitive entries, a process that wastes both time and computer space. Moreover, the table will be extremely difficult to use and update.

The alternative to storing all attribute data in a single table is to manage attribute data in separate tables and to use a **database management system (DBMS).** A **DBMS** is a set of computer programs for managing an integrated and shared database (Laurini and Thompson 1992). A DBMS provides tools for data input, search, retrieval, manipulation, and output.

Label-ID	pH	Depth	Fertility
1	6.8	12	High
2	4.5	4.8	Low

→ Row

↓ Column

Figure 6.2

A feature attribute table consists of rows and columns. Each row represents a map feature, and each column represents a property or characteristic of the map feature.

Most commercial GIS packages include database management tools for local or internal databases. ArcGIS uses Microsoft Access for the geodatabase model, ArcInfo Workstation uses INFO, AutoCAD Map uses VISION, and MGE uses database management tools from Oracle and Informix. These GIS packages also provide access to external databases managed using Oracle, Informix, SYBASE, SQL Server, IBM DB2, or other DBMSs.

The use of a DBMS has other advantages beyond its GIS applications. Often a GIS is part of an enterprisewide information system, and attribute data needed for the GIS may reside in various departments of the same organization. Therefore, the GIS must function within the overall information system and interact with other information technologies. One effect of the object-oriented data model is the total immersion of GIS in a DBMS.

6.2.2 Type of Attribute Data

One method for classifying attribute data is the data type. A data type determines the kind of data a GIS package can store. Common data types include character, integer, floating, and date. Each common type may have subtypes. The integer type can be short or long, and the floating type can be single or double. Each field in an attribute table must be defined with a data type.

Another method is to define attribute data by measurement scale. The measurement scale concept groups attribute data into nominal, ordinal, interval, and ratio data (Stevens 1946; Chang 1978). **Nominal data** describe different kinds or different categories of data such as land-use types or soil types. **Ordinal data** differentiate data by a ranking relationship. For example, soil erosion may be ordered from severe, moderate, to light. **Interval data** have known intervals between values. For example, a temperature reading of 70°F is warmer than 60°F by 10°F. **Ratio data** are the same as interval data except that ratio data are based on a meaningful, or absolute, zero value. Population densities are an example of ratio data, because a density of 0 is an absolute zero. In GIS applications the four measurement scales may be grouped into two general categories: **categorical data** include nominal and ordinal scales, and **numeric data** include interval and ratio scales.

Measurement scales are important to data display and data analysis in GIS. The choice of map symbols depends on the data to be displayed. For example, different-sized symbols are not appropriate for displaying nominal data. Chapter 8 discusses in more detail the data–symbol relationship.

GIS analysis often involves computation, which is limited to numeric data. For example, descriptive statistics such as mean and standard deviation can only be derived from numeric data. But GIS projects, especially those dealing with suitability analysis, commonly assign scores to nominal or ordinal data and use these scores in computation (Chrisman 2001). For example, a suitability analysis for a housing development may assign different scores to different soil types and then combine these scores with scores from other variables in computing the total suitability score. Assigning scores to categorical data requires information that is not in the base data. Scores in this case represent interpreted data based on expert judgments.

Data type and measurement scale are obviously related. Character strings are appropriate for nominal and ordinal data. Integers and real numbers are appropriate for interval and ratio data, depending on whether decimal digits are included or not. But there are exceptions. For example, a study may classify the potential for groundwater contamination as high, medium, and low, but enter the information as numeric data using a lookup table. The lookup table may show 1 for low, 2 for medium, and 3 for high. It would be wrong to say that the medium potential is twice as high as the low potential because numbers in this case are just numeric codes. GIS packages also have functionalities to convert between data types (e.g., from character strings to numbers, or from numbers to character strings). Therefore, GIS users must pay attention to the nature of attribute data before using them in analysis.

6.3 THE RELATIONAL DATABASE MODEL

A database is a collection of interrelated tables in digital format. There are four types of database design: flat file, hierarchical, network, and relational (Figure 6.3) (Laurini and Thompson 1992; Worboys 1995). A **flat file** contains all data in a large table. An extended feature attribute table is like a flat file. Another common example of a flat file is a spreadsheet with data only. A **hierarchical database** organizes its data at different levels and uses only the one-to-many association between levels. The simple example in Figure 6.3 shows the hierarchical levels of zoning, parcel, and owner. Based on the one-to-many association, each level is divided into different branches. A **network database** builds connections across tables, as shown by the linkages between the tables in Figure 6.3. One problem with both the hierarchical and the network database designs is that the linkages between tables must be known in advance and built into the computer code. This requirement tends to make a complicated and inflexible database.

GIS vendors typically use the relational database model for data management (Codd 1970; Codd 1990; Date 1995). A **relational database** is a collection of tables, also called relations, which can be connected to each other by keys. A **primary key** represents one or more attributes whose values can uniquely identify a record in a table. Its counterpart in another table for the purpose of linkage is called a **foreign key**. Therefore, a key common to two tables can establish connections between corresponding records in the tables. In Figure 6.3, the key connecting zoning and parcel is the zone code and the key connecting parcel and owner is the PIN (parcel ID number). When used together, the keys can relate zoning and owner.

Compared to other database designs, a relational database is simple and flexible. It has two distinctive advantages. First, each table in the database can be prepared, maintained, and edited separately from other tables. This feature is important because, with the increased popularity of GIS technology, more data are being recorded and

(a) Flat file

PIN	Owner	Zoning
P101	Wang	Residential (1)
P101	Chang	Residential (1)
P102	Smith	Commercial (2)
P102	Jones	Commercial (2)
P103	Costello	Commercial (2)
P104	Smith	Residential (1)

(b) Hierarchical

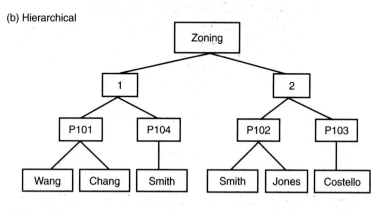

(c) Network

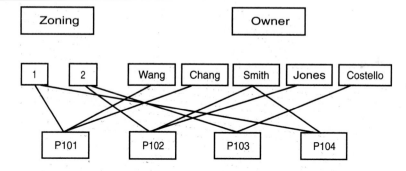

(d) Relational

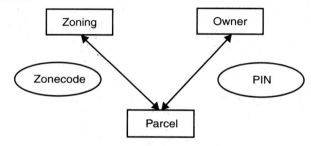

Figure 6.3
Four types of database design: (a) flat file, (b) hierarchical, (c) network, and (d) relational.

managed in spatial units. Second, the tables can remain separate until a query or an analysis requires that attribute data from different tables be linked together. Because the need for linking tables is often temporary, a relational database is efficient for both data management and data processing.

6.3.1 MUIR: An Example of Relational Database

The Natural Resources Conservation Service (NRCS), formerly the Soil Conservation Service, produces the **Soil Survey Geographic (SSURGO) database** nationwide. The NRCS collects SSURGO data from field mapping and archives the data in 7.5-minute quadrangle units for a soil survey area. A soil survey area may consist of a county, multiple counties, or parts of multiple counties. The SSURGO database represents the most detailed level of soil mapping by the NRCS in the United States. Linked to SSURGO data is a **Map Unit Interpretation Record (MUIR)** attribute database, which contains about 88 estimated soil physical and chemical properties, interpretations, and performance data in a series of tables. This comprehensive attribute database is available on the Internet (**http://www.statlab.iastate.edu/soils/muir/**).

The NRCS structures the MUIR database by soil survey area, map units within the survey area, and map unit components within the map unit (Figure 6.4). Four keys are essential in relating tables in the database: state code, soil survey area ID, map unit symbol, and sequence number. The first two keys define the state and the area of the soil survey. The map unit symbol is a numeric code for each soil polygon in the survey. The sequence number key indicates the components of a soil map unit, because a unit may consist of a single dominant soil or two or more soil components. Additionally, various other keys are used to relate soil tables to lookup tables. Examples of lookup tables include soil classification and plants' scientific and common names.

GIS users tend to be overwhelmed by the sheer size of the MUIR database. Actually, the database is not difficult to use if one has a proper

SSarea	Keys: state code, soil survey area ID

Map unit	Keys: state code, soil survey area ID, map unit symbol

Comp	Keys: state code, soil survey area ID, map unit symbol, sequence number

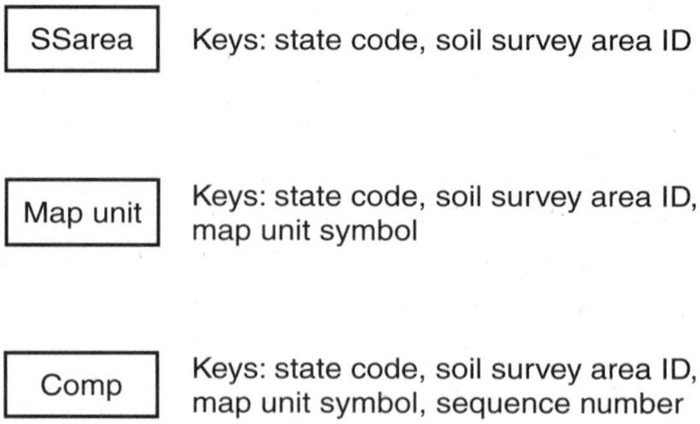

Figure 6.4
The three most important tables in the MUIR database. Ssarea is the soil survey area table, map unit is the soil survey map unit table, and comp is the map unit components table. Each table has keys that can relate to other tables in the database.

understanding of the relational database model and the keys used in relating MUIR tables. This chapter uses the MUIR database to illustrate the type of relationship between tables. Chapter 9 uses the database as an example in data exploration.

6.3.2 Normalization

Preparing a relational database such as MUIR must follow certain rules. One of the rules is called normalization. **Normalization** is a process of decomposition, taking a table with all the attribute data and breaking it down to small tables while maintaining the necessary linkages between them (Vetter 1987). Normalization is designed to achieve several objectives:

- To avoid redundant data in tables that waste space in the database and may cause data integrity problems.
- To ensure that attribute data in separate tables can be maintained and updated separately and can be linked whenever necessary.
- To facilitate a distributed database.

Table 6.1 shows attribute data for a parcel map. Table 6.1 is unnormalized because it contains redundant data. Owner addresses are repeated for Smith and residential and commercial zoning are

TABLE 6.1	An Unnormalized Table					
PIN	**Owner**	**Owner Address**	**Sale Date**	**Acres**	**Zone Code**	**Zoning**
P101	Wang	101 Oak St	1-10-98	1.0	1	Residential
	Chang	200 Maple St				
P102	Smith	300 Spruce Rd	10-6-68	3.0	2	Commercial
	Jones	105 Ash St				
P103	Costello	206 Elm St	3-7-97	2.5	2	Commercial
P104	Smith	300 Spruce Rd	7-30-78	1.0	1	Residential

TABLE 6.2	First Step in Normalization					
PIN	**Owner**	**Owner Address**	**Sale Date**	**Acres**	**Zone Code**	**Zoning**
P101	Wang	101 Oak St	1-10-98	1.0	1	Residential
P101	Chang	200 Maple St	1-10-98	1.0	1	Residential
P102	Smith	300 Spruce Rd	10-6-68	3.0	2	Commercial
P102	Jones	105 Ash St	10-6-68	3.0	2	Commercial
P103	Costello	206 Elm St	3-7-97	2.5	2	Commercial
P104	Smith	300 Spruce Rd	7-30-78	1.0	1	Residential

entered twice. An unnormalized table can be difficult to prepare and manage. For example, the number of values for the fields of owner and owner address in Table 6.1 varies from record to record, making it difficult to define the fields and to store field values. Also, if ownership changes, the table must be updated with all the attribute data. The same difficulty applies to operations in which values are added and deleted.

Table 6.2 represents the first step in normalization. Often called the first normal form, Table 6.2 no longer has multiple values in its cells but the problem of data redundancy has increased. The parcels P101 and P102 are repeated twice except for their owners and owner addresses. Smith's address is repeated twice. And the zoning descriptions of residential and commercial are repeated three times each. Also, identification of the owner address is not possible with PIN alone but requires a compound key of PIN and owner.

Figure 6.5 represents the second step in normalization. In place of Table 6.2 are three small tables for parcel, owner, and address tables. PIN is the key relating the parcel and owner tables. Owner name is the key relating the address and owner tables. The relationship between the parcel and address tables can be established through the keys of PIN and owner name. The only problem with the second normal form is data redundancy with the fields of zone code and zoning.

The final step in normalization with the above example is summarized in Figure 6.6. A new table, zone, is created to take care of the remaining data redundancy problem with zoning. Zone code is the key relating the parcel and zone tables. Unnormalized data in Table 6.1 are now fully normalized.

Although it can achieve objectives consistent with the relational database model, normalization does have a major drawback of slowing down data access. To find the addresses of parcel owners, for

	PIN	Sale date	Acres	Zone code	Zoning
Parcel table	P101	1-10-98	1.0	1	Residential
	P102	10-6-68	3.0	2	Commercial
	P103	3-7-97	2.5	2	Commercial
	P104	7-30-78	1.0	1	Residential

	PIN	Owner name
Owner table	P101	Wang
	P101	Chang
	P102	Smith
	P102	Jones
	P103	Costello
	P104	Smith

	Owner name	Owner address
Address table	Wang	101 Oak St
	Chang	200 Maple St
	Jones	105 Ash St
	Smith	300 Spruce Rd
	Costello	206 Elm St

Figure 6.5

Separate tables from the second step in normalization.

	PIN	Sale date	Acres	Zone code
Parcel table	P101	1-10-98	1.0	1
	P102	10-6-68	3.0	2
	P103	3-7-97	2.5	2
	P104	7-30-78	1.0	1

	Owner name	Owner address
Address table	Wang	101 Oak St
	Chang	200 Maple St
	Jones	105 Ash St
	Smith	300 Spruce Rd
	Costello	206 Elm St

	PIN	Owner name
Owner table	P101	Wang
	P101	Chang
	P102	Smith
	P102	Jones
	P103	Costello
	P104	Smith

	Zone code	Zoning
Zone table	1	Residential
	2	Commercial

Figure 6.6

Separate tables after normalization.

example, one must link three tables (parcel, owner, and address) and employ two keys (PIN and owner name). One way to increase the performance in data access is to reduce the level of normalization by, for example, removing the address table and including the addresses in the owner table. Therefore, normalization should be maintained in the conceptual design of a database but performance and other factors should be considered in its physical design.

6.3.3 Type of Relationship

A relational database may contain four types of relationship between tables: one-to-one, one-to-many, many-to-one, and many-to-many (Figure 6.7). The **one-to-one relationship** means that one and only one record in a table is related to one and only one record in another table. The **one-to-many relationship** means that one record in a table may be related to many records in another table. For example, the street address of an apartment complex may include several households. The **many-to-one relationship** means that many records in a table may be related to one record in another table. For example, several households may share the same street address. The **many-to-many relationship** means that many records in a table may be related to many records in another table. For example, a vegetation stand can grow more than one species and a species can grow in more than one stand. The many-to-many relationship usually requires special handling in GIS (Laurini and Thompson 1992).

To explain these relationships, especially one-to-many and many-to-one, the designation of the tables can be helpful. For example, if the purpose is to join attribute data from another table to the feature attribute table, then the feature attribute table is the base table and the other table is the table to be joined. The feature attribute table has the primary key and the other table has the foreign key. Often the designation of tables depends on the data stored in the tables and the information sought. This is illustrated in the following two examples.

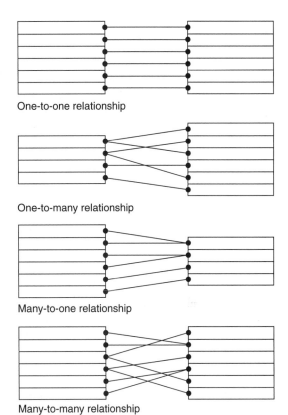

One-to-one relationship

One-to-many relationship

Many-to-one relationship

Many-to-many relationship

Figure 6.7
Four types of data relationship between tables: one-to-one, one-to-many, many-to-one, and many-to-many.

6.3.3.1 Example 1: Type of Relationship between Normalized Tables

Figure 6.6 shows four tables from normalization: parcel, owner, address, and zone. Suppose the question is to find who owns a selected parcel. To answer this question, one can treat the parcel table as the base table and the owner table as the table to be linked. The relationship between the tables is one-to-many: one record in the parcel table may correspond to more than one record in the owner table.

Suppose the question is now changed to find land parcels owned by a selected owner. The owner table becomes the base table and the parcel table is the table to be linked. The relationship is many-to-one: more than one record in the owner table may correspond to one record in the parcel table. The same is true between the parcel table and the zone table. If the question is to find the zoning code for a selected parcel, it is a many-to-one relationship. If the question is to find land parcels that are zoned commercial, it is a one-to-many relationship.

6.3.3.2 Example 2: Type of Relationship in the MUIR Database

One approach to better understanding the MUIR database is to sort out the relationship between tables. Many relationships are either one-to-many or many-to-one. For example, the many-to-one relationship exists between the map unit components table and the soil classification table because different soil components may be grouped into the same soil class (Figure 6.8). On the other hand, the one-to-many relationship applies to the map unit components table and the soil profile layers table because a soil component may have multiple profile layers (Figure 6.9).

6.3.3.3 Type of Relationship and Data Display

The type of relationship can influence how data are displayed. To display the parcel ownership would be simple if the relationship between the parcel and ownerships tables is one-to-one or many-to-one. The one-to-one relationship would have one unique symbol for each parcel, and the many-to-one relationship would have one symbol for either one or more than one parcel.

The one-to-many relationship would create a problem in symbolization. Because a parcel may have more than one owner, it would not be right to show a parcel with multiple owners with an owner who happens to be the first on the list of owners. One solution to the problem is to treat multiple owners as a group or to use a different symbol design for multiple owners.

6.3.4 Working with Attribute Data Management in GIS

Attribute data management is an important part of a GIS project. Most GIS packages provide tools for connecting tables with different relationships. A Windows-based GIS package is clearly better than a GIS package using command lines for attribute data management, because a Windows-based GIS package allows the user to open two or more tables and view the linking between tables simultaneously.

As examples, Boxes 6.1 and 6.2 describe attribute data management using ArcView and ArcInfo Workstation, respectively.

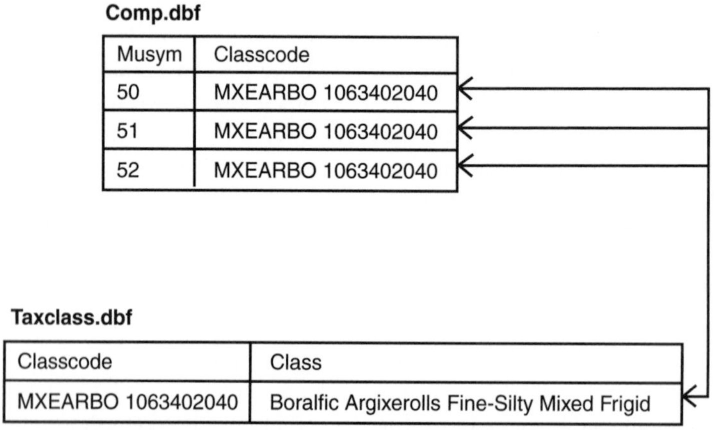

Figure 6.8
An example of the many-to-one relationship in the MUIR database. Three map unit components in comp.dbf are related to the same soil class in the taxclass.dbf.

6.4 ATTRIBUTE DATA ENTRY

6.4.1 Field Definition

The first step in attribute data entry is to define each field in the table. A field definition usually includes data width, data type, and number of decimal places. The width refers to the number of spaces to be reserved for a field. The width should be large enough for the largest number or the longest string in the data. Spaces for the negative sign and the decimal point should also be included in the width. The data type must follow data types allowed in the GIS package. The number of decimal places is part of the definition for the real numeric data type.

Character and numeric type definitions can be confusing at times. For example, the soil map unit in the MUIR database is defined as character, although the map unit is coded as numbers such as 610001, 610002, and so on. These are ID numbers to identify soil map units; therefore, these unit numbers are nominal data rather than interval or ratio data. In other words, computations such as subtracting 610001 from 610002 are meaningless with soil map unit numbers. But treating soil map unit numbers as numeric may be useful in certain types of data query, such as use of a numeric range from 610001 to 610010 in a query.

6.4.2 Methods of Data Entry

Attribute data entry is like digitizing for spatial data entry. One must enter attribute data by typing. A map with 4000 polygons, each with 50 fields of attribute data, could require entering 200,000

Comp.dbf

Muid	Seqnum	Musym
610001	1	1

Layer.dbf

Muid	Seqnum	Layernum	Laydepl	Laydeph
610001	1	1	0	7
610001	1	2	7	18
610001	1	3	18	58
610001	1	4	58	62

Figure 6.9

An example of the one-to-many relationship in the MUIR database. One map unit component in comp.dbf is related to four layers in layer.dbf.

 Box 6.1 **Attribute Data Management Using ArcView**

The ArcMap application of ArcView has two commands for managing tables: join and relate. Join brings together two tables, given that each table has a common key. The keys, called the relate field, do not have to have the same name but must be of the same data type. Join is usually recommended for the one-to-one or many-to-one relationship. Given the one-to-one relationship, two tables are joined by record. Given the many-to-one relationship, many records in the base table have the same value from a record in the other table. Join is inappropriate with the one-to-many relationship, because only the first matching record value from the other table is assigned to a record in the base table.

Relate temporarily connects two tables but keeps the tables separate. Relate is therefore appropriate for all four relationships. It can also work with three or more tables simultaneously by first establishing connections between these tables in pairs. One disadvantage of relate is that it tends to slow down data access.

The geodatabase model allows ArcEditor and ArcInfo users to set up relationship classes in ArcCatalog. A relationship class establishes a relationship between feature classes or tables and is functionally similar to relate in ArcMap. After a relationship class is defined, it is automatically activated whenever the feature classes or tables that participate in the relationship are added to a map in ArcMap.

Box 6.2 | Attribute Data Management Using ArcInfo Workstation

ArcInfo Workstation has the same attribute data management capabilities as ArcView, but requires use of command lines. JOINITEM joins two tables permanently with a shared item. The shared item must be defined exactly the same, including the name, in both tables. JOINITEM can work with the one-to-one or many-to-one relationship. RELATE temporarily connects two tables. Use of two or more relates can link three or more tables. In the case of a one-to-many relationship, the CURSOR command can be used to highlight records that correspond to a single record in the base table. CURSOR then allows the user to step through the highlighted (selected) records one record at a time.

values. How to reduce time and effort in attribute data entry is of interest to any GIS user.

As in finding existing data for GIS projects, it is best to determine if a government agency or an organization has already entered attribute data in digital format. If the answer is yes, then one can simply import the digital data file into a GIS. The data format is important in importing. Most GIS packages can import dBASE and ASCII files as well as data from database servers such as Oracle, Access, Sybase, IBM DB2, and Informix.

If attribute data files do not exist, then typing is the option. But the amount of typing can vary depending on which method or command is used. The ArcInfo Workstation command UPDATE, for example, can be used for attribute data entry but requires the typing of values one record at a time.

One way to save time in typing is to take advantage of the relational database model by having a lookup table. For example, instead of entering attribute data for each polygon in a landform map, one can use a key and a lookup table. First, each landform polygon is assigned a map unit symbol. Next, a lookup table is prepared to list landform attributes for each map unit symbol. By using the map unit symbol as the key, one can then join or relate data from the lookup table to the landform map. The same lookup table can be used for other maps as well.

Attribute data entry using a GIS package is usually more cumbersome than using a wordprocessing or spreadsheet package. A wordprocessing package offers cut-and-paste, find-and-replace, and other functions that are helpful to a typist, especially a poor typist. A GIS package typically does not offer these wordprocessing options. Therefore, it is wise to enter attribute data using a "typist friendly" package, export the data file in a format compatible with the GIS package, and import the file.

6.5 ATTRIBUTE DATA VERIFICATION

Attribute data verification consists of two parts. The first is to make sure that attribute data are properly linked to spatial data: the label or object ID should be unique and should not contain null (empty) values. The second is to verify the accuracy of attribute data. Data verification is difficult because inaccuracies may be attributed to a large number of factors including observation errors, out-of-date data, and data entry errors.

There are at least two methods for checking data entry errors. First, attribute data can be printed for manual verification. This is like using check plots to check the accuracy of spatial data. Second, computer programs can be written to verify data accuracy. For example, the NRCS has developed automated data validation procedures to catch obvious errors in soil attribute data such as data type, data length, data range, and key data field. The procedures also test attribute data against a master set of valid attributes to catch logical errors such as having incompatible components in the same soil map unit. Automated data validation procedures require expert knowledge about attribute data and a well-designed computer program that can catch all possible errors.

One advantage of using the geodatabase model is its validation rules (Zeiler 1999). Directly related to attribute data verification is the rule about attribute domains, which groups objects into classes by a valid range of values or a valid set of values for an attribute. Suppose the field zoning has the value of 1 for residential, 2 for commercial, and 3 for industrial for a land parcel feature class. This set of zoning values can be enforced whenever the field zoning is edited. Therefore, if a zoning value of 9 is entered, the value will be flagged or rejected because it is outside the valid set of values. Similar constraint using a valid numeric range instead of a valid set of values can be applied to lot size or height of building.

6.6 CREATING NEW ATTRIBUTE DATA FROM EXISTING DATA

6.6.1 Attribute Data Classification

New attribute data can be created from existing data through data classification. Data classification reduces a data set to a small number of classes on the basis of the values of an attribute or attributes. For example, elevations may be grouped into lower than 500 meters, 500 to 1000 meters, and so on. Another example is to classify by elevation and slope: class 1 has lower than 500 meters in elevation and 0 to 10% slope, class 2 has lower than 500 meters in elevation and 10 to 20% slope, and so on.

Operationally, creating new attribute data by classification involves three steps: defining a new field for saving the classification result, selecting a data subset through query, and assigning a value to the selected data subset. The second and third steps are repeated until all records are classified and assigned new field values, unless a computer code is written to automate the procedure (see Task 6 in the Applications section). The main advantage of data classification is that it reduces or simplifies a data set so that the new data set can be more easily used in GIS analysis or modeling.

6.6.2 Attribute Data Computation

New attribute data can also be created from existing data through computation. Operationally, it involves two steps: defining a new field, and computing the new field values from the values of an existing attribute or attributes.

A simple example is to create a new attribute called feet for a trail map that is measured in meters. This new attribute can be computed by length $\times$ 3.28, where length is an existing attribute.

Another example is to create a new attribute that measures the quality of wildlife habitat by evaluating the existing attributes of slope, aspect, and elevation. The first step in completing this task is to develop a scoring system for each variable. One can then compute the total score that measures the quality of wildlife habitat by summing the scores for slope, aspect, and elevation. In some cases, different weights may be assigned to different variables. For example, if elevation is three times as important as slope and aspect, then one can compute the total score by using the following equation: slope score + aspect score + 3 $\times$ elevation score.

KEY CONCEPTS AND TERMS

Categorical data: Data that are measured at a nominal or an ordinal scale.

Database Management System (DBMS): A set of computer programs for managing an integrated and shared database for such tasks as data input, search, retrieval, manipulation, and output.

Field: A column in a table that describes an attribute of a map feature. Also called *column* or *item*.

Flat file: A database that contains all data in a large table.

Foreign key: One or more attributes that can uniquely identify a record in another table.

Georelational data model: A common model used in a GIS package that stores spatial data in binary files and attribute data in tables, and links the two data components by the IDs of map features.

Hierarchical database: A database that is organized at different levels and uses the one-to-many association between levels.

Interval data: Data with known intervals between values, such as temperature readings.

Many-to-many relationship: One type of data relationship, in which many records in a table are related to many records in another table.

Many-to-one relationship: One type of data relationship, in which many records in a table are related to one record in another table.

Map Unit Interpretation Record (MUIR):   An attribute database linked to SSURGO data.

Network database: A database that is based on the built-in connections across tables.

Nominal data: Data that show different kinds or different categories, such as land-use types or soil types.

Normalization: The process of taking a table with all the attribute data and breaking it down to small tables while maintaining the necessary linkages between them in a relational database.

Numeric data: Data that are measured at an interval or a ratio scale.

Object-oriented data model: A data model that stores spatial data and attribute data in a single database. An example is ArcGIS' geodatabase model.

One-to-many relationship: One type of data relationship, in which one record in a table is related to many records in another table.

One-to-one relationship: One type of data relationship, in which one record in a table is related to one and only one record in another table.

Ordinal data: Data that are ranked, such as large, medium, and small cities.

Primary key: One or more attributes that can uniquely identify a record in a table.

Ratio data: Data with known intervals between values and are based on a meaningful zero value, such as population densities.

Record: A row in a table that represents a map feature. Also called *row* or *tuple*.

Relational database: A database that consists of a collection of tables and uses keys to connect the tables.

Soil Survey Geographic (SSURGO) database: A database maintained by the Natural Resources Conservation Service (NRCS), formerly the Soil Conservation Service, which archives soil survey data in 7.5-minute quadrangle units.

APPLICATIONS: ATTRIBUTE DATA ENTRY AND MANAGEMENT

This applications section has seven tasks. Tasks 1 and 2 cover attribute data entry. Task 1 uses a geodatabase feature class and Task 2 uses a shapefile. Tasks 3 and 4 cover relating tables and joining tables respectively. Tasks 5 and 6 show how to create new attributes through data classification. Task 5 uses the conventional method of repeatedly selecting a data subset and assigning a class value. Task 6, on the other hand, uses a Visual Basic script to automate the procedure. Task 7 shows how to create new attributes through data computation.

Task 1: Enter Attribute Data of a Geodatabase Feature Class

What you need: *landat.shp,* a polygon shapefile with 19 records.

In Task 1, you will learn how to enter attribute data using a geodatabase feature class and a domain. The domain and its coded values can restrict values to be entered, thus avoiding data entry errors.

1. Start ArcCatalog, and make connection to the Chapter 6 database. You will first create a personal geodatabase. Right-click the Chapter 6 database in the Catalog tree, point to New, and select Personal Geodatabase. Rename the new personal geodatabase *land.mdb*. (If the extension .mdb does not appear, do the following: select Options from the Tools menu, click the General tab, and uncheck the box next to Hide file extensions at the bottom of the Options dialog.)

2. This step is to add *landat.shp* as a feature class to *land.mdb*. Right-click *land.mdb*, point to Import, and select Shapefile to Geodatabase. Use the browse button or the drag-and-drop method to add *landat.shp* as the input shapefile. Click OK and dismiss the Shapefile to Geodatabase dialog. Click the plus sign next to *land.mdb*. *Landat* is a feature class in the geodatabase.

3. Now you will create a domain for the geodatabase. Select Properties from the context menu of *land.mdb*. The Database Properties dialog uses three frames: Domains, Domain Properties, and Coded Values. You need to work with all three frames. Click the first cell under Domain Name, and enter lucodevalue. Click the cell next to Field Type, and select Short Integer. Click the cell next to Domain Type, and select Coded Values. Click the first cell under Code and enter 100. Click the cell next to 100 under Description, and enter urban. Enter 200, 300, 400, 500, 600, and 700 following 100 under Code and enter their respective descriptions of agriculture, brushland, forestland, water, wetland, and barren under Description. Click Apply and OK to dismiss the Database Properties dialog.

4. This step is to add a new field to *landat* and to specify the field's domain. Right-click

landat and select Properties. Click the Fields tab. Click the first empty cell under Field Name and enter lucode. Click the cell next to lucode and select Short Integer. Click the cell next to Domain in the Field Properties frame and select lucodevalue. Click Apply and OK to dismiss the Properties dialog.

5. Launch ArcMap. Add *landat* to Layers, and rename the data frame Tasks 1&2. Select Open Attribute Table from the context menu of *landat*. Lucode appears with Null values in the last field of the table.

6. Click the Editor Toolbar button. Click the Editor dropdown arrow and select Start Editing. Right-click the field of LANDAT_ID and select Sort Ascending. Now you are ready to enter the lucode values. Click the first cell under lucode and select forestland (400). Enter the rest of the lucode values according to the table below.

Landat-ID	Lucode	Landat-ID	Lucode
59	400	69	300
60	200	70	200
61	400	71	300
62	200	72	300
63	200	73	300
64	300	74	300
65	200	75	200
66	300	76	300
67	300	77	300
68	200		

7. When you finish entering the lucode values, select Stop Editing from the Editor dropdown list. Save the edits.

Task 2: Enter Attribute Data of a Shapefile

What you need: *landat.shp*, same as in Task 1.

Task 2 is the same as Task 1 except that you will use a shapefile. No coded values can be predefined for a shapefile. Therefore, data entry in Task 2 will be slower and more prone to errors.

1. Make sure that ArcCatalog is still connected to the Chapter 6 database. You must add a new field to *landat.shp* first. Select Properties from the context menu of *landat.shp*. Under the Fields tab, click the empty cell below LANDAT_ID in the Field Name column and enter lucode. Click the cell next to lucode in the Data Type column, and select Short Integer. Click OK to dismiss the Properties dialog. Now preview the table of *landat.shp*. LUCODE has been added to the table and populated with 0s.

2. Activate ArcMap, and add *landat.shp* to Tasks 1&2. Select Open Attribute Table from the context menu of *landat* (the one on top). Click the Editor Toolbar, and select Start Editing from the Editor dropdown list.

3. In the Attributes of *landat* table, the fields in white such as lucode can be edited. Right-click LANDAT_ID and select Sort Ascending. Click the first cell under lucode and enter 400. Enter the other lucode values according to the table in Task 1.

4. Select Stop Editing from the Editor dropdown list, and save edits.

Task 3: Relate Tables in ArcMap

What you need: *wp.shp,* a forest stand shapefile; *wpdata.dbf* and *wpact.dbf,* two attribute data files that can be linked to *wp.shp*. *Wpdata.dbf* includes vegetation and land-type data, and *wpact.dbf* includes activity records.

Task 3 lets you relate two tables in ArcMap.

1. Select Data Frame from the Insert menu in ArcMap. Rename the new data frame Task 3. Add *wp.shp*, *wpdata.dbf*, and *wpact.dbf* to Task 3.

2. Check the key that can be used in relating tables. Right-click *wp* and select Open Attribute Table. The field ID will be used in linking. Open *wpact* and *wpdata*. Both tables also have the field ID. Close the tables.

3. The first relate is between *wp* and *wpdata*. Right-click *wp*, point to Joins and Relates,

and select Relate. In the Relate dialog, select ID in the first dropdown list, *wpdata* in the second list, and ID in the third list, and accept Relate1 as the relate name.

4. The second relate is between *wpdata* and *wpact*. Right-click *wpdata*, point to Joins and Relates, and select Relate. In the Relate dialog, select ID in the first dropdown list, *wpact* in the second list, and ID in the third list, and enter Relate2 as the relate name.

5. The three tables are now related in pairs by two relates. Right-click *wpdata* and select Open. Click the Options dropdown arrow in the *wpdata* table and choose Select by Attributes. In the next dialog, create a new selection by entering the following SQL statement in the expression box: "ORIGIN" > 0 AND "ORIGIN" <= 1900. Click Apply. Click Selected at the bottom of the table so that only selected records are shown.

6. To see which records in the *wp* attribute table are related to the selected records in *wpdata*, go through the following steps. Click the Options dropdown arrow in the *wpdata* table, point to Related Tables, and click Relate1. The Selected Attributes of *wp* table shows the related records. And the *wp* map shows where those selected records are located.

7. You can follow the same procedure as in the previous step to see which records in *wpact* are related to those selected polygons in *wp*.

Task 4: Join Tables in ArcMap

What you need: *wp.shp* and *wpdata.dbf,* same as in Task 3.

Task 4 asks you to join a dBASE file to a feature attribute table. Join combines attribute data from different tables into a single table, making it possible to use all attribute data in query, classification, or computation.

1. Select Data Frame from the Insert menu in ArcMap. Rename the new data frame Task 4. Add *wp.shp* and *wpdata.dbf* to Task 4.

2. Right-click *wp* and select Open Attribute Table. The field ID will be used in joining tables. Open *wpdata*. *Wpdata* contains attributes such as Origin, As, and Elev for *wp*. The field ID will be used in joining tables. Close both the *wp* attribute table and *wpdata*.

3. This step is to join *wpdata* to the attribute table of *wp*. Right-click *wp*, point to Joins and Relates, and select Join. At the top of the Join Data dialog, opt for joining attributes from a table. Then, select ID in the first dropdown list, *wpdata* in the second list, and ID in the third list. Click OK to dismiss the dialog. You can open the attribute table of *wp* to see the expanded table.

Task 5: Create New Attribute by Data Classification

What you need: *wpdata.dbf*.

Task 5 demonstrates how the existing attribute data can be used for data classification and creation of a new attribute.

1. Exit ArcMap , and go back to ArcCatalog. Select Properties from the context menu of *wpdata.dbf* in the Catalog tree. Under the Fields tab, click the first empty cell in the Field Name column and enter ELEVZONE. Click the cell next to ELEVZONE in the Data Type column, and select Short Integer from the dropdown list. Click Apply and then OK.

2. Launch ArcMap. Rename the data frame Tasks 5&6. Add *wpdata.dbf* to Tasks 5&6.

3. Select Open from the context menu of *wpdata*. ELEVZONE should appear in the table with 0s. Click the Options dropdown arrow and choose Select by Attributes. Make sure that the selection method is to create a new selection. Enter the following SQL statement in the expression box: "ELEV" > 0 AND "ELEV" <= 40. Click Apply. Click Selected at the bottom of the table so that only selected records are shown. These

selected records fall within the first class of ELEVZONE.

4. Right-click the field ELEVZONE and select Calculate Values. A Field Calculator message box appears with information about calculating values outside an edit session. Click Yes to proceed. Enter 1 in the expression box of the Field Calculator dialog, and click OK. The selected records in *wpdata* are now populated with the value of 1, meaning that they all fall within class 1.

5. Go back to the Select by Attributes dialog. Enter the SQL statement: "ELEV" > 40 AND "ELEV" <= 45. Click Apply. Follow the same procedure above to calculate the ELEVZONE value of 2 for the selected records.

6. Repeat the same procedure to select the remaining two classes of 46–50 and > 50, and to calculate their ELEVZONE values of 3 and 4, respectively.

Task 6: Use Advanced Method for Attribute Data Classification

What you need: *wpdata.dbf*.

You have classified ELEVZONE in *wpdata.dbf* by repeating the procedure of selecting a data subset and calculating the class value in Task 5. This task shows you how you can use a Visual Basic code and the advanced option to calculate the ELEVZONE values all at once.

1. Select Open from the context menu of *wpdata*. ELEVZONE should appear in the table with values calculated in Task 5. If necessary, clear selected records in *wpdata* by selecting Clear Selection from Options' dropdown menu. Right-click ELEVZONE and select Calculate Values. A Field Calculator message box appears with information about calculating values outside an edit session. Click Yes to proceed.

2. Check Advanced in the Field Calculator dialog. Next you will create a Visual Basic script to perform the classification. The first

line of your script declares an integer variable called intclass, which stores the classification value. Enter the following statements in the first box under "Pre-Logic VBA Script Code":

```
Dim intclass as integer
If [ELEV] > 0 and [ELEV] <= 40 Then
Intclass = 1
ElseIf [ELEV] > 40 and [ELEV] <= 45 Then
Intclass = 2
ElseIf [ELEV] > 45 and [ELEV] <= 50 Then
Intclass = 3
ElseIf [ELEV] > 50 Then
Intclass = 4
End If
```

3. Enter intclass in the second box under "ELEVZONE =". In other words, ELEVZONE will be assigned the value from intclass. Click OK to dismiss the Field Calculator. ELEVZONE is now populated with values calculated by the Visual Basic code. They should be the same as those in Task 5.

Task 7: Create New Attribute by Data Computation

What you need: *wp.shp* and *wpdata.dbf*.

You have created a new field from data classification in Tasks 5 and 6. Another common method for creating new fields is computation. Task 7 shows how a new field can be created and computed from existing attribute data.

1. Close ArcMap if it is open with *wp*. Go back to ArcCatalog. Select Properties from the context menu of *wp.shp* in the Catalog tree. Under the Fields tab, click the first empty cell in the Field Name column, and enter ACRES. Click the cell next to ACRES in the Data Type column, and select Float from the dropdown list. Click Apply and OK to dismiss the Properties dialog.

2. Activate ArcMap. Rename the new data frame Task 7. Add *wp.shp* to Task 7. Open the attribute table of *wp*. The new field ACRES appears in the table with 0s.

3. Right-click ACRES to select Calculate Values. Click Yes in the message box about calculating outside an edit session. In the Field Calculator dialog, enter the following expression in the box below ACRES =: [AREA] / 1000000 * 247.11. Click OK. The field ACRES now shows the polygons in acres.

REFERENCES

Chang, K. 1978. Measurement Scales in Cartography. *The American Cartographer* 5: 57–64.

Chrisman, N. 2001. *Exploring Geographic Information Systems*, 2d ed. New York: Wiley.

Codd, E. F. 1970. A Relational Model for Large Shared Data Banks. *Communications of the Association for Computing Machinery* 13: 377–87.

Codd, E. F. 1990. *The Relational Model for Database Management,* Version 2. Reading, MA: Addison-Wesley.

Date, C. J. 1995. *An Introduction to Database Systems*. Reading, MA: Addison-Wesley.

Laurini, R., and D. Thompson. 1992. *Fundamentals of Spatial Information Systems*. London: Academic Press.

Stevens, S. S. 1946. On the Theory of Scales of Measurement. *Science* 103: 677–80.

Vetter, M. 1987. *Strategy for Data Modelling*. New York: Wiley.

Worboys, M. F. 1995. *GIS: A Computing Perspective*. London: Taylor & Francis.

Zeiler, M. 1999. *Modeling Our World: The ESRI Guide to Geodatabase Design*. Redlands, CA: ESRI Press.

CHAPTER

7

RASTER DATA

7.1 INTRODUCTION

The vector data model uses the geometric objects of point, line, and area to represent spatial features. Although ideal for discrete features with well-defined locations and shapes, the vector data

133

model does not work well with spatial phenomena that vary continuously over the space such as precipitation, elevation, and soil erosion. A better option for representing continuous phenomena is the raster data model.

The raster data model uses a regular grid to cover the space and the value in each grid cell to correspond to the characteristic of a spatial phenomenon at the cell location. Conceptually, the variation of the spatial phenomenon is reflected by the changes in the cell value. Raster data have been described as field-based, as opposed to object-based vector data (Burrough and McDonnell 1998).

A wide variety of data used in GIS are encoded in raster format. They include digital elevation data, satellite images, digital orthophotos, scanned maps, and graphic files. Commercial GIS packages can display raster and vector data simultaneously, and can convert from raster to vector data or from vector to raster data. Raster data also introduce a large set of data analysis functions to GIS. In many ways, raster data complement vector data in GIS applications. Integration of both types of data has therefore become a common and desirable feature in a GIS project.

Because the data model determines how GIS data are structured, stored, and processed, raster data are covered in this chapter, separate from vector data. This chapter is divided into the following six sections. Section 7.2 discusses the basic elements of raster data. Section 7.3 describes different types of raster data. Section 7.4 provides an overview of data structure, data compression, and data files. Section 7.5 explains projection and geometric transformation of raster data. Section 7.6 discusses conversion between raster and vector data. Section 7.7 offers examples of integration of raster data and vector data in GIS.

7.2 ELEMENTS OF THE RASTER DATA MODEL

A raster data model is variously called a grid, a raster map, a surface cover, or an image in GIS.

Grid is adopted in this chapter. A grid consists of rows, columns, and cells. The origin of rows and columns is at the upper-left corner of the grid. Rows have the same function as y-coordinates and columns as x-coordinates in a plane coordinate system. A fundamental quality of a cell or pixel is that its location is explicitly defined by the row and column position in the grid.

Raster data represent points by single cells, lines by sequences of neighboring cells, and areas by collections of contiguous cells (Figure 7.1). Each cell in a grid carries a value, either an **integer** value (a value with no decimal digits) or a **floating-point** value (a value with decimal digits). Integer cell values typically represent categorical (nominal or ordinal) data. For example, a land cover model may use 1 for urban land use, 2 for forested land, 3 for water bodies, and so on. Floating-point cell values represent continuous, numeric (interval or ratio) data. For example, a precipitation model may have precipitation values of 20.15, 12.23, and so forth. A floating-point grid requires more computer memory than an integer grid—an important factor for a GIS project that covers a large area. Also, data query and display of a floating-point grid should be based on value ranges such as 12.0-19.9, rather than individual values because the chance of finding a specific value is small.

Each cell value in a grid represents the characteristic of a spatial phenomenon at the location denoted by its row and column. The raster data model does not clearly separate spatial and attribute data and, unlike vector data, raster data have little use for database management. Although an integer grid can have a value attribute table (similar to a feature attribute table for a vector-based map), the table is used primarily for summarizing the cell values and their frequencies.

One option to store multiple attributes in a grid is to use some kind of cell ID as the cell value. For example, a grid representing counties in the Pacific Northwest may use the county FIPS (U.S. Federal Information Processing Standards) code as the cell value and store county population, income data, and so forth, in separate fields. The primary

---

Box 7.1 | **Rules in Determining a Categorical Cell Value**

If a large cell does cover forest, pasture, and water on the ground, which category should be entered for the cell? Probably the most common method is to enter the category that occupies the largest area percentage of the cell. But in some situations the majority rule may not be the best option. For example, studies of endangered species are more inclined to use the presence/absence rule than the majority rule (Chris-

man 2001). As long as an endangered species has been recorded in a cell, no matter how much of the cell the species occupies, the cell will be coded with a value for presence. Similarly, the determination of cell values may be based on a ranking of the importance of spatial features to the study. If pasture is deemed to be more important than forest and water, a mixed cell of all three features will be coded as pasture.

---

function of such a value attribute table is data storage. The grid can be used for displaying population or income data, but any query or analysis with the grid is limited to the FIPS codes. The strong association between a grid and its cell values is an important difference between the raster data model and the vector data model.

The cell size determines the resolution of the raster data model. A cell size of 30 meters means that each cell measures 900 square meters (30×30 meters). A large cell size cannot represent the precise location of spatial features, thus increasing the chance of having mixed features such as forest, pasture, and water in a cell (Box 7.1). These problems lessen when a grid uses a smaller cell size. But a small cell size increases the data volume and the data processing time.

A grid is normally projected onto a plane coordinate system such as the UTM (Universal Transverse Mercator) coordinate system. Therefore, raster data can be displayed with vector data if they are based on the same coordinate system. The area extent of a grid can also be determined by reading the real-world coordinates of its lower-left corner and upper-right corner points. One should note, however, that a plane coordinate system has its origin at the lower-left corner, different from the origin of a grid (the upper-left corner).

Although the raster data model has its weakness in representing the precise location of spatial features, it has the distinctive advantage in having

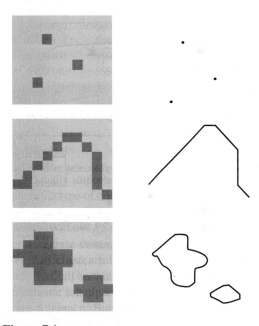

Figure 7.1
Representation of point, line, and area features in raster format (left) and vector format (right).

fixed cell locations (Tomlin 1990). In computing algorithms, a grid can be treated as a matrix with rows and columns, and its cell values can be stored into a two-dimensional array. All commonly used programming languages can easily handle arrayed variables. Raster data are therefore much easier for data manipulation, aggregation, and analysis than vector data.

7.3 TYPES OF RASTER DATA

7.3.1 Satellite Imagery

Remotely sensed satellite data are recorded in raster format. The spatial resolution of a satellite image relates to the ground pixel size. For example, a spatial resolution of 30 meters means that each pixel on the satellite image corresponds to a ground pixel of 30 meters by 30 meters. The pixel value in a satellite image represents light energy reflected or emitted from the Earth's surface (Lillesand and Kiefer 2000). The measurement of light energy is based on spectral bands from a continuum of wavelengths known as the electromagnetic spectrum. Panchromatic images are comprised of a single spectral band, whereas multispectral images are comprised of multiple bands.

The U.S. **Landsat** satellites, started by the National Aeronautics and Space Administration (NASA) with the cooperation of the U.S. Geologic Survey (USGS) in 1972, have produced the most widely used imagery worldwide. Landsat 1, 2, and 3 acquired images by the Multispectral Scanner (MSS) with a spatial resolution of about 79 meters. When launched in 1982, Landsat 4 carried a new sensor, the Thematic Mapper (TM) scanner. TM images have a spatial resolution of 30 meters and seven spectral bands (blue, green, red, near infrared, mid-infrared I, thermal infrared, and mid-infrared II). A second TM was launched aboard Landsat 5 in 1984. Landsat 6 failed to reach its orbit after launch in 1993.

Landsat 7 was launched successfully in April 1999, carrying an Enhanced Thematic Mapper-Plus (ETM+) sensor (**http://landsat7.usgs.gov/**). The enhanced sensor is designed to seasonally monitor small-scale processes on a global scale such as cycles of vegetation growth, deforestation, agricultural land use, erosion and other forms of land degradation, snow accumulation and melt, and urbanization. The spatial resolution of Landsat 7 imagery is 15 meters in the panchromatic band; 30 meters in the six visible, near, and shortwave infrared bands; and 60 meters in the thermal infrared band.

In December 1999, NASA's Earth Observing System launched the Terra spacecraft to study the interactions among the Earth's atmosphere, lands, oceans, life, and radiant energy (heat and light) (**http://terra.nasa.gov/About/**). Terra carries five sensors, of which ASTER (Advanced Spaceborne Thermal Emission and Reflection Radiometer) is the only high-spatial-resolution instrument. ASTER's spatial resolution is 15 meters in the visible and near infrared range, 30 meters in the shortwave infrared band, and 90 meters in the thermal infrared band. A major application of ASTER data products is land cover classification and change detection.

The U.S. National Oceanic and Atmospheric Administration (NOAA) uses weather satellites as an aid to weather prediction and monitoring. NOAA's Polar Orbiting Environmental Satellites (POES) carry the **AVHRR (Advanced Very High Resolution Radiometer)** scanner, which provides data useful for large-area land cover and vegetation mapping (**http://edcdaac.usgs.gov/1KM/1kmhomepage.html/**). AVHRR data have a spatial resolution of 1.1 kilometers, which may be too coarse for some GIS projects. But the coarse spatial resolution is offset by the daily coverage and reduced volume of AVHRR data.

The USGS has recently initiated AmericaView, a new program designed to make satellite data from the U.S. government more accessible to the public through a network of state consortia (**http://americaview.usgs.gov/index.html/**). The pilot consortium, OhioView, offers Landsat 7 and ASTER data for the State of Ohio and elsewhere (**http://www.ohioview.org/**). The USGS expects to expand the program to all 50 states in the next several years.

A number of countries have programs similar to Landsat in the United States. The French **SPOT** satellite series began in 1986 (**http://www.spot.com/**). Each SPOT satellite carries two types of sensors. Spot 1 to 4 acquire single-band imagery with a 10-meter spatial resolution and multiband imagery with a 20-meter resolution. Spot 5, launched successfully in May 2002, sends back higher-resolution imagery: 5 and 2.5 meters in single-band, and 10 meters in multiband. High spatial resolution makes SPOT images good spatial data sources for GIS projects. Other important satellite programs have also been established since the late 1980s in India (**http://www.isro. org/**) and Japan (**http://www.nasda.go.jp/**).

The privatization of the Landsat program in the United States in 1985 has opened the door for private companies to gather and market remotely sensed data using various platforms and sensors. (The French space agency has also handed over responsibilities of commercial operation of Spot 5 to Spot Image in 2002.) Ikonos, a satellite designed by Space Imaging (**http://www.spaceimaging.com/**) to acquire panchromatic images with a 1-meter spatial resolution and multi-spectral images with a 4-meter resolution, was successfully launched in September 1999. A resolution of 1 meter is high enough to detect ground objects such as cars, small houses, fires, and troop deployments. DigitalGlobe's QuickBird reached its orbit successfully in October 2001 (**http://www.digitalglobe.com/**). QuickBird collects panchromatic images with a 61-centimeter spatial resolution and multispectral images with a 2.44-meter resolution.

An image processing system must be able to access and process multiband imagery. This is usually the basis for separating image processing from GIS packages, even though image processing packages such as ERDAS (**http://www.erdas. com/**) and ER Mapper (**http://www.ermapper. com/**) have the same data display and processing functions as a GIS package. One can display satellite images in black and white or in color, and simulate color photographs or color infrared photographs using pixel values from the blue, green, and red spectral bands. By analyzing the pixel values, one can extract a variety of themes from satellite images, such as land cover, hydrography, water quality, and areas of eroded soils. For example, the 1990s National Land Cover Data Set for the conterminous United States is based primarily on Landsat 5 TM data (Vogelmann et al. 2001).

7.3.2 Digital Elevation Models

A **digital elevation model** (**DEM**) consists of an array of uniformly spaced elevation data. A DEM is point-based, but it can easily be converted to raster data by placing each elevation point at the center of a cell. Most GIS users in the United States use DEMs from the USGS. The USGS has a website that updates the status of DEM data by state (**http://mcmcweb.er.usgs.gov/status/dem_ stat.html/**). USGS DEMs include the 7.5-minute DEM, 30-minute DEM, 1-degree DEM, and Alaska DEM. Each DEM file contains the header information on measurement units, minimum and maximum elevations, projection parameters, and data accuracy statistics.

7.3.2.1 The 7.5-Minute DEM

The 7.5-minute DEMs provide elevation data at a spacing of 30 meters or 10 meters on a grid measured in UTM coordinates. Each DEM covers a 7.5-by-7.5-minute block that corresponds to a USGS 1:24,000 scale quadrangle (Box 7.2). The USGS groups the 7.5-minute DEMs into three levels by data accuracy and production method, with level 1 having the least accurate data. Only level 1 and level 2 DEMs are currently available for most states. Level 1 DEMs have the vertical accuracy of 15 meters or better, whereas level 2 DEMs have the vertical accuracy of 7 meters. Level 2 DEMs, which are derived from contour lines, can have resolutions at 30 meters or 10 meters (Kumler 1994).

7.3.2.2 The 30-Minute DEM

The 30-minute DEMs provide elevation data at a spacing of 2 arc-seconds on the geographic grid (about 60 meters in the midlatitudes). Each DEM covers a 30-by-30-minute block, corresponding to the east or west half of a USGS 30-by-60-minute 1:100,000 quadrangle. The vertical accuracy of the 30-minute DEMs is equal to or better than 25 meters, one-half a contour interval of the 1:100,000 quadrangle.

7.3.2.3 The 1-Degree DEM

The 1-degree DEMs provide elevation data at a spacing of three arc-seconds on the geographic grid (about 100 meters in the midlatitudes). Each

Box 7.2 **No-Data Slivers in 7.5-Minute DEM**

A 7.5-minute DEM is supposed to correspond to a USGS 1:24,000 scale quadrangle, but a sliver of no data often shows up along the border of the DEM. This is because the bounding coordinates of the DEM do not form a rectangle. The 7.5-minute DEM data are stored as a series of profiles, with a constant spacing of 30 meters along and between each profile. When DEMs are cast on the UTM coordinate system, the bounding coordinates of a 7.5-minute DEM form a quadrilateral, rather than a rectangle. Thus it cannot match the quadrangle perfectly. The no-data value is used to fill the uneven rows and columns.

DEM covers a 1-by-1-degree block, corresponding to the east or west half of a USGS 1-by-2-degree 1:250,000 quadrangle. The Defense Mapping Agency (DMA, now the National Imagery and Mapping Agency or NIMA) originally produced the 1-degree DEMs by interpolation from digitized contour lines. The vertical accuracy of elevation data is about 30 meters.

7.3.2.4 Alaska DEMs

The USGS provides the 7.5-minute and 15-minute Alaska DEMs, with spacing referenced to latitude and longitude. The spacing for the 7.5-minute Alaska DEMs is 1 arc-second of latitude by 2 arc-seconds of longitude, and the spacing for the 15-minute Alaska DEMs is 2 arc-seconds of latitude by 3 arc-seconds of longitude.

7.3.2.5 Non–USGS DEMs

A basic method for producing DEMs is to use a stereoplotter and aerial photographs with overlapped areas. The stereoplotter creates a three-dimensional model, which allows the operator to compile elevation data as well as orthophotos. Although this method can produce highly accurate DEM data at a finer resolution than USGS DEMs, it is expensive for coverage of large areas.

There are alternatives to the use of a stereoplotter. A popular alternative with GIS users is to generate DEMs from satellite imagery such as the SPOT stereo model. Software packages for extracting elevation data from SPOT images on the personal computer are commercially available. Besides imagery data, the data extraction process requires ground control points, which can be measured in the field by GPS (global positioning system) with differential correction. The quality of such DEMs depends on the software package and the quality of the inputs.

Stereo radar data can also be the source for producing DEMs. An active remote sensor, radar emits microwave pulses and measures returned energy from ground objects. Radar can penetrate cloud cover, which usually presents a major problem with aerial photography. As an example, Intermap Technologies produces elevation data from stereo radar data at a spatial resolution of 5 to 15 meters and with a vertical resolution of 2 to 5 meters (**http://www.intermaptechnologies.com/**).

LIDAR (light detection and ranging) is a new technology for producing DEMs (Turner 2000). The basic components of a LIDAR system include a laser scanner mounted in an aircraft, GPS, and an Inertial Measurement Unit (IMU). LIDAR uses laser to measure distance. At the same time, the location and orientation of the laser source are determined by GPS and IMU, respectively. One application of LIDAR technology is the creation of high-resolution DEMs, with a vertical accuracy of about 15 centimeters. These DEMs are already georeferenced and can have different levels such as ground elevation and canopy elevation (Figure 7.2).

7.3.2.6 Global DEMs

DEMs at different resolutions are now available on the global scale. ETOPO5 (Earth Topography–5 Minute) data cover both the land surface and

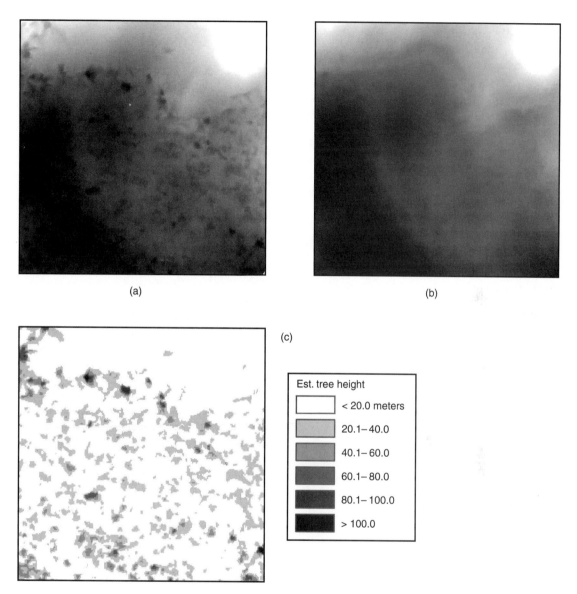

(a)

(b)

(c)

Est. tree height

	< 20.0 meters
	20.1– 40.0
	40.1– 60.0
	60.1– 80.0
	80.1– 100.0
	> 100.0

Figure 7.2
This illustration shows elevation grids derived from LIDAR data at the canopy level (a) and the ground level (b). The difference between (a) and (b) is the estimated tree height (c). The process of deriving an elevation grid (or DEM) from LIDAR data is as follows: convert LIDAR data to a TIN (triangulated irregular network), and use linear interpolation to convert the TIN to a grid.

ocean floor of the Earth, with a grid spacing of 5 minutes of latitude by 5 minutes of longitude (**http://edcwww.cr.usgs.gov/glis/hyper/guide/eto po5).** Both GTOPO30 (**http://edcdaac.usgs.gov/** gtopo30/gtopo30.html/)** and GLOBE (**http:// www.ngdc.noaa.gov/seg/topo/globe.shtml/)** offer global DEMs with a horizontal grid spacing of 30 arc-seconds or approximately 1 kilometer.

7.3.3 Digital Orthophotos

A **digital orthophoto quad** (**DOQ**) is a digitized image prepared from an aerial photograph or other remotely sensed data, in which the displacement caused by camera tilt and terrain relief has been removed. A digital orthophoto is georeferenced and can be registered with topographic and other maps. The USGS began producing DOQs in 1991 from 1:40,000-scale aerial photographs of the National Aerial Photography Program and expects to have complete coverage of the conterminous United States by 2004.

The standard USGS DOQ format is either a 3.75-minute quarter quadrangle or 7.5-minute quadrangle in the form of a black-and-white, color infrared, or natural color image. Most GIS users use 3.75-minute quarter quadrangles in black and white. These quarter quadrangles have a 1-meter ground resolution (i.e., each pixel in the image measures 1-by-1 meter on the ground) and have pixel values representing 256 gray levels, similar to a single-band satellite image.

For decades, mapmakers have produced photomaps by superimposing map symbols on aerial photographs, either rectified or unrectified. Continuing with this tradition, DOQs combine the image characteristics of a photograph with the geometric qualities of a map. Easily integrated in a GIS, DOQs are the ideal background for data display and data editing.

7.3.4. Binary Scanned Files

A **binary scanned file** is a scanned image containing values of 1 or 0. Scanned files for tracing are examples of the binary scanned files (Chapter 4). Maps to be digitized are typically scanned at 300 or 400 dots per inch (dpi) to produce the scanned files.

7.3.5 Digital Raster Graphics (DRG)

A **digital raster graphic** (**DRG**) is a scanned image of a USGS topographic map. The USGS scans the 7.5-minute topographic map at 250 dots per inch, thus producing a DRG with a ground resolu-

tion of 2.4 meters. The USGS uses up to 13 colors on each 7.5-minute DRG. The DRG is georeferenced to the UTM coordinate system by using 2.5-minute grid ticks as control points.

7.3.6 Graphic Files

Maps, photographs, and images can be stored as digital graphic files. Many popular graphic files are in raster format, such as TIFF [tag(ged) image file format], GIF (graphics interchange format), and JPEG (Joint Photographic Experts Group). The USGS distributes DOQs in TIFF or GeoTIFF format. GeoTIFF is a georeferenced version of TIFF. By having the geographic reference information of the image, DOQs can be readily used with other GIS data.

7.3.7 GIS Software-Specific Raster Data

GIS packages use raster data that are imported from DEMs, satellite images, scanned images, graphic files, and ASCII files or are converted from vector data. These raster data are named differently. For example, ArcGIS and MGE name raster data grids, GRASS raster map layers, IDRISI images, and PCI surface covers.

ESRI grids are based on a proprietary format from ESRI Inc. ESRI grids are stored as either integer or floating-point grids. An integer grid has a value attribute table that stores cell values. A floating-point grid does not, because of its potentially large number of records. ArcGIS can convert a floating-point grid to an integer grid, and vice versa. ArcGIS can also convert vector-based coverages and shapefiles to grids, and vice versa.

7.4 Data Structure, Data Compression, and Header File

7.4.1 Data Structure

Data structure refers to the storage of raster data so that they can be used and processed by the computer. The simplest data structure is called the **cell-by-cell encoding** method: a raster model is stored as a matrix, and its cell values are written into a file by row and column (Figure 7.3). Functioning at the cell level, this method is an ideal choice if the cell values of a raster model change continuously.

DEMs use the cell-by-cell data structure because the neighboring elevation values are rarely the same. Satellite images also use this method for data storage. With multiple spectral bands, however, each pixel in a satellite image has more than one value. Multiband imagery is typically stored in the following three formats. The band interleaved by line (.bil) method stores the first value of every row sequentially, followed by the second value of every row, and so on, in one image file. The band sequential (.bsq) method stores the values for each band of satellite data sequentially in one image file. The band interleaved by pixel (.bip) method stores each row of an image sequentially: row 1 all bands, row 2 all bands, and so on.

The cell-by-cell encoding method becomes inefficient if a raster model contains many redundant cell values. For example, a binary scanned file from a soil map would have many 0s representing non-inked areas and only occasional 1s representing the inked soil lines. Raster models such as binary scanned files can be more efficiently stored using the **run length encoding** (RLE) method, which records the cell values by row and by group (Figure 7.4). Each group includes a cell value and the number of cells with that value. If all cells in a row contain the same value, only one group is recorded, thus saving the computer memory.

Working along one row at a time, the RLE method is not efficient for recording two-dimensional features, often called regions in the literature. Three methods are designed for two-dimensional regions: chain codes, block codes, and quad trees (Worboys 1995; Burrough and McDonnell 1998).

The **chain code** method represents the boundary of a region by using a series of cardinal directions and cells. For example, N1 means moving north by 1 cell and S4 means moving south by 4 cells. Figure 7.5 illustrates the chain coding of a region. A variation of the method is to

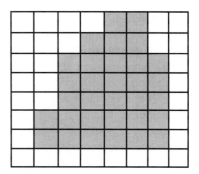

Row 1: 0 0 0 0 1 1 0 0
Row 2: 0 0 0 1 1 1 0 0
Row 3: 0 0 1 1 1 1 1 0
Row 4: 0 0 1 1 1 1 1 0
Row 5: 0 0 1 1 1 1 1 0
Row 6: 0 1 1 1 1 1 1 0
Row 7: 0 1 1 1 1 1 1 0
Row 8: 0 0 0 0 0 0 0 0

Figure 7.3
The cell-by-cell data structure records each cell value by row and column.

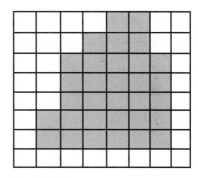

Row 1: 5 6
Row 2: 4 6
Row 3: 3 7
Row 4: 3 7
Row 5: 3 7
Row 6: 2 7
Row 7: 2 7

Figure 7.4
The run length encoding method records the cell values in runs. Row 1, for example, has two adjacent cells in columns 5 and 6 that are gray or have the value of 1. Row 1 is therefore encoded with one run, beginning in column 5 and ending in column 6. The same method is used to record other rows.

code the cardinal directions numerically: for example, 0 for east, 1 for north, 2 for west, and 3 for south.

The **block code** method uses square blocks to represent a region. A unit square represents a cell, a 4-square block represents 2-by-2 cells, a 9-square block represents 3-by-3 cells, and so on. Using the medial axis transform (Rosenfeld 1980), each square block is coded only with the location of a cell (e.g., the lower left of the block), and the side length of the block. Figure 7.6 shows the block encoding of a region.

Regional **quad tree** uses recursive decomposition to divide a grid into a hierarchy of quadrants (Samet 1990) (Figure 7.7). A quadrant having cells with the same value will not be subdivided, and it is stored as a leaf node. Leaf nodes are coded with the value of the homogeneous quadrant (gray and white in Figure 7.7). A quadrant having different cell values will be subdivided until a quadrant at the finer level contains only one value. Recursive decomposition refers to this process of continuous subdivision. After the subdivision is complete, cells of interest can be coded using a spatial indexing method and a hierarchical structure. For example, the level-1 NW quadrant (with the spatial index of 0) in Figure 7.7 has two gray leaf nodes. The first, 02, refers to the level-2 SE quadrant, and the second, 032, refers to the level-3 SE quadrant of the level-2 NE quadrant.

Regional quad tree is an efficient method for storing area data, especially if the data contain few categories. The method is also efficient for data processing (Samet 1990). Use of quad tree in GIS extends beyond simple raster data. Researchers have proposed using a hierarchical quad tree structure for storing global data (Tobler and Chen 1986; Dutton 1999; Ottoson and Hausak 2002). This kind of data structure applies directly to the spherical surface of the Earth, thus circumventing the need for map projection.

GIS packages use different raster data structures. Both GRASS and IDRISI use raster data that are stored using either the cell-by-cell or RLE method. SPANS uses a quad tree data structure.

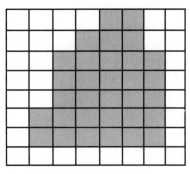

N1 E1 N3 E1 N1 E1 N1 E1 S2 E1 S4 W5

Figure 7.5
Starting at the lower left cell of the region, the chain codes method records the region's boundary by using the principal direction and the number of cells. In this example, the recording follows a clockwise direction.

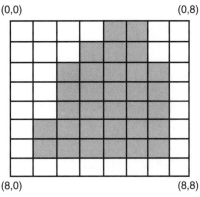

Three unit squares: (7,1; 6,1; 2,3)

One 4-squares : (2,4)

One 25-squares : (7,2)

Figure 7.6
The block codes method works with square blocks. The region in the diagram is divided into three unit square blocks, one 4-square block, and one 25-square block. Each block is encoded with the coordinates of its lower left corner. The origin of the coordinates is at the upper left corner of the grid.

ESRI grids use a hierarchical tile-block-cell data structure: a tile consists of a series of rectangular

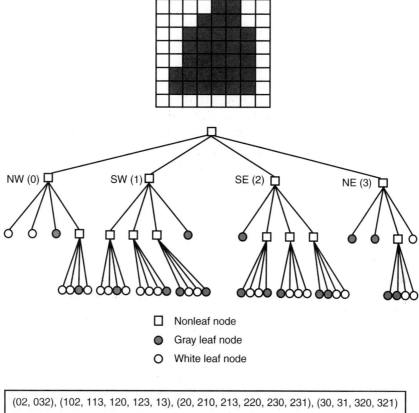

Figure 7.7
The regional quad tree method divides a grid into a hierarchy of quadrants. The division stops when a quadrant is made of cells of the same value (gray or white). A quadrant that cannot be subdivided is called a leaf node. In the diagram, the quadrants are indexed spatially: 0 for NW, 1 for SW, 2 for SE, and 3 for NE. Using the spatial indexing method and the hierarchical quad tree structure, the gray cells can be coded as 02, 032, and so on. See text for more explanation.

blocks, and a block consists of square cells. Cell values are stored by block. Each block is stored as one variable-length record using either the cell-by-cell or RLE method.

7.4.2 Data Compression

Data compression refers to the reduction of data volumes, a topic closely related to raster data structure. For instance, a raster data file stored using the RLE method is called the run length compressed (RLC) file because the storage method by group saves the computer memory.

A binary scanned file of a 7.5-minute soil quadrangle map, scanned at 300 dots per inch, can be over 8 megabytes (MB). A quarter DOQ from the USGS is about 40 to 50 MB in size. Data compression is therefore important for the sharing of image files. When compressed using the RLE method, the 8-MB binary scanned file is reduced to about 800,000 bytes at a 10:1 compression ratio. Some raster data such as DEMs and satellite images are difficult to compress because of their continuously changing values.

Graphic files such as TIFF, GIF, and JPEG files are regularly compressed using a variety of image compression algorithms. TIFF and GIF files use **lossless compression**, which allows the original image to be precisely reconstructed, whereas JPEG files use **lossy compression**, which can achieve high-compression ratios but cannot reconstruct fully the original image. Image degradation through lossy compression can affect a GIS-related task such as extracting ground control points from air photographs or satellite images (Section 7.5). A recent study, however, suggests that this should not be a concern if JPEG compression is under a 10:1 ratio (Li et al. 2002).

MrSID (Multi-resolution Seamless Image Database) is a compression technology originally developed at the Los Alamos National Laboratory and later awarded to LizardTech Inc. (**http://www.lizardtech.com/**). Multiresolution means that MrSID has the capability of recalling the image data at different resolutions or scales. Seamless means that MrSID can compress a large image such as a DOQ with subblocks and eliminates the artificial block boundaries during the compression process. MrSID assumes that the level of detail in an image is not constant. Therefore, MrSID encodes the detailed parts of the image at high resolution and the less detailed parts at low resolution.

Data compression programs, such as WinZip for PCs and gzip for the UNIX environment, can work with any type of data files. Given simple binary scanned files, these programs can achieve extremely high compression ratios.

7.4.3 Header File

To import raster data for use, a GIS package must have information about the data structure and, if applicable, the compression method. This information is contained in the header file, often denoted by the extension .hdr, which is like metadata in function. A header file also contains information

on numbers of rows and columns, number of spectral bands, number of bits per pixel, value for no data, x- and y-coordinates of the origin, and pixel size (Box 7.3).

Some raster data use other files besides the header file or organize the data information into different files. Satellite images, for example, may have two optional files. The statistics file, often with the extension .stx, describes statistics such as minimum, maximum, mean, and standard deviation for each spectral band in an image. The color file, often with the extension .clr, associates colors with different pixel values in an image. An ESRI grid has its own separate directory, which contains at least six files. The first three files describe the area extent of the grid, the statistical data about the cell values, and, in the case of an integer grid, the cell value and the number of cells that have the same value. Of the next three files, the header file stores information about the grid's cell size, blocking factor, and compression technique. The other two files store information on the data and spatial indexing for the first tile in the grid.

7.5 PROJECTION AND GEOMETRIC TRANSFORMATION OF RASTER DATA

Map projection and geometric transformation are just as important to raster data as to vector data. Unless they are projected onto the same coordinate system, raster data such as satellite imagery and DEMs cannot be registered spatially with vector data such as roads, streams, and other map features in a GIS database.

Projected raster data are still based on rows and columns but the rows and columns are measured in real-world coordinates. For example, an elevation grid may be defined as follows:

- Rows: 463, columns: 318, cell size: 30 m
- UTM coordinates at the lower-left corner: 499995, 5177175
- UTM coordinates at the upper-right corner: 509535, 5191065

Box 7.3   A Header File Example

The following is a header file example for a GTOPO30 DEM. The explanation of each entry in the file is given after /*.

BYTEORDER   M   /* byte order in which image pixel values are stored. M = Motorola byte order.

LAYOUT   BIL   /* organization of the bands in the file. BIL = band interleaved by line.

NROWS   6000   /* number of rows in the image.

NCOLS   4800   /* number of columns in the image.

NBANDS   1   /* number of spectral bands in the image. 1 = single band.

NBITS   16   /* number of bits per pixel.

BANDROWBYTES   9600   /* number of bytes per band per row.

TOTALROWBYTES   9600   /* total number of bytes of data per row.

BANDGAPBYTES   0   /* number of bytes between bands in a BSQ format image.

NODATA   -9999   /* value used for masking purpose.

ULXMAP   -99.9958333333334   /* longitude of the center of the upper-left pixel (decimal degrees).

ULYMAP   39.99583333333333   /* latitude of the center of the upper-left pixel (decimal degrees).

XDIM   0.00833333333333   /* x dimension of a pixel in geographic units (decimal degrees).

YDIM   0.00833333333333   /* y dimension of a pixel in geographic units (decimal degrees).

The numbers of rows and columns can be derived from the bounding UTM coordinates of the grid and the cell size. In the above example, 463 = (5191065 − 5177175) / 30, and 318 = (509535 − 499995) / 30. One can also derive the UTM coordinates defining each grid cell. For example, the cell in row 1, column 1 has the UTM coordinates of 499995, 5191035 (5191065 − 30) at the lower-left corner and 500025 (499995 + 30), 5191065 at the upper-right corner.

Geometric transformation of satellite imagery is often called **georeferencing** in image processing (Verbyla and Chang 1997; Lillesand and Kiefer 2000). Two common methods for georeferencing are affine transformation and polynomial equations. An affine transformation can georeference an image through rotation, translation, and scaling. For example, a rotation adjustment can correct a satellite image with the northeast orientation, rather than the north orientation, because the satellite orbit is from northeast to southwest.

An affine transformation of image data uses the same transformation equations as for vector data (Chapter 4):

(7.1)
$$x' = Ax + By + C$$

(7.2)
$$y' = Dx + Ey + F$$

The differences are that (1) x and y represent column number and row number and (2) the coefficient E is negative because of the different origins of an image and a coordinate system.

Transformation of image data, however, requires the additional work of selecting control points. Because they are not defined in an image, control points, also called **ground control points**, must be chosen. Ideally, control points are those features that show up clearly as single distinct cells on the satellite image. For example, one may select control points at road intersections, rock outcrops, small ponds, or distinctive features along shore-lines. After control points are identified, their real-world coordinates can then be obtained from digital maps or GPS readings.

Because ground control points are selected by the user and are subject to errors in interpretation, the affine transformation of satellite images is usually an iterative process. The affine transformation is run with an initial set of control points. The root mean square (RMS) error is examined (Chapter 4). If the RMS error exceeds the required tolerance value, for example, 1 pixel, then the control points that contribute most to the RMS error are removed, and a different set of control points is entered for the next affine transformation. This process continues until a satisfactory RMS error is obtained.

A file must be created after an affine transformation to store the values of the six coefficients in the order of A, D, B, E, C, and F. ArcInfo Workstation, for example, use a world file (an ASCII file) and has the letter w in the file extension for easy recognition. Using a world file (e.g., tmrect.blw), an image file (e.g., tmrect.bil) can be converted to real-world coordinates and registered with vector-based map features.

The use of polynomial equations for georeferencing is also called **warping** or *rubber sheeting*. A polynomial equation provides a mathematical model for differential scaling and rotating across an image. The degree of complexity of the model is expressed by the order of the polynomial, which may range from 2 to 5. The 2nd order polynomial, for example, uses the following equations for transformation:

(7.3)

$$x' = a_0 + a_1x + a_2y + a_3xy + a_4x^2 + a_5y^2$$

(7.4)

$$y' = b_0 + b_1x + b_2y + b_3xy + b_4x^2 + b_5y^2$$

GIS packages that can integrate image data with vector and raster data are usually capable of performing geometric transformation. As an example, the spatial adjustment tools of ArcEditor include geometric transformation and rubber sheeting.

The result of geometric transformation is a new grid. Data **resampling** is therefore required to fill each cell of the new grid with the value of the corresponding cell or cells in the original grid. Three common resampling methods for filling the new raster grid are nearest neighbor, bilinear interpolation, and cubic convolution (Figure 7.8).

The **nearest neighbor** resampling method fills each cell of the new grid with the nearest cell value from the original grid. The **bilinear interpolation** method fills each cell of the new grid with the weighted average of the four nearest cell values from the original grid. The **cubic convolution**

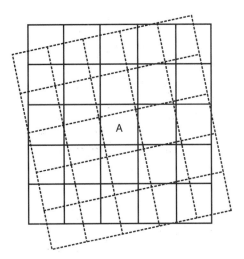

Figure 7.8
The grid in solid line is a new grid created by transforming the grid in dashed lines. Resampling is required to fill the value of each cell in the new grid such as cell A.

method fills each cell of the new grid with the weighted average of the 16 nearest cell values from the original grid. Nearest neighbor is primarily used for categorical data such as land cover types because the method will not change the cell values. All three methods can be used for numeric data such as elevation or satellite imagery. The overall smoothness of the resampled output—at the expense of the processing time—increases from nearest neighbor to bilinear interpolation to cubic convolution.

7.6 DATA CONVERSION

The conversion of vector data to raster data is called **rasterization**, and the conversion of raster data to vector data is called **vectorization** (Figure 7.9). These two types of conversion use different computer algorithms (Piwowar et al. 1990).

The simpler of the two conversion methods, rasterization involves three basic steps (Clarke 1995). The first step is to set up a grid with a specified cell size to cover the area extent of the coverage and to assign initially all cell values as zeros. The second step is to change the values of those cells that correspond to points, lines, or polygon boundaries. The cell value is set to 1 for a point, the line's value for a line, and the polygon's value for a polygon boundary. The third step is to fill the interior of the polygon outline with the polygon value. Errors from rasterization depend on the design of the computer algorithm and the size of raster cell (Bregt et al. 1991).

Chapter 4 has already covered vectorization as part of the scanning method for spatial data entry. In short, the process involves the following steps: raster line thinning, vector line extraction, topological reconstruction, and line smoothing.

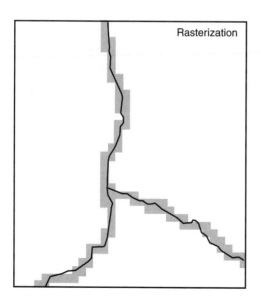
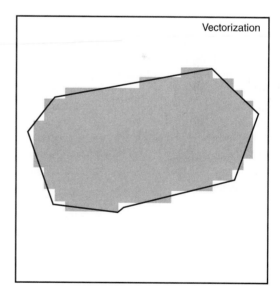

Figure 7.9
On the left is an example of conversion from vector to raster data, or rasterization. On the right is an example of conversion from raster to vector data, or vectorization.

Box 7.4   Linking Vector Data with Images

The hyperlink tool in ArcMap can link a point, line, or polygon feature in vector format with an image. To use the tool, one must first use a field in the feature attribute table to specify the path to the image to be linked to the feature. An example of the hyper-link application is to link a map showing houses for sale with pictures of the houses so that, when a house (point) is clicked, it opens a picture of the house in a graphic file.

Data conversion is a standard feature in a GIS package. Using ArcInfo, for example, one can convert a polygon coverage to a grid or convert a grid to a polygon coverage. Similar commands are available for points and lines. Using ArcView, one can convert a shapefile to a grid and a grid to a shapefile.

7.7 INTEGRATION OF RASTER AND VECTOR DATA

Integration of raster data and vector data in GIS can take place in different forms. DOQs, DRGs, and graphic files are essentially pictures. They are useful as the background for data display, or as the source for spatial data input or revision (Box 7.4). Binary scanned files can be used for tracing and digitizing (Chapter 4).

DEMs are input data, which can be processed to extract topographic features such as contour, slope, aspect, drainage network, and watersheds (Jenson and Domingue 1988; Band 1993; Dymond et al. 1995). One can process a DEM in its original raster format or convert it to a vector-based TIN (triangulated irregular network) before processing (Chapter 12). Topographic features extracted from DEMs may also be represented in either raster or vector format.

Satellite images are pictures like DOQs, but contain quantitative spectral data that can be analyzed to create new map information. When used as pictures, georeferenced satellite images can be displayed with other spatial features. When used in image processing, satellite images can create maps such as land cover, vegetation, urbanization, snow accumulation, and environmental degradation (Mesev et al. 1995; Hinton 1996). These new maps can be stored in either raster or vector format.

Use of vector data as ancillary information in image processing is well documented in the literature (Ehlers et al. 1989; Ehlers et al. 1991; Wilkinson 1996; Hinton 1996). Image stratification is a good example. This method uses vector data to divide the landscape into major areas of different characteristics or different elevation zones and then treats these areas or zones separately in image processing and classification. Another example is the use of vector data in selecting control points for the georeferencing of remotely sensed data (Couloigner et al. 2002).

Finally, the technologies of GIS and remote sensing can be integrated at different levels (Ehlers et al. 1989). At the simple level, GIS and remote sensing reside in two separate software packages and the linkage between them is through data exchange. At the next level GIS and remote sensing software packages still exist separately but share a common user interface, and are capable of simultaneous display. The highest level is total integration: GIS and remote sensing reside in one software unit with combined processing. The current status of data integration is near the second level. Commercial GIS packages allow simultaneous displays of raster data and vector data, and use a common user interface to access remote sensing and GIS data. But one must perform data conversion before analyzing raster and vector data separately.

KEY CONCEPTS AND TERMS

AVHRR (Advanced Very High Resolution Radiometer) scanner: A weather satellite that provides large-area land cover data at a 1-kilometer spatial resolution.

Bilinear interpolation: A resampling method that fills each cell of the new raster grid with the weighted average of the four nearest cell values from the original grid.

Binary scanned file: A scanned file containing values of 1 or 0.

Block code: A raster data structure that uses square blocks to represent a region.

Cell-by-cell encoding: A raster data structure that stores cell values in a matrix by row and column.

Chain code: A raster data structure that represents the boundary of a region by using a series of cardinal directions and cells.

Cubic convolution: A resampling method that fills each cell of the new raster grid with the weighted average of the 16 nearest cell values from the original grid.

Data compression: Reduction of data volumes, especially for raster data.

Digital elevation model (DEM): A digital model with an array of uniformly spaced elevation data in raster format.

Digital orthophoto quad (DOQ): A digitized image prepared from an aerial photograph or other remotely sensed data, in which the displacement caused by camera tilt and terrain relief has been removed.

Digital raster graphic (DRG): A scanned image of a USGS topographic map.

ESRI grid: A proprietary ESRI format for raster data.

Floating-point grid: A grid that contains cells of continuous values.

Georeferencing: The process of using a set of control points to convert images from image coordinates to real-world coordinates.

Ground control points: Control points used in georeferencing an image.

Integer grid: A grid that contains cell values of integers.

Landsat: An orbiting satellite that provides repeat images of the Earth's surface. Landsat 7 was launched in April 1999.

Lossless compression: One type of data compression that allows the original image to be precisely reconstructed.

Lossy compression: One type of data compression that can achieve high-compression ratios but cannot reconstruct fully the original image.

Nearest neighbor: A resampling method that fills each cell of the new raster grid with the nearest cell value from the original grid.

Quad tree: A raster data structure that divides a raster model into a hierarchy of quadrants.

Rasterization: Conversion of vector data to raster data.

Resampling: The process of filling each cell of a new grid with the value of the corresponding cell or cells in an original grid.

Run Length Encoding (RLE): A raster data structure that records the cell values by row and by group. A run length encoded file is also called run length compressed (RLC) file.

SPOT: A French satellite that provides repeat images of the Earth's surface. SPOT 5 was launched in May 2002.

Vectorization: Conversion of raster data to vector data.

Warping: The process of using polynomial equations for georeferencing an image. Also called *rubber sheeting*.

APPLICATIONS: RASTER DATA

The first two tasks of this applications section let you view two types of raster data: DEM and Landsat TM imagery. Task 3 lets you convert a line shapefile and a polygon shapefile to grids.

Task 1: View USGS DEM Data

What you need: *filer.dem*, a USGS 7.5-minute DEM.

In Task 1, you will use ArcToolbox to import a USGS 7.5-minute DEM to a grid and use ArcCatalog to examine the grid's properties.

1. Start ArcCatalog and make connection to the Chapter 7 database. Launch ArcToolbox. Double-click on Conversion Tools. Double-click on Import to Raster, and select DEM to Grid.

2. In the DEM to Grid dialog, use the browse button to select *filer.dem* for USGS DEM. Click the radio button next to Integer. Save the output grid as *filergrd*. Click OK to dismiss the dialog.

3. This step is to examine *filergrd* in ArcCatalog. Right-click *filergrd* in the Catalog tree and select Properties. The General tab shows the number of rows, number of columns, cell size, and elevation statistics of *filergrd*. The Spatial Reference tab shows the area extent and the projection. *Filergrd* is projected onto Zone 11 (with the central meridian at 117°W) of the UTM coordinate system. The map units are in meters and the datum is NAD27.

Task 2: View a Satellite Image in ArcMap

What you need: *tmrect.bil*, a Landsat TM image comprised of the first five bands.

Task 2 lets you view a Landsat TM image with five bands. By changing the color assignment to each of the bands, you can alter the view of the image.

1. Make sure that ArcCatalog is still connected to the Chapter 7 database. Select Properties from the context menu of *tmrect.bil*. The General tab shows that *tmrect.bil* has 366 rows, 651 columns, 5 bands, and a cell size of 25 (meters).

2. Launch ArcMap. Rename the data frame from Layers to Task 2. Add *tmrect.bil* to Task 2. The Table of Contents shows *tmrect.bil* as a RGB Composite with Red for Band_1, Green for Band_2, and Blue for Band_3.

3. Select Properties from the context menu of *temrect.bil*. Click the Symbology tab. Use the dropdown lists to change the RGB composite: Red for Band_3, Green for Band_2, and Blue for Band_1. Click OK. You should see the image as a color photograph.

4. Next, use the following RGB composite: Red for Band_4, Green for Band_3, and Blue for Band_2. You should see the image as a color infrared photograph.

Task 3: Convert Vector Data to Raster Data

What you need: *nwroads.shp* and *nwcounties.shp*, shapefiles showing major highways and counties in the Pacific Northwest, respectively.

As mentioned in the chapter, integration of vector and raster data in a GIS project often requires conversion of vector to raster data and vice versa. In Task 3, you will convert a line shapefile (*nwroads.shp*) and a polygon shapefile (*nwcounties.shp*) to grids. Covering Idaho, Washington, and Oregon, both shapefiles are projected onto a Lambert conformal conic projection and are measured in meters.

1. Select Data Frame from the Insert menu in ArcMap. Rename the new data frame Task 3. Add *nwroads.shp* and *nwcounties.shp* to Task 3. Select Extensions from the Tools menu,

and make sure that Spatial Analyst is checked in the Extensions dialog. Select Toolbars from the View menu and make sure that Spatial Analyst is checked.

2. Click the Spatial Analyst dropdown arrow, point to Convert, and select Features to Raster. In the Features to Raster dialog, select *nwroads* for the Input features and RTE_NUM1 for the Field. Enter 5000 for the Output cell size. Save the Output raster as *nwroads_gd*. Click OK and dismiss the Features to Raster dialog. *Nwroads_gd* appears in the map in different colors. Each color represents a numbered highway. The highways look blocky because of the large cell size (5000 meters).

3. Click the Spatial Analyst dropdown arrow, point to Convert, and select Features to Raster. In the Features to Raster dialog, select *nwcounties* for the Input features and FIPS for the Field. Enter 5000 for the Output cell size. Save the Output raster as *nwcounties_gd*. Click OK. *Nwcounties_gd* appears in the map with symbols representing the classified values. Double-click *nwcounties_gd* in the Table of Contents. Click the Symbology tab. Select Unique Values in the Show box, and click OK. Now the map shows *nwcounties_gd* with a unique symbol for each county.

REFERENCES

Band, L. E. 1993. Extraction of Channel Networks and Topographic Parameters from Digital Elevation Data. In M. J. Kirkby and K. Beven, eds., *Channel Network Hydrology*, pp. 13–42. Chichester, NY: Wiley.

Bregt, A. K., J. Denneboom, H. J. Gesink, and Y. Van Randen. 1991. Determination of Rasterizing Error: A Case Study with the Soil Map of the Netherlands. *International Journal of Geographical Information Systems* 5: 361–67.

Burrough, P. A., and R. A. McDonnell. 1998. *Principles of Geographical Information Systems*. Oxford: Oxford University Press.

Chrisman, N. 2001. *Exploring Geographic Information Systems,* 2d ed. New York: Wiley.

Clarke, K. C. 1995. *Analytical and Computer Cartography,* 2d ed.

Englewood Cliffs, NJ: Prentice Hall.

Couloigner, I., K. P. B. Thomson, Y. Bedard, B. Moulin, E. LeBlanc, C. Djima, C. Latouche, and N. Spicher. 2002. Towards Automating the Selection of Ground Control Points in Radarsat Images Using a Topographic Database and Vector-Based Data Matching. *Photogrammetric Engineering and Remote Sensing* 68: 433–40.

Dutton, G. H. 1999. *A Hierarchical Coordinate System for Geoprocessing and Cartography*. Berlin: Springer-Verlag.

Dymond, J. R., R. C. Derose, and G. R. Harmsworth. 1995. Automated Mapping of Land Components from Digital Elevation Data. *Earth Surface Processes and Landforms* 20: 131–37.

Ehlers, M., G. Edwards, and Y. Bedard. 1989. Integration of Remote Sensing with Geographical Information Systems: A Necessary Evolution. *Photogrammetric Engineering and Remote Sensing* 55: 1619–27.

Ehlers, M., D. Greenlee, T. Smith, and T. Star. 1991. Integration of Remote Sensing and GIS: Data and Data Access. *Photogrammetric Engineering and Remote Sensing* 57: 669–75.

Hinton, J. E. 1996. GIS and Remote Sensing Integration for Environmental Applications. *International Journal of Geographical Information Systems* 10: 877–90.

Jenson, S. K., and J. O. Domingue. 1988. Extracting Topographic Structure from Digital Elevation Data for Geographical Information System Analysis.

Photogrammetric Engineering and Remote Sensing 54: 1593–1600.

Kumler, M. P. 1994. An Intensive Comparison of Triangulated Irregular Networks (TINs) and Digital Elevation Models (DEMs). *Cartographica* 31 (2): 1–99.

Li, Z., X. Yuan, and K. W. K. Lam. 2002. Effects of JPEG Compression on the Accuracy of Photogrammetric Point Determination. *Photogrammetric Engineering and Remote Sensing* 68: 847–53.

Lillesand, T. M., and R. W. Kiefer. 2000. *Remote Sensing and Image Interpretation,* 4th ed. New York: Wiley.

Mesev, T. V., P. A. Longley, M. Batty, and Y. Xie. 1995. Morphology from Imagery— Detecting and Measuring the Density of Urban Land-Use. *Environment and Planning A* 27: 759–80.

Ottonson, P., and H. Hauska. 2002. Ellipsoidal Quadtrees for Indexing of Global Geographical Data. *International Journal of Geographical Information Science* 16: 213–26.

Piwowar, J. M., E. F. LeDraw, and D. J. Dudycha. 1990. Integration of Spatial Data in Vector and Raster Formats in a Geographic Information System Environment. *International Journal of Geographical Information Systems* 4: 429–44.

Rosenfeld, A. 1980. Tree Structures for Region Representation. In H. Freeman and G. G. Pieroni (eds.), *Map Data Processing,* pp. 137-50. New York: Academic Press.

Samet, H. 1990. *The Design and Analysis of Spatial Data Structures*. Reading, MA: Addison-Wesley.

Tobler, W., and Z. Chen. 1986. A Quadtree for Global Information Storage. *Geographical Analysis* 18: 360–71.

Tomlin, C. D. 1990. *Geographic Information Systems and Cartographic Modeling*.

Englewood Cliffs, NJ: Prentice Hall.

Turner, A. K. 2000. LIDAR Provides Better DEM Data. *GEOWorld* 13(11): 30–31.

Verbyla, D. L., and Chang, K. 1997. *Processing Digital Images in GIS*. Santa Fe, NM: OnWord Press.

Vogelmann, J. E., S. M. Howard, L. Yang, C. R. Larson, B. K. Wylie, and N. Van Driel. 2001. Completion of the 1990s National Land Cover Data Set for the Conterminous United States from Landsat Thematic Mapper Data and Ancillary Data Sources. *Photogrammetric Engineering and Remote Sensing* 67: 650–62.

Wilkinson, G. G. 1996. A Review of Current Issues in the Integration of GIS and Remote Sensing Data. *International Journal of Geographical Information Systems* 10: 85–101.

Worboys, M. F. 1995. *GIS: A Computing Perspective*. London: Taylor & Francis.

```
001    52000736E1211 36000                    35.00N
INTRO TO GEOGRAPHIC INFORMATION SYSTEMS
730     0703723/0307                             1.35
HIGH 1576 1061 1 UPKT TURKSENT
708     0471003554N0                             1.45
TWIN PKT FOLDER 33246 N/NB
         5  3 0.29

                      SUB TOTAL     38.94
                      SALES TAX      0.21
                          TOTAL     39.15
                          VISA      39.15


         ACCT XXXXXXXXXXXX 7469
         Auth 009205                    Ref

SA   R    006 09:13:23 01/12/04 1/99
  YOUR SALES ASSOCIATE WAS:  JUDY C
```

THE CO-OP STORE
979 PRATT DRIVE
INDIANA, PA 724-357-9143

LAST DAY TO RETURN TEXTBOOKS IS JAN 23, 2004

WWW.IUPSTORE.COM

002	97800713097173 N0008		$5.50N
	INTRO TO GEOGRAPHIC INFORMATION SYSTEMS		
150	0109132509B1		1.95
	HIGH SCHOOL EQUIV. PREP DIGESTBL		
200	093710033X00		1.15
	HIGH PTL LIBRARY CLEAN MEMO		

$ 8.0.23

SUB TOTAL	38.94
SALES TAX	0.21
TOTAL	03.15
VISA	04.15

ACCT XXXXXXXXXX 7808
AUTH 003205   REF:

SA	T	008 03113 29 0112/2/01 5799
YOUR BILLS WITH BILL ONE: JODY L		

I CARDHOLDER ACKNOWLEDGE(S) RECEIPT OF
GOODS/SERVICES IN THE TOTAL AMOUNT SHOWN
ABOVE AND AGREES TO PAY THE INDEBTEDNESS
ACCORDING TO ITS CURRENT TERMS.

****
THANK YOU FOR SHOPPING WITH YOUR IUP CAMPUS PROVIDER
****
PLEASE SAVE YOUR RECEIPT FOR RETURNS

DATA DISPLAY AND CARTOGRAPHY

8.1 INTRODUCTION

Maps are an interface to GIS (Kraak and Ormeling 1996). We view, query, and analyze maps. We plot maps to examine results of query and analysis. And we produce maps for presentation and reports. As a visual tool maps are most effective in communicating spatial data, whether the emphasis is on the location or the distribution pattern of spatial data.

Common map elements are the title, body, legend, north arrow, scale, acknowledgment, and neatline/map border (Figure 8.1). Other elements include the graticule or grid, name of map projection, inset or location map, and data quality information. Together, these elements bring spatial

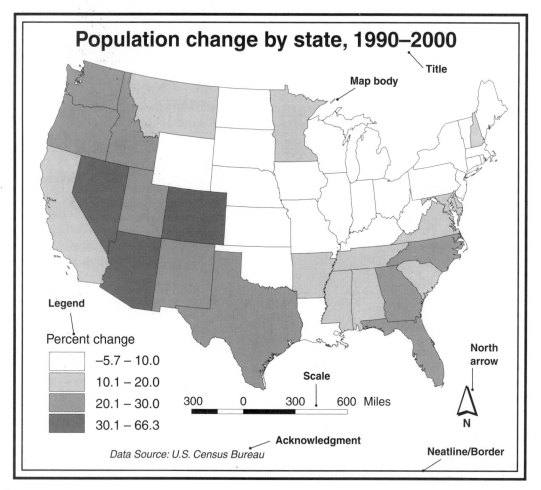

Figure 8.1
Common map elements.

information to the map reader. The map body is the most important part of a map because it contains the map information. Other elements of the map support the communication process. For example, the title suggests the subject matter, and the legend relates map symbols to spatial data. In practical terms, mapmaking may be described as a process of assembling map elements.

Data display is one area that commercial GIS packages have greatly improved on in recent years. Desktop GIS packages with the graphical user interface (GUI) are excellent for data display

for two reasons. First, the mapmaker can simply point and click the graphic icons to construct a map. In comparison, a command-line driven package requires the user to become familiar with many commands and their parameters before making a map. Second, these packages have incorporated some design options such as symbol choices and color schemes into menu selections.

For a novice mapmaker, these easy-to-use GIS packages and their "default options" can result in maps of questionable quality. Mapmaking should be guided by a clear idea of map design and map communication. A well-designed map can help the mapmaker communicate spatial information to the map reader, whereas a poorly designed map can confuse the map reader and even distort the information intended by the mapmaker.

This chapter is divided into the following five sections. Section 8.2 discusses cartographic symbolization including the data–symbol relationship and use of color. Section 8.3 considers different types of maps by function and by map symbol. Section 8.4 provides an overview of typography, selection of type variations, and placement of text. Section 8.5 covers map design and the design elements of layout and visual hierarchy. Section 8.6 examines issues related to map production.

8.2 CARTOGRAPHIC SYMBOLIZATION

8.2.1 Spatial Features and Map Symbols

Spatial features are characterized by their locations and attributes. To display a spatial feature on a map, we use a map symbol to indicate the feature's location and a visual variable, or visual variables, with the symbol to show the feature's attribute data. For example, a thick line in red may represent an interstate highway and a thin line in black may represent a state highway. The line symbol shows the location of the highway in both cases, but the line width and color—two visual variables with the line symbol—separate the interstate from the state highway. Choosing the appropriate map symbol and visual variables is therefore the main concern for data display (Robinson et al. 1995; Dent 1999; Slocum 1999).

Choice of map symbol is simple for raster data: the map symbol applies to cells whether the spatial feature to be depicted is a point, line, or area. Choice of map symbol for vector data depends on the feature type (Figure 8.2). The general rule is to use point symbols for point features, line symbols for line features, and area symbols for area features. But this general rule does not apply to volumetric data or aggregate data. There are no volumetric symbols for data such as elevation, temperature, and precipitation. Instead, isolines are

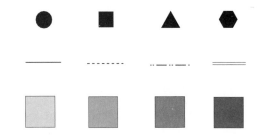

Figure 8.2

Type of map symbol. Top to bottom: point, line, and area symbols.

often used to map volumetric data. Aggregate data such as county populations are data reported at an aggregate level. A common approach is to assign aggregate data to the center of each county and display the data using point symbols.

Visual variables for data display include hue, value, chroma, size, texture, shape, and pattern (Figure 8.3). Choice of visual variable depends on the type of data to be displayed. Hue, value, and chroma relate to color, which is covered in the next section. Size and texture (the spacing of symbol markings) are more appropriate for displaying quantitative (ordinal, interval, and ratio) data. For example, a map may use different-sized circles to represent different-sized cities. Shape and pattern (the type of symbol markings) are more appropriate for displaying qualitative (nominal) data. For example, a map may use different area patterns to show different land-use types. Most GIS packages organize the selection of visual variables into palettes so that the user can easily select the variables that make up the map symbols. Some packages also allow custom pattern designs.

The choice of visual variables is limited in the case of raster data. The visual variables of shape and size do not apply to raster data because of the use of cells. Using texture or pattern is possible for low-resolution raster data but difficult with very small cells. Display of raster data is therefore limited to different colors or different shades of colors. Raster data display can benefit from a tool called **transparency,** which controls the percentage of a map layer that is transparent. By "toning

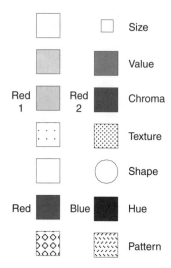

Figure 8.3
Visual variables in cartographic symbolization.

down" a raster layer, transparency makes it possible to superimpose other map symbols.

8.2.2 Use of Color

Because color adds a special appeal to a map, mapmakers will choose color maps over black and white maps whenever possible. But color is probably the most misused visual variable according to published critiques of computer-generated maps (Monmonier 1996). Use of color in mapmaking must begin with an understanding of the color dimensions of hue, value, and chroma.

Hue is the quality that distinguishes one color from another, such as red from blue. Hue can also be defined as the dominant wavelength of light making up a color. We tend to relate different hues with different kinds of data. **Value** is the lightness or darkness of a color, with black at the lower end and white at the higher end. We generally perceive darker symbols on a map as being more important, or more in terms of magnitude (Robinson et al. 1995). Also called saturation or intensity, **chroma** refers to the richness, or brilliance, of a color. A fully saturated color is pure, whereas a low saturation approaches gray. We generally as-

sociate higher-intensity symbols with greater visual importance.

The first rule of thumb in use of color is simple: hue is a visual variable better suited for qualitative data, whereas value and chroma are better suited for quantitative data.

Qualitative mapping is relatively simple. It is not difficult to find 12 or 15 distinctive hues for a qualitative map. If a map requires more symbols, one can add pattern or text to hue to make up more map symbols.

Quantitative mapping has received much more attention than qualitative mapping. Over the years, cartographers have suggested general color schemes that combine value and chroma for displaying quantitative data (Cuff 1972; Mersey 1990; Brewer 1994; Robinson et al. 1995). A basic premise among these color schemes is that the map reader can easily perceive the progression from low to high values (Antes and Chang 1990). The following is a summary of these color schemes.

- **The single hue scheme.** This color scheme uses a single hue but varies the combination of value and chroma to produce a sequential color scheme such as from light red to dark red. It is a simple but effective option for displaying quantitative data (Cuff 1972).
- **The hue and value scheme.** This color scheme progresses from a light value of one hue to a darker value of a different hue. Examples are yellow to dark red and yellow to dark blue. In an experimental study, Mersey (1990) found that color sequences incorporating both regular hue and value variations outperformed other color schemes on the recall or recognition of general map information.
- **The diverging or double-ended scheme.** This color scheme uses graduated colors between two dominant colors. For example, a diverging scheme may progress from dark blue to light blue and then from light red to dark red. The diverging color scheme is a natural choice for displaying data with positive and negative values, or increases and decreases. But Brewer et al. (1997) reported

Box 8.1 Locating Dots on a Dot Map

The most difficult part of making a dot map is locating dots. Suppose a dot represents 500 persons and a county has a population of 5000. This means that the county should have 10 dots within its border. The next question is where to place the 10 dots. Ideally the dots should be placed at the locations of towns and villages, rather than in unsettled areas. Most computer mapping packages including ArcMap, however, use a random method in placing dots. The random method can create an unrealistic dot map. One way to improve the accuracy of dot maps is to base them on the smallest administrative unit possible. Instead of placing dots at the county level, consult a census block group or block map with its population statistics to locate dots. Another way is to exclude areas such as water bodies that should not have dots. ArcMap allows use of mask areas for this purpose.

that the diverging scheme was in fact better than other color schemes for the map reader to retrieve information from quantitative maps that do not necessarily include positive and negative values. To further support her finding, Brewer (2001) used the diverging scheme to map various demographic data from Census 2000 (**http://www.census.gov/population/www/cen2000/atlas.html/**).

- **The part spectral scheme.** This color scheme uses adjacent colors of the visible spectrum to show variations in magnitude. Examples of this color scheme include yellow to orange to red and yellow to green to blue.
- **The full spectral scheme.** This color scheme uses all colors in the visible spectrum. A conventional use of the full spectral scheme is found in elevation maps. Cartographers usually do not recommend this option for mapping other quantitative data because there is no logical sequence between hues.

8.3 TYPE OF MAP

Cartographers classify maps by function and by symbolization. By function, maps can be general reference or thematic. The **general reference map** is used for general purposes. An example would be a USGS quadrangle map, which shows a variety of spatial features, including boundaries, hydrology, transportation, contour lines, settlements, and land covers. The **thematic map** is also called the special purpose map, because its main objective is to show the distribution pattern of a theme, such as the distribution of population densities by county in a state.

By map symbol, maps can be qualitative or quantitative. A qualitative map uses visual variables that are appropriate for portraying qualitative data, whereas a quantitative map uses visual variables that are appropriate for communicating quantitative data. The following describes several common types of quantitative maps (Figure 8.4) (Robinson et al. 1995; Dent 1999; Slocum 1999).

- The **dot map** uses uniform point symbols to show spatial data, with each symbol representing a unit value. One-to-one dot mapping uses the unit value of one such as one dot representing one crime location. But in most cases, it is one-to-many data mapping and the unit value is greater than one. The placement of dots becomes a major consideration in one-to-many dot mapping (Box 8.1).
- The **choropleth map** symbolizes, with shading, derived data based on administrative units (Box 8.2). An example is a map showing average household income by county. The derived data are usually

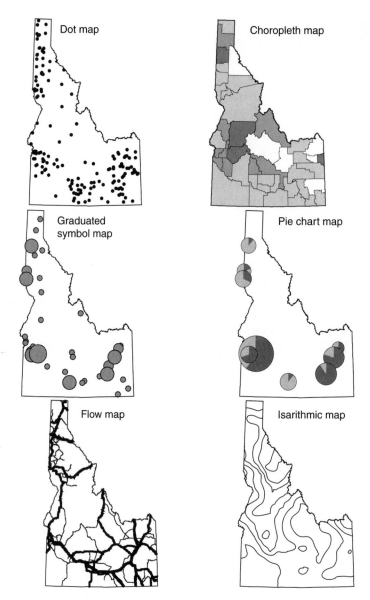

Figure 8.4
Six common types of quantitative maps.

classified prior to mapping and are symbolized using a color scheme for quantitative data. The **dasymetric map** is a variation of the simple choropleth map. By using statistics and additional information, the dasymetric map delineates areas of

homogeneous values, rather than follows administrative boundaries (Robinson et al. 1995) (Figure 8.5). Dasymetric mapping used to be a time-consuming task, but the analytical functions of a GIS have simplified the mapping procedure (Holloway et al.

Box 8.2 | Mapping Derived and Absolute Values

Cartographers distinguish between absolute and derived values in mapping (Chang 1978). Absolute values are magnitudes or raw data such as county population, whereas derived values are normalized values such as county population densities (derived from dividing the county population by the area of the county). County population densities are independent of the county size. Therefore, two counties with equal populations but different sizes will have different population densities, and thus different symbols, on a choropleth map. If choropleth maps are used for mapping absolute values such as county populations, size differences among counties can severely distort map comparisons (Monmonier 1996). Cartographers recommend the graduated symbols for mapping absolute values.

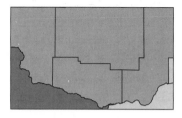 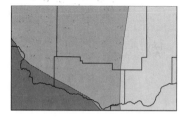

Figure 8.5
Choropleth map (left) versus dasymetric map (right). Map symbols follow the boundaries in the choropleth map but not the dasymetric map.

1999). ArcGIS uses the term **graduated color map** to cover the choropleth and dasymetric maps because both map types use a graduated color scheme to show the variation in spatial data.

- The **graduated symbol map** uses different-sized symbols such as circles, squares, or triangles to represent different ranges of values. For example, one circle size may represent a population of 10,000 to 30,000, another 30,000 to 50,000, and so on. Two important issues to this map type are the range of sizes and the discernible difference between sizes. Both issues are obviously related to the number of graduated symbols on a map. A **proportional symbol map** is a map that uses a specific symbol size for each numeric value rather than a range of values.

Therefore, one circle size may represent a population of 10,000, another 15,000, and so on.
- The **chart map** uses either pie charts or bar charts. A variation of the graduated circle, the pie chart can display two sets of quantitative data: its size can be made proportional to a value such as a county population, and its subdivisions can show the makeup of the value, such as the racial composition of the county population. Bar charts use vertical bars and their height to represent quantitative data. Bar charts are particularly useful for comparing data side by side.
- The **flow map** displays different quantities of flow data such as traffic volume and stream flow by varying the width of line symbols. Similar to the graduated symbols, the flow symbols usually represent ranges of values.

- The **isarithmic map** uses a system of isolines to represent a surface. Each isoline connects points of equal value. GIS users often use the isarithmic map to display the statistical surface created by spatial interpolation (Chapter 13).

GIS has introduced a new classification of maps based on vector and raster data. Maps prepared from vector data are the same as traditional maps using point, line, and area symbols. Most of this chapter applies to the display of vector data. Maps prepared from raster data, although they may look like traditional maps, are cell-based. But raster data can also be classified as either quantitative or qualitative. The color schemes for qualitative and quantitative maps described in this section can therefore be applied to mapping raster data as well.

8.4 TYPOGRAPHY

A map cannot be understood without lettering or type on it. Lettering is needed for almost every map element. Mapmakers treat type as a map symbol because, like point, line, or area symbols, type has many variations. Using type variations to create a pleasing and coherent map is a major challenge to mapmakers.

8.4.1 Type Variations

Type varies in typeface, form, size, and color (Box 8.3). **Typeface** refers to the design character of the type. There are two main groups of typefaces: **serif** (with serif) and **sans serif** (without serif) (Figure 8.6). Serifs are small, finishing touches at the ends of line strokes, which tend to make it easier to read running text in newspapers and books. Compared to serif types, sans serif types appear simpler and bolder. Although rarely used in books or other text-intensive materials, sans serif type stands up well on maps with complex map symbols and remains legible even in small sizes. Sans serif types have an additional advantage in mapmaking because many of them come in a wide range of type variations.

Times New Roman

Tahoma

Figure 8.6
Times New Roman is a serif typeface, and Tahoma is a sans serif typeface.

Helvetica Normal

Helvetica Italic

Helvetica Bold

Helvetica Bold-Italic

Times Roman Normal

Times Roman Italic

Times Roman Bold

Times Roman Bold-Italic

Figure 8.7
Type variations in weight and roman/italic.

Type form refers to the variant of letterform. Type form variations include **type weight** (bold, regular, or light), **type width** (condensed or extended), upright versus slanted (or roman versus italic), and uppercase versus lowercase (Figure 8.7). A **font** is a complete set of all variants of a given typeface and size.

Type size measures the height of a letter in **points,** with 72 points to an inch. Printed letters look smaller than what their point sizes suggest. The point size is supposed to be measured from a metal type block, which must accommodate the lowest point of the descender (such as p or g) to the highest part of the ascender (such as d or b). But no letters extend to the very edge of the block.

Type color is the color of letters. Color, drop shadow, halo, and fill pattern provide the type effects that are separate from the type form variations.

Box 8.3 **Terms for Type Variations**

Except for type size and type color, terms for type variations are not standardized. This lack of standardization can create confusion when GIS users turn to cartography textbooks for help in type design. The design character of the type may be referred to as type style, typeface, or font. Type variations such as weight, width, uppercase versus lowercase, and upright versus italic may collectively be referred to as type form, letterforms, or type style. ArcMap uses the term text symbol to cover type variations including font, font style, color, effects, and size.

8.4.2 Selection of Type Variations

Type variations can function in the same way as visual variables for map symbols. Differences in typeface, type color, and upright versus italic are more appropriate for qualitative data, whereas differences in type size, weight, and uppercase versus lowercase are more appropriate for quantitative data. For example, a reference map showing different sizes of cities typically shows the largest cities in a large type size and in bold and capital letters, and the smallest cities in a small type size and in thin and lowercase letters.

Cartographers also recommend legibility, harmony, and conventions in type selection (Dent 1999). Legibility is difficult to control on a map because it can be influenced by the choice of typeface, type size, placement of lettering, letter spacing, as well as contrast between type and the background symbol. GIS users have the additional problem of having to design a map on the computer monitor and to print it on a much larger plot. GIS users therefore need to experiment to ensure type legibility in all parts of the map.

Type legibility, however, should be balanced with harmony. The function of type is to communicate the map content. Thus type should be legible but should not draw too much attention. Mapmakers can generally achieve harmony by adopting only one or two typefaces on a map and can use other type variations for different elements or symbols (Figure 8.8). For example, many mapmakers use a sans serif type in the body of a map

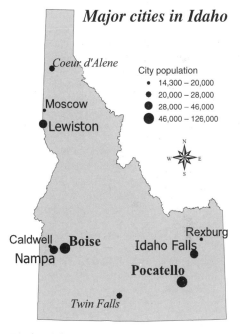

Figure 8.8
The look of the map is not harmonious because of the use of too many typefaces.

and a serif type for the map's title and legend. Use of conventions may also lend support for harmony. Conventions in type selection include italic for names of water features, uppercase and letter spacing for names of administrative units, and variations in type size and form for names of cities that are ranked in a hierarchical order.

Most GIS users have to work with fonts that have been loaded on the computer from the printer manufacturer or other software packages. These fonts are used for wordprocessing, graphics, and mapping. Therefore, not all of them are suitable for mapmaking. Some of the fonts installed by ArcGIS such as geology 24k and survey are designed for special types of maps. ArcGIS users should be able to select from the preloaded fonts enough type variations for their maps. If not, they can use other fonts by importing them into ArcGIS.

8.4.3 Placement of Text

The placing of lettering, or labeling, on a map is just as important as selecting type variations. As a general rule, lettering should be placed to show the location or the area extent of the named spatial feature. Cartographers recommend placing the name of a point feature to the upper right of its symbol, the name of a line feature in a block and parallel to the course of the feature, and the name of an area feature to indicate its area extent. Other general rules suggest aligning names with either the map border or lines of latitude, and placing names entirely on land or on water.

Implementing labeling algorithms in a GIS package is no easy task (Mower 1993; Chirié 2000). Automated name placement presents several difficult problems to the computer programmer: names must be legible, names cannot overlap other names, names must be clear as to their intended referent symbols, and name placement must follow cartographic conventions. These problems worsen at smaller map scales as competition for map space intensifies between names. GIS users should not expect that labeling be completely automated. Usually, some interactive editing is needed to improve the final map's appearance. For this reason many GIS packages offer more than one method of labeling.

As an example, ArcGIS offers interactive and dynamic labeling. Interactive labeling works with one label at a time. The label can be placed along a horizontal line or a curved line. If the placement does not work out well, the label can be moved immediately. Interactive labeling is ideal for adding data source, acknowledgment, or explanatory notes on a map.

Dynamic labeling is probably the method of choice for most users because it can automatically label all or selected features with an attribute value. Using dynamic labeling, one must first define placement options for point and line features. A label can be placed in different positions around a point feature. A label can be placed to follow the curve of a line feature, and above, below, or on the line. Additionally, rules can be set to prioritize labels competing for the same map space. By default, ArcGIS does not allow overlapped labels. This constraint can impact the placement of labels and may require adjustment of some labels (Figure 8.9).

Dynamic labels cannot be selected or adjusted individually. But they can be first converted to text elements so that these elements can be individually moved and changed in the same way as interactive labels (Figure 8.10). Task 3 in the Applications section shows how dynamic labeling and interactive labeling can be used together for placement of text. Another way to take care of labels in a congested area is to use a leader line to link a label to the map feature (Figure 8.11).

Perhaps the most difficult task in labeling is to label names of streams. The general rule states that the name should follow the course of the stream, either above or below it. Both interactive labeling and dynamic labeling can curve the name if it appears along a curvy part of the stream. But the appearance of the curved name depends on the smoothness of the corresponding stream segment, the length of the stream segment, and the length of the name. Placing every name in its correct position the first time is nearly impossible (Figure 8.12). Problem names must be removed and relabeled using the **spline text** tool, which can align a text string along a curved line that is digitized on-screen (Figure 8.13).

Figure 8.9

Dynamic labeling of major cities in the United States. The initial result from ArcMap is good but not totally satisfactory. Philadelphia is missing. Labels of San Antonio, Indianapolis, and Baltimore overlap slightly with point symbols. San Francisco is too close to San Jose.

Figure 8.10

A revised version of Figure 8.9. Philadelphia is added to the map, and several city names are moved individually to be closer to their point symbols.

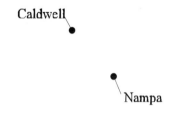

Figure 8.11
A leader line connects a point symbol to its label.

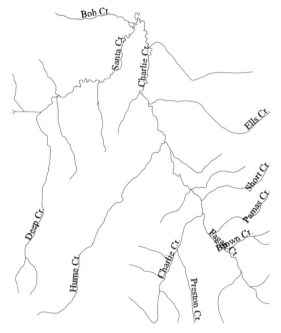

Figure 8.12
The Find Best Placement option in ArcMap may not work with every label. Brown Cr. overlaps with Fagan Cr., and Pamas Cr. and Short Cr. do not follow the course of the creek.

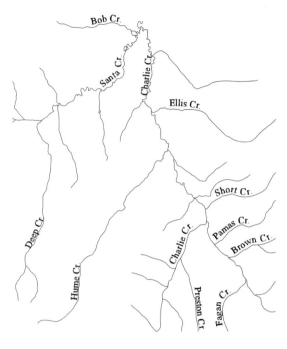

Figure 8.13
Problem labels in Figure 8.12 are redrawn with the Spline Text tool in ArcMap.

fusing and disoriented (Antes et al. 1985). Map design is both an art and science. There may not be clear-cut right or wrong designs for maps, but there are better or more effective maps and worse or less effective maps.

Map design overlaps with the field of graphic arts, and many map design principles have their origin in visual perception. Cartographers usually study map design from the perspectives of layout and visual hierarchy.

8.5 Map Design

Like graphic design, **map design** is a visual plan to achieve a goal. The purpose of map design is to enhance map communication, which is particularly important for thematic maps. A well-designed map is balanced, coherent, ordered, and interesting to look at, whereas a poorly designed map is con-

8.5.1 Layout

Layout, or planar organization, deals with the arrangement and composition of various map elements on a map. Major concerns with layout are focus, order, and balance. A thematic map should have a clear focus, which is usually the map body or a part of the map body. To draw the map

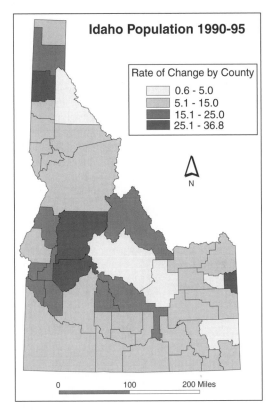

Figure 8.14
Use a box around the legend to draw the map reader's attention to it.

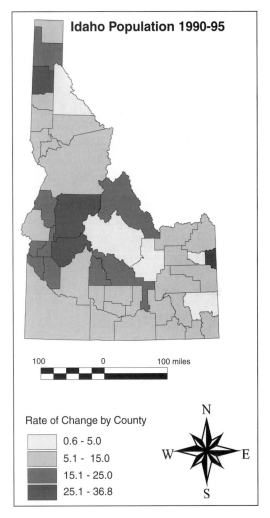

Figure 8.15
A poorly balanced map.

reader's attention, the focal element should be placed near the optical center, just above the map's geometric center. The focal element should be differentiated from other map elements by contrast in line width, texture, value, detail, and color.

After viewing the focal element, the reader should be directed to the rest of the map in an ordered fashion. For example, the legend and the title are probably the next elements that the viewer needs to look at after the map body. To smooth the transition, the mapmaker should clearly place the legend and the title on the map, with perhaps a box around the legend to draw attention to it (Figure 8.14).

A finished map should look balanced. It should not give the map reader an impression that

the map "looks" heavier on the top, bottom, or the side (Figure 8.15). But balance does not suggest the breaking down of the map elements and placing them, almost mechanically, in every part of the map. Although in that case the elements would be in balance, the map would be disorganized and confusing. Mapmakers therefore should deal with balance within the context of organization and map communication.

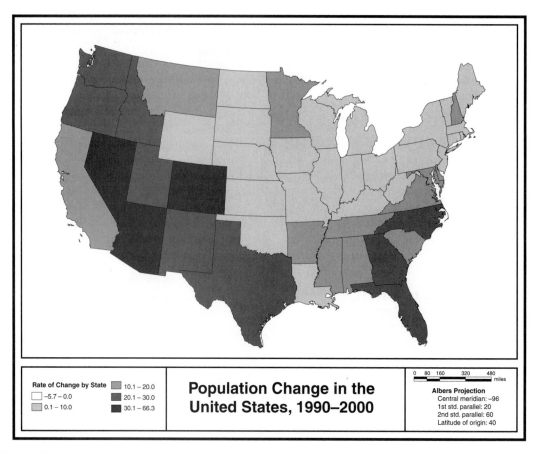

Figure 8.16
The basic structure of the conterminous USA layout template in ArcMap.

Cartographers used to use thumbnail sketches to experiment with balance on a map. Now they use computers to manipulate map elements on a layout page. ArcGIS, for example, offers two basic methods for layout design. The first method is to use a layout template. These templates are grouped into general, industry, USA, and world. Each group has a list of choices. For example, the layout templates for the United States include USA, conterminous USA, and five different regions of the country. Figure 8.16 shows the basic structure of the conterminous USA layout template. The idea of using a layout template is to use a built-in design option to quickly compose a map.

The second option is to open a layout page and build on it different map elements one at a time. ArcGIS offers the elements of map (called data frame), title, text, neatline, legend, north arrow, scale bar, scale text, and the graticule (Box 8.4). The user can choose one of these map elements, place it on the layout page, and manipulate it graphically. For example, the user can change the type design of the title, enlarge or reduce it, and move it around the layout. A layout design created using the second option can be saved as a template for future use.

Regardless of the method used in layout design, the legend deserves special consideration. The legend includes descriptions of all the layers that make up a map. For example, a map showing different classes of cities and roads in a state requires a minimum of three layers: one for the

Box 8.4   Wizards for Adding Map Elements

ArcMap provides wizards for adding map elements to a map. The graticule/grid wizard is one of them. The graticule refers to lines or tics of longitude and latitude. The grid refers to lines or tics of coordinates of a plane coordinate system. The graticule/grid wizard guides the user through a series of dialogs to choose the intervals, symbol design, and placement option. Wizards are particularly useful for the beginner. But for experienced users, wizards can be tedious and limiting. One can choose to turn off the wizard and to work directly with the main property dialog of a map element.

cities, one for the roads, and one for the state boundary. As default, these descriptions are placed together as a single graphic element, which can become quite lengthy with multiple layers. A lengthy legend presents a problem in balancing a layout design (Figure 8.17a). The solution is to break the legend into two or more columns and to remove useless legend descriptions such as the description of an outline symbol (Figure 8.17b).

8.5.2 Visual Hierarchy

Visual hierarchy is the process of developing a visual plan to introduce the 3-D effect or depth to maps (Figure 8.18). Mapmakers create the visual hierarchy by placing map elements at different visual levels according to their importance to the map's purpose. The most important element should be at the top of the hierarchy and should appear closest to the map reader. The least important element should be at the bottom. A thematic map may consist of three or more levels in a visual hierarchy.

The concept of visual hierarchy is an extension of the **figure–ground relationship** in visual perception (Arnheim 1965). The figure is more important visually, appears closer to the viewer, has form and shape, has more impressive color, and has meaning. The ground is the background. Cartographers have adopted the depth cues for developing the figure–ground relationship in map design.

Probably the simplest and yet most effective principle in creating a visual hierarchy is called interposition or superimposition (Dent 1999). **Interposition** uses the incomplete outline of an object to make it appear behind another. Examples of interposition abound in maps, especially in newspapers and magazines. Continents on a map look more important or occupy a higher level in visual hierarchy if the lines of longitude and latitude stop at the coast. A map title, a legend, or an inset map looks more prominent if it lies within a neatline (box), with or without the drop shadow. When the map body is deliberately placed on top of the neatline around a map, the map body will stand out more (Figure 8.19). Because interposition is so easy to use, it can be overused or misused. A map looks confusing if several of its elements compete for the map reader's attention (Figure 8.20).

Subdivisional organization is a map design principle that groups map symbols at the primary and secondary levels according to the intended visual hierarchy (Robinson et al. 1995). Each primary symbol is given a distinctive hue, and the differentiation among the secondary symbols is based on color variation, pattern, or texture. For example, all tropical climates on a climate map are shown in red, and different tropical climates (e.g., wet equatorial climate, monsoon climate, and wet–dry tropical climate) are distinguished by different shades of red. Subdivisional organization is most useful for maps with many map symbols, such as climate, soils, geology, and vegetation maps.

Contrast is a basic element in map design, important to layout as well as to visual hierarchy. Contrast in size or width can make a state outline look more important than county boundaries and larger cities look more important than smaller ones (Figure 8.21). Contrast in color can separate the figure

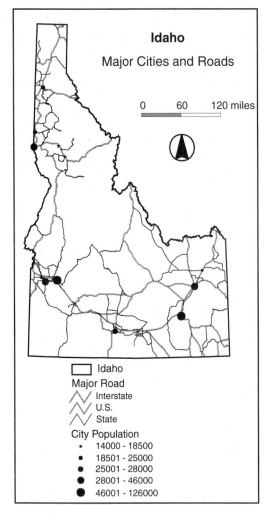

Figure 8.17a
A lengthy legend is confusing and can create a problem
in layout design.

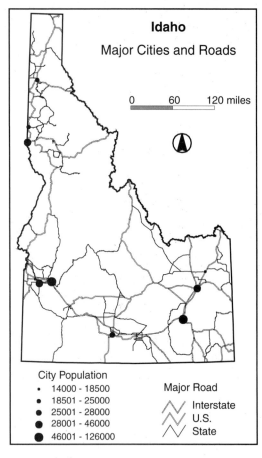

Figure 8.17b
The lengthy legend in Figure 8.17a is separated into
two parts. Also, the unnecessary outline symbol for
Idaho is removed from the legend.

from the ground. Cartographers often use a warm
color (e.g., orange to red) for the figure and a cool
color (e.g., blue) for the ground. Contrast in texture
can also differentiate between the figure and the
ground because the area containing more details or
a greater amount of texture tends to stand out on a
map. Like the use of interposition, too much contrast can create a confusing map appearance. For instance, if bright red and green are used side by side
as area symbols on a map, they appear to vibrate.

8.6 MAP PRODUCTION

GIS users design and make maps on the computer
screen. These soft-copy maps can be used in a variety of ways. They can be printed, exported for
use on the Internet, used in overhead computer
projection systems, exported to other software
packages, or further processed for publishing (Box
8.5). Map production is a complex topic. Many
GIS users probably have had the experience of

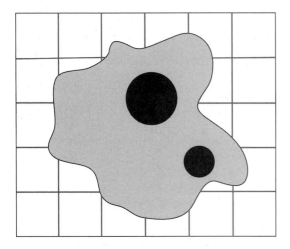

Figure 8.18
A visual hierarchy example. The two black circles are
on top (closest to the map reader), followed by the gray
polygon and the grid.

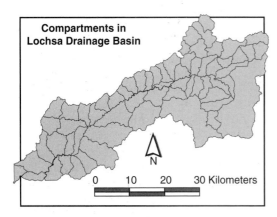

Figure 8.19
The interposition effect in map design.

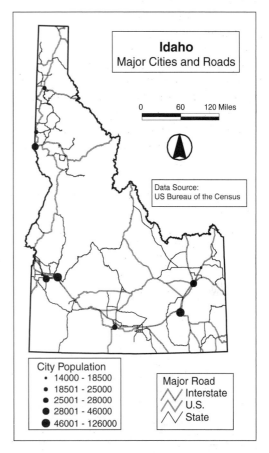

Figure 8.20
A map looks confusing if it has too many boxes to
highlight individual map elements.

plotting a blank or totally black map from a supposedly perfect color map on the computer screen.

GIS users are often surprised that color symbols from their color printers do not exactly match
those on the computer screen. This discrepancy results from the use of different media and color
models. Most computer screens use **Cathode-Ray
Tubes** (CRTs) for data display. A CRT screen has a
built-in fine mesh of pixels, and each pixel has
three phosphors, one red, one green, and one blue.
These phosphors emit light when struck by electrons from three electron guns with the **RGB** (red,
green, and blue) designation. Therefore, a color
symbol we see on the computer screen is made of
pixels, and the color of each pixel is actually a
mixture of RGB.

The intensity of each primary color in a mixture determines its color. The number of intensity
levels each primary color can have depends on the
number of bit-planes assigned to the electron gun.
Typically, a color display system has 8 bit-planes

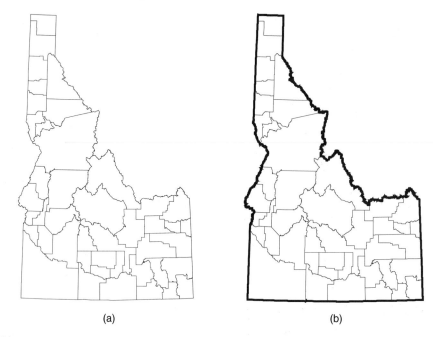

(a)   (b)

Figure 8.21
Contrast is missing in (a), whereas line contrast makes the state outline look more important than the county boundaries in (b).

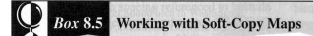

Box 8.5　Working with Soft-Copy Maps

When completed, a soft-copy map can be either printed or exported. Printing a map from the computer screen requires the software interface (often called the driver) between the operating system and the print device. ArcMap uses Enhanced Metafiles (EMF) as the default and offers two additional choices of PostScript (PS) and ArcPress. EMF files are native to the Windows operating system for printing graphics. PS, developed by Adobe Systems Inc., in the 1980s, is an industry standard for high-quality printing. Developed by ESRI Inc., ArcPress is PS-based and useful for printing maps containing raster datasets, images, or complex map symbols.

One must specify an export format to export a computer-generated map for other uses. ArcMap, for example, offers JPEG (Joint Photographic Experts Group) for the Internet, TIFF (tagged image file format), AI (Adobe Illustrator format), BMP (bitmap), and CGM (Computer Graphics Metafile) for other graphics packages, and EMF, EPS (Encapsulated PostScript), AI, and PDF (portable document format) for printing and publishing.

Offset printing is the standard method for printing a large number of copies of a map. A typical procedure to prepare a computer-generated map for offset printing involves the following steps. First, the map is exported to separate CMYK PostScript files in a process called color separation. CMYK stands for the four process colors of cyan, magenta, yellow, and black, commonly used in offset printing (see text for more explanation). Second, these files are processed through an image setter to produce high-resolution plates or film negatives. Third, if the output from the image setter consists of film negatives, the negatives are used to make plates. Finally, offset printing runs the plates through the press to print color maps.

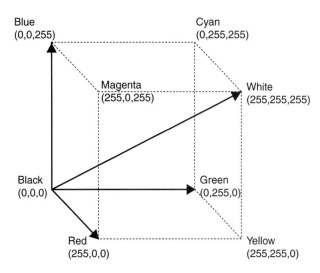

Figure 8.22
The RGB (red, green, and blue) color model.

per gun, meaning that the system can generate 16.8 million colors ($2^8 * 2^8 * 2^8$, or 256 * 256 * 256).

Many GIS packages offer the RGB color model for color specification. ArcGIS, for example, uses the RGB color model for multiband image data. But color mixtures of RGB are not intuitive (Figure 8.22). It is difficult to perceive that a mixture of red and green at full intensity is a yellow color. This is why other color models have been developed to specify colors that are based on the perception of hue, value, and chroma. ArcGIS, for example, has the **HSV** (hue/saturation/value) color model for specifying custom colors.

Printed color maps differ from CRT color displays in two ways: color maps reflect rather than emit light; and the creation of colors on color maps is a subtractive rather than additive process. The three primary subtractive colors are cyan, magenta, and yellow. In printing, these three primary colors plus black form the four process colors of **CMYK.**

Color symbols are produced on a printed map in much the same way as on a computer screen. In place of phosphors and pixels are color dots, and in place of varying light intensities are percentages of area covered by color dots. A deep orange color on a printed map may represent a combination of 60% magenta and 80% yellow, whereas a light orange color may represent a combination of 30% magenta and 90% yellow. To match a color symbol on the computer screen with a color symbol on the printed map requires a translation from the RGB color model to the CMYK color model. As yet there is no exact translation between the two, and therefore the color map looks different when printed (Slocum 1999). The International Color Consortium, a consortium of over 70 companies and organizations worldwide, has been working since 1993 on a color management system that can be used across different platforms and media (**http://www.color.org/).** Until such a color management system is developed, one must experiment with colors on different media.

Map production, especially production of color maps, can be a challenging task to GIS users. Box 8.6 describes a free Web tool that can help GIS users choose color symbols that are appropriate for a particular mode of map production.

Box 8.6 **A Web Tool for Map Production**

Afree Web tool for map production is available at **http://www.colorbrewer.org**. Funded initially by the National Science Foundation's Digital Government Program, the website is maintained by Cindy Brewer and Mark Harrower of Pennsylvania State University. The tool offers three main types of color schemes: sequential, diverging, and qualitative. One can select a color scheme and see how the color scheme looks on a sample choropleth map. One can also add point and line symbols to the sample map and change the colors for the map border and background. Then, for each color selected, the tool shows its color values in terms of CMYK, RGB, and other models. And, for each color scheme selected, the tool rates its potential uses including photocopy in black and white, CRT display, color printing, LCD (liquid crystal display) projector, and laptop LCD display.

KEY CONCEPTS AND TERMS

Cathode-Ray Tube (CRT) screen: A display device with a built-in fine mesh of pixels, each of which has three phosphors (red, green, and blue) and can emit light when struck by electrons from electron guns.

Chart map: A map that uses charts such as pie charts or bar charts as map symbols.

Choropleth map: A map that applies shading symbols to data or statistics collected for administrative units such as counties or states.

Chroma: The richness or brilliance of a color. Also called *saturation* or *intensity*.

CMYK: A color model in which colors are specified by the four process colors of cyan (C), magenta (M), yellow (Y), and black (K).

Contrast: A basic element in map design that enhances the look of a map or the figure–ground relationship by varying the design of size, width, color, and texture.

Dasymetric map: A map that uses statistics and additional information to delineate areas of homogeneous values, rather than following administrative boundaries.

Dot map: A map that uses uniform point symbols to show spatial data, with each symbol representing a unit value.

Figure–ground relationship: A tendency in visual perception to separate more important objects (figures) in a visual field from the background (ground).

Flow map: A map that displays different quantities of flow data by varying the width of line symbols.

Font: A complete set of all variants of a given typeface and size.

General reference map: One type of map used for general purposes such as the USGS topographic map.

Graduated color map: A map that uses a progressive color scheme such as light red to dark red to show the variation in spatial data.

Graduated symbol map: A map that uses different-sized symbols such as circles, squares, or triangles to represent different magnitudes.

HSV: A color model in which colors are specified by their hue (H), saturation (S), and value (V).

Hue: The quality that distinguishes one color from another, such as red from blue. Hue is the dominant wavelength of light.

Interposition: A tendency for an object to appear behind another because of its incomplete outline.

Isarithmic map: A map that uses a system of isolines to represent a surface.

Layout: The arrangement and composition of map elements on a map.

Map design: The process of developing a visual plan to achieve the map's purpose.

Point: Measurement unit of type, with 72 points to an inch.

Proportional symbol map: A map that uses a specific-sized symbol for each numeric value.

RGB: A color model in which colors are specified by their red (R), green (G), and blue (B) components.

Sans serif: Without serifs.

Serif: Small, finishing touches added to the ends of line strokes.

Spline text: A text string aligned along a curved line.

Subdivisional organization: A map design principle that groups map symbols at the primary and secondary levels according to the intended visual hierarchy.

Thematic map: One type of map used to emphasize the spatial distribution of a theme, such as a map that shows the distribution of population densities by county in a state.

Transparency: A display tool that controls the percentage of a map layer that is transparent.

Typeface: A particular style or design of type.

Type weight: Relative blackness of a type such as bold, regular, or light.

Type width: Relative width of a type such as condensed or extended.

Value: The lightness or darkness of a color.

Visual hierarchy: The process of developing a visual plan to introduce the 3-D effect or depth to maps.

APPLICATIONS: DATA DISPLAY AND CARTOGRAPHY

This applications section consists of three tasks. Task 1 guides you through the process of making a choropleth map. Task 2 involves type design, graduated symbols, color combination, and the highway shield symbols. Task 3 focuses on the placement of text. Making maps, especially maps for presentation, is tedious. You must be patient and willing to experiment.

Task 1: Make a Choropleth Map

What you need: *us.shp,* a shapefile showing population change by state in the United States between 1990 and 2000. The shapefile is projected onto the Albers conic equal-area projection and is measured in meters.

Choropleth maps display statistics by administrative unit. For this task you will map the rate of population change between 1990 and 2000 by state. The map consists of the following elements: a map of the conterminous United States and a bar scale, a map of Alaska and a bar scale, a map of Hawaii and a bar scale, a title, a legend, a north arrow, a data source statement, a map projection statement, and a neatline around all elements. The basic layout of the map is as follows. One-third of the map on the left, from top to bottom, has the title, map of Alaska, and map of Hawaii. Two-thirds of the map on the right, from top to bottom, has the map of the conterminous United States and all the other elements.

1. Start ArcCatalog, and make connection to the Chapter 8 database. Launch ArcMap. Maximize the view window of ArcMap. Add *us.shp* to Layers, and rename the data frame from Layers to Conterminous. Use the Zoom In tool to zoom in the conterminous U.S.

2. You are ready to symbolize the rate of population change by state. Right-click *us* and select Properties. Click the Symbology tab in the Layer Properties dialog. Click Quantities in the Show box and select Graduated colors. In the Fields frame, click the Value dropdown arrow and choose ZCHANGE (rate of population change from 1990 to 2000). Cartographers recommend use of round numbers and logical breaks such as 0 in data classification. Click the first cell under Range and enter 0. The new range should read −5.7 – 0.0. Enter 10, 20, and 30 for the next three cells, and click the empty space below the cells to unselect. Next, change the color scheme for ZCHANGE. Right-click the Color Ramp box and uncheck Graphic View. Click the dropdown arrow and choose Yellow to Green to Dark Blue. The first class from −5.7 to 0.0 is shown in yellow, and the other classes are shown in green to dark blue. Click OK to dismiss the Layer Properties dialog.

3. The Alaska map is next. Select Data Frame from the Insert menu. Rename the new data frame Alaska. Add *us.shp* to Alaska. Use the Zoom In tool to zoom in Alaska. Follow the same procedure above to display ZCHANGE.

4. The Hawaii map is next. Select Data Frame from the Insert menu. Rename the new data frame Hawaii. Add *us.shp* to Hawaii. Use the Zoom In tool to zoom in Hawaii. Follow the same procedure to display ZCHANGE.

5. The Table of Contents in ArcMap now has three data frames: Conterminous, Alaska, and Hawaii. Select Layout View from the View menu. Click the Zoom Whole Page button. Select Page Setup from the File menu. Click the radio button next to Landscape for Page Orientation in the Map Size frame, and click OK.

6. The three data frames are stacked up on the layout. You will rearrange the data frames according to the basic layout plan. Click the Select Elements button. Click the Conterminous data frame. When the handles around the data frame appear, move the data frame to the upper-right side of the layout so that it occupies about two thirds of the layout in both width and height and shows only the lower 48 states. Click the Alaska data frame, and move it to the center left of the layout. Click the Hawaii data frame, and place it below the Alaska data frame.

7. Now you will add a scale bar to each data frame. Begin with the Conterminous data frame by clicking on it. Select Scale Bar from the Insert menu. Click the selection of Alternating Scale Bar 1, and then Properties. The Scale Bar dialog has three tabs: Scale and Units, Numbers and Marks, and Format. Click the Scale and Units tab. Start with the bottom half of the dialog: select Adjust width when resizing, choose Kilometers for Division Units, and enter km for Label. Now work with the upper half of the dialog: enter the division value of 1000 (km) for the Division value, select 2 for Number of divisions, select 0 for Number of subdivision, opt not to Show one division before zero. Click the Numbers and Marks tab, and select divisions from the Frequency dropdown list. Click the Format tab, and select Times New Roman from the Font dropdown list. Click OK in both the Scale Bar and Scale Bar Selector dialogs. The scale bar appears in the map with the handles. Move the scale bar to the lower-left corner of the Conterminous data frame. The scale bar should have the divisions of 1000 and 2000 kilometers. Separate bar scales are needed for the other two data frames. Click the Alaska data frame. Follow the same

procedure above to prepare its bar scale but use 500 kilometers for the division value. Click the Hawaii data frame. Add its scale bar by using 100 kilometers for the division value.

8. So far, you have completed the data frames of the map. The map must also have the title, legend, and other elements. Select Title from the Insert menu. An Enter Map Title box appears on the layout. Click outside the box. When the outline of the box is shown in cyan, double-click it. A Properties dialog appears with two tabs. Under the Text tab, enter two lines in the Text box: "Population Change" in the first line, and "by State, 1990–2000" in the second line. Click Change Symbol. The Symbol Selector dialog lets you choose type color, size, font, and style. Use the dropdown lists to choose 20 for Size and Bookman Old Style (or another serif type). Click B for boldface. Click OK in the Symbol Selector dialog and then in the Properties dialog. Move the title to the upper left of the layout above the Alaska data frame.

9. The legend is next. Because all three data frames use the same legend, it does not matter which data frame is used. The active data frame has boldface in the Table of Contents. Click ZCHANGE of the active data frame, and click it one more time so that ZCHANGE is shown in a box. Highlight ZCHANGE and delete it. (Unless ZCHANGE is removed in the Table of Contents, it will show up in the layout as a confusing legend descriptor.) Select Legend from the Insert menu. The Legend Wizard uses five panels. In the first panel, make sure *us* is the layer to be included in the legend. The second panel lets you enter the legend title and its type design. Delete Legend in the Legend Title box, and enter 'Rate of Population Change (%).' Then choose 14 for the Size, choose Times New Roman for the Font, and uncheck B (Bold). Skip the third and fourth panels, and click Finish in the

fifth panel. Move the Legend to the right of the Hawaii data frame.

10. A north arrow is next. Select North Arrow from the Insert menu. Choose ESRI North 6, a simple north arrow from the selector, and click OK. Move the north arrow to the upper right of the legend.

11. Next is the data source. Select Text from the Insert menu. An Enter Text box appears on the layout. Click outside the box. When the outline of the box is shown in cyan, double-click it. A Properties dialog appears with two tabs. Under the Text tab, enter 'Data Source: US Census 2000' in the Text box. Click on Change Symbol. Select 14 for the Size and Times New Roman in the Symbol Selector dialog. Click OK in both dialogs. Move the data source statement below the north arrow in the lower right of the layout.

12. Follow the same procedure as for the data source to add a text statement about the map projection. Enter Albers Conic Equal-area Projection in the Text box by using the Size of 10 and the font of Times New Roman. Move the projection statement below the data source.

13. Finally, add a neatline to the layout. Select Neatline from the Insert menu. Make sure that the radio button next to Place around all elements is checked in the Neatline dialog. Select Double, Graded from the dropdown list for the Border, and Sand for the Background color. Click OK.

14. The layout is now complete. If you want to rearrange a map element, select the element and then move it to a new location. You can also enlarge or reduce a map element by using its handles.

15. If your PC is connected to a color printer, you can print the map directly by selecting Print in the File menu. Or you can save the map as an ArcMap document or export the map as a graphic file in PostScript, JPG, and other formats. Both options are accessible through the File menu. Exit ArcMap.

Task 2: Use Graduated Symbols, Line Symbols, Highway Shield Symbols, and Text Symbols

What you need: *idlcity.shp,* a shapefile showing the 10 largest cities in Idaho; *idhwy.shp,* a shapefile showing interstate and U.S. highways in Idaho; and *idoutl.shp,* an outline map of Idaho.

Task 2 lets you experiment with type design, use graduated symbols, and try the highway shield symbols in ArcMap. The task also involves choice of color symbols and map design in general.

1. Make sure that ArcCatalog is still connected to the Chapter 8 database. Launch ArcMap. Rename the data frame Task 2, and add *idlcity.shp, idhwy.shp,* and *idoutl.shp* to Task 2. Select Layout View from the View menu. Maximize the view window, and click the Zoom Whole Page button. Select Page Setup from the File menu. Make sure that the page has a width of 8.5 (inches), a height of 11 (inches), and a portrait orientation.

2. The only decision to be made about *idoutl* is its color, the background color for the state, which should be contrasted with the point symbols for *idlcity* and the highway symbols. Select Properties from the context menu of *idoutl.* Under the Symbology tab, click the Symbol box and select Grey for the symbol color.

3. Select Properties from the context menu of *idhwy.* Click the Symbology tab. Click Categories in the Show frame and select ROUTE_DESC from the Value Field dropdown list. Click Add All Values at the bottom. Interstate and U.S. appear as the values. Uncheck the box for all other values. Double-click the Symbol next to Interstate and select the Freeway symbol in the Symbol Selector box. Double-click the Symbol next to U.S. Select the Major Road symbol but change its color to Ginger Pink in the Options frame. Click OK in both dialogs.

4. Select Properties from the context menu of *idlcity.* Click the Symbology tab. Click Quantities and then Graduated Symbols in the Show frame. Select Population in the Value dropdown list. The default classification groups the cities into 5 classes by population, and a default symbol scheme is assigned. First, change the number of classes from 5 to 3. Change the Range values by entering 28000 in the first class and 46000 in the second class. Click the empty space below the cells to unselect. Second, change the symbol color. Click Template. Click the Color option and choose Solar Yellow.

5. Labeling the cities is next. Each of the three classes of cities will be labeled with a different type design. Start with the class of large cities. Select Data View from the View menu. Right-click *idlcity* and select Properties. Click the Labels tab. First, make sure that the Label Field is CITY_NAME. Then, click Symbol. Use the dropdown lists to select 12 for Size and Century Gothic (or another sans serif type) and click B (Bold) in the Symbol Selector dialog. Click OK in the Symbol Selector and Layer Properties dialogs. On the Draw toolbar, click the New Text (A) arrow and select the Label tool. In the Labeling Options dialog, choose to place label at position clicked and to use properties set for the feature layer. Place the Label tool at the center of the circle for Boise. When the name Boise appears in a box, left-click on the mouse. Follow the same procedure to label Pocatello.

6. The labels of Boise and Pocatello appear to be too large and too close to the circles in Data View. But you must switch to Layout View to see how the labels will appear on a plot. Select Layout View from the View menu. Select 100% from the Zoom Control list. Use the Layout Pan tool to see how the labels of Boise and Pocatello will appear on an 8.5-by-11-inch plot. If the label is too far away from the circle, switch to Data View. Use the Select Elements tool to select the label before making any necessary adjustment.

7. Next, label the cities in the medium class. Select Properties from the context menu of *idlcity*. Click the Labels tab. Make sure that CITY_NAME is the Label Field, and click Symbol. In the Symbol Selector dialog, choose a size of 12 and Century Gothic (regular, not bold). Label Idaho Falls, Nampa, and Lewiston. Again, switch to Layout View to view how the labels will appear on the plot. Make adjustment if necessary.

8. Follow the same procedure but a size of 10 and Century Gothic (regular) to label cities in the small class: Coeur d'Alene, Moscow, Caldwell, and Rexburg.

9. The last part of this task is to label the interstates and U.S. highways with the highway shield symbols. Switch to Data View. Right-click *idhwy,* and select Properties. Click the Labels tab. Make sure that the Label Field is MINOR1, which lists the highway number. Then click Symbol, and select the U.S. Interstate HWY shield. Click OK to dismiss the Symbol Selector dialog. Click Label Placement Options in the Layer Properties dialog. In the Constrain placement frame, check Centered on line and uncheck the others. Choose Horizontal from the Angle dropdown list. Click OK to dismiss the Placement Properties dialog. Click OK to dismiss the Layer Properties dialog. Click the Label tool. Opt to place label at position clicked. Move the Label tool over an interstate in the map. When a number appears in a box, left-click on the mouse. (The highway number may vary along the same interstate because the interstate has multiple numbers such as 90 and 10, or 80 and 30.)

10. Follow the same procedure but use the U.S. Route HWY shield to label U.S. highways. Switch to Layout View and make sure that the highway shield symbols are labeled appropriately.

11. To complete the layout, you must add the title, legend, and other map elements. Switch

to Layout View. Start with the title. Select Title from the Insert menu. When Enter Map Title appears in a box, click outside the box. When the outline of the box is shown in cyan, double-click it. A Properties dialog appears with two tabs. Click the Text tab. Enter in the Text box: Idaho Cities and Highways. Then click on Change Symbol. In the Symbol Selector dialog, choose size 18, Bookman Old Style (or another serif type), and B (Bold). Click OK in the Symbol Selector and Properties dialogs. Move the title to the upper right of the layout.

12. Next is the legend. The default legend subtitles are the field names of Population and Route_Desc. You can use more descriptive subtitles. Click Population in the Table of Contents. Click it one more time so that Population appears in a box. Change Population to City Population. Follow the same procedure to change Route_Desc to Highway Type.

13. Select Legend from the Insert menu. By default, the legend includes all layers from the map. *Idoutl* shows the outline of Idaho and does not have to be included in the legend. You can remove *idoutl* from the legend by clicking *idoutl* in the Legend Items box and then the left arrow button. Click Next. In the second panel, highlight Legend in the Legend Title box and delete it. (If you want to keep the word Legend on the map, do not delete it.) Skip the next two panels, and click Finish in the fifth panel. Move the legend to the upper right of the layout below the title.

14. A scale bar is next. Select Scale Bar from the Insert menu. Click Alternating Scale Bar 1, and then click Properties. Click the Scale and Units tab. Start with the bottom half of the dialog: select Adjust width when resizing and choose Miles for Division Units. Now work with the upper half of the dialog: enter the division value of 50 (miles), select 2 for the number of divisions, select 0 for the number

of subdivisions, opt not to Show one division before zero. Click the Numbers and Marks tab, and select divisions from the Frequency dropdown list. Click the Format tab, and select Times New Roman from the Font dropdown list. Click OK in both the Scale Bar and Scale Bar Selector dialogs. The scale bar appears in the map. Use the handles to place the scale bar below the legend.

15. A north arrow is next. Select North Arrow from the Insert menu. Choose ESRI North 6, a simple north arrow from the selector, and click OK. Place the north arrow below the scale bar.

16. Finally, change the design of the data frame. Right-click Task 2, and select Properties. Click the Frame tab. Select Double, Graded from the Border dropdown list. Click OK.

17. If your PC is connected to a color printer, you can print the map directly by selecting Print in the File menu. Or you can save the map as an ArcMap document or export the map as a graphic file in PostScript, JPG, or other formats. Both options are accessible through the File menu. Exit ArcMap.

Task 3: Label Streams

What you need: *charlie.shp,* a shapefile showing Santa Creek and its tributaries in north Idaho.

Task 3 lets you try the automatic labeling method in ArcMap. Although the method can label all features on a map and remove duplicate names, it requires adjustments on some individual labels and overlapped labels. Therefore, Task 3 also requires you to use the Spline Text tool.

1. Launch ArcMap. Rename the data frame Task 3, and add *charlie.shp* to Task 3. Select Layout View from the View menu. Maximize the view window and click the Zoom Whole Page button.

2. Select Page Setup from the File menu. Enter 5 (inches) for Width and 7 (inches) for Height. Click OK to dismiss the Page Setup dialog.

3. Right-click *charlie* and select Properties. Click the Labels tab. Check to label features in this layer. Opt for the method of labeling all the features the same way. Make sure that the Label Field is NAME. Click Symbol. On the right side of the Symbol Selector dialog, select black for the Color, 10 for the Size, Times New Roman for the font, and I (italic). Click OK to dismiss the Symbol Selector dialog.

4. Click Label Placement Options in the Layer Properties dialog. Under the Placement tab, check the box to produce labels that follow the curve of the line. Click OK in the Placement Properties dialog and the Layer Properties dialog.

5. Names of streams appear in the layout. Select 100% from the Zoom Control dropdown list. Use the Layout Pan tool to check the labeling of stream names. The result is generally satisfactory except for Fagan Cr. and Charlie Cr. (on top).

6. Dynamic labeling, which is what you have done up to this point, does not allow individual labels to be selected and modified. To fix the placement of Fagan Cr. and Charlie Cr., you can convert labels to annotation. Right-click *charlie,* and select Convert Labels to Annotation. Accept the defaults in the Convert Labels to Annotation dialog, and click OK.

7. The Overflow dialog shows labels that could not be placed on the map. (You may see Pamas Cr., a tributary north of Brown Cr., in the Overflow dialog. If so, you will add Pamas Cr. after Fagan Cr. is moved.)

8. To make sure that the annotation you add to the map has the same look as other labels, you must specify the drawing symbol options. Click the Drawing dropdown. Point to Active Annotation Target, and check charlie anno. Click the Drawing arrow again, and select Default Symbol Properties. Click Text Symbol. In the Symbol Selector dialog, select 10 for the Font Size, Times New Roman, and I (Italic). Then click Properties

in the Symbol Selector dialog. Click the Formatted Text tab in the Properties frame, and change Character Spacing from 0 to 1. Click OK to dismiss the dialogs.

9. Switch to Data View. Click Fagan Cr. and delete it. Click the New Text (A) dropdown arrow and choose the New Splined Text tool. Move the mouse pointer to just below the junction of Brown Cr. and Fagan Cr., and click along the course of the stream. Double-click to end the spline. Enter Fagan Cr. in the Text box. The Fagan Cr. annotation appears

in its default options. If Pamas Cr. is listed in the Overflow dialog, you can follow the same procedure to add Pamas Cr. to the map.

10. Next, select Charlie Cr. at the top of the map and delete it. Follow the same procedure to place Charlie Cr. below its original location and on the left side of the stream.

11. If the added names have narrower character spacing than 1 on the map that you print directly from Layout View, you can export the layout in PostScript format and then print the PostScript file.

REFERENCES

Antes, J. R., and K. Chang. 1990. An Empirical Analysis of the Design Principles for Quantitative and Qualitative Symbols. *Cartography and Geographic Information Systems* 17: 271–77.

Antes, J. R., K. Chang, and C. Mullis. 1985. The Visual Effects of Map Design: An Eye Movement Analysis. *The American Cartographer* 12: 143–55.

Arnheim, R. 1965. *Art and Visual Perception.* Berkeley: University of California Press.

Brewer, C. A. 1994. Color Use Guidelines for Mapping and Visualization. In A. M. MacEachren and D. R. F. Taylor, eds., *Visualization in Modern Cartography,* pp. 123–47. Oxford: Pergamon Press.

Brewer, C. A. 2001. Reflections on Mapping Census 2000. *Cartography and Geographic Information Science* 28: 213–36.

Brewer, C. A., A. M. MacEachren, L. W. Pickle, and D. Herrmann. 1997. Mapping Mortality:

Evaluating Color Schemes for Choropleth Maps. *Annals of the Association of American Geographers* 87: 411–38.

Chang, K. 1978. Measurement Scales in Cartography. *The American Cartographer* 5: 57–64.

Chirié, F. 2000. Automated Name Placement with High Cartographic Quality: City Street Maps. *Cartography and Geographic Information Science* 27: 101–10.

Cuff, D. J. 1972. Value versus Chroma in Color Schemes on Quantitative Maps. *Canadian Cartographer* 9: 134–40.

Dent, B. D. 1999. *Cartography: Thematic Map Design,* 5th ed. Dubuque, IA: WCB/McGraw-Hill.

Holloway, S., J. Schumacher, and R. L. Redmond. 1999. People and Place: Dasymetric Mapping Using ARC/INFO. In S. Morain, ed., *GIS Solutions in Natural Resource Management: Balancing the Technical–Political Equation,*

pp. 283–91, Santa Fe, NM: OnWord Press.

Kraak, M. J., and F. J. Ormeling. 1996. *Cartography: Visualization of Spatial Data.* Harlow, England: Longman.

Mersey, J. E. 1990. Colour and Thematic Map Design: The Role of Colour Scheme and Map Complexity in Choropleth Map Communication. *Cartographica* 27(3): 1–157.

Monmonier, M. 1996. *How to Lie with Maps,* 2d ed. Chicago: Chicago University Press.

Mower, J. E. 1993. Automated Feature and Name Placement on Parallel Computers. *Cartography and Geographic Information Systems* 20: 69–82.

Robinson, A. H., J. L. Morrison, P. C. Muehrcke, A. J. Kimerling, and S. C. Guptill. 1995. *Elements of Cartography,* 6th ed. New York: Wiley.

Slocum, T. A. 1999. *Thematic Cartography and Visualization.* Upper Saddle River, NJ: Prentice Hall.

C H A P T E R

9

DATA EXPLORATION

9.1 INTRODUCTION

Starting data analysis in a GIS project can be overwhelming. The GIS database may have dozens of map layers and hundreds of attributes. Where do you begin? What attributes do you look for? What data relationships are there? One way to ease into the analysis phase is data exploration. Broadly defined, **data exploration** is data-centered query and analysis. It allows the user to examine the general trends in the data, to take a close look at data subsets, and to focus on possible relationships between data sets. The purpose of data exploration is to better understand the data and to provide a starting point in formulating research questions and hypotheses.

An important component of data exploration consists of interactive and dynamically linked visual tools. Maps (both vector- and raster-based), graphs, and tables are displayed in multiple windows and dynamically linked so that selecting a record or records from a table will automatically highlight the corresponding features in a graph and a map. Data exploration allows data to be viewed from different perspectives, making it easier for information processing and synthesis. Windows-based GIS packages, which can work with maps, graphs, and tables in different windows simultaneously, are well suited for data exploration.

Chapter 9 is organized into the following five sections. Section 9.2 discusses interactive data exploration. Section 9.3 covers vector data query using attribute data, spatial data, or both. Raster data query using the cell value or graphic method is covered in Section 9.4. Section 9.5 briefly explains the use of charts and statistics in data exploration. And Section 9.6 deals with geographic visualization and different methods for manipulating and displaying map-oriented data. The emphasis of Chapter 9 is the dynamic linkage between spatial data and attribute data in GIS.

9.2 DATA EXPLORATION

Statisticians have traditionally used descriptive statistics and graphs to explore data structure and to discover data patterns. More recently, data exploration has expanded to include exploratory data analysis and dynamic graphics. **Exploratory data analysis** advocates the use of a variety of techniques for examining data more effectively as the first step in statistical analysis and as a precursor to more formal and structured data analysis (Tukey 1977; Tufte 1983). **Dynamic graphics** enhances exploratory data analysis by using multiple and dynamically linked windows and by letting the user directly manipulate data points in charts and diagrams (Cleveland and McGill 1988; Cleveland 1993). This section examines these tools for data exploration.

9.2.1 Descriptive Statistics

Descriptive statistics such as minimum, maximum, range, mean, variance, and standard deviation summarize the values of a data set. The minimum is the lowest value, the maximum is the highest value, and the **range** is the difference between the minimum and maximum values. The **mean** is the average of data values, the **variance** is the average of the squared deviations of each data value about the mean, and the **standard deviation** is the square root of the variance. Whereas the mean is a measure of central tendency, the variance and standard deviation are measures of dispersion. The computational formulas for the mean and standard deviation are as follows:

(9.1)

$$\text{Mean} = \sum_{i=1}^{n} x_i / n$$

(9.2)

$$\text{Standard deviation} = \sqrt{\sum_{i=1}^{n} (x_i - \text{mean})^2 / n}$$

where x_i is the ith value and n is the number of values in the data set.

9.2.2 Graphs

Common graphs for data exploration include line graphs, bar charts, scatterplots, and their variations (Figure 9.1). A line graph displays data as a line. The line graph example in Figure 9.1 shows state

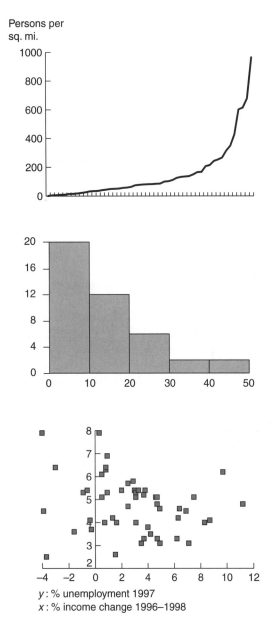

Figure 9.1

Graphs for data exploration: a line graph (top), a bar chart (middle), and a scatterplot (bottom).

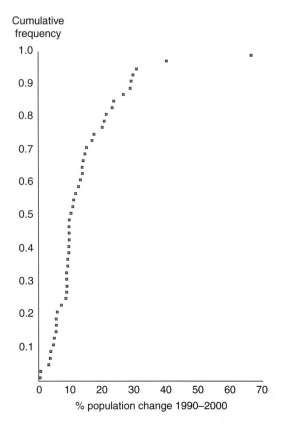

Figure 9.2

This cumulative distribution graph plots percent population change, 1990–2000, by state in the United States. The y-axis shows the cumulative frequencies from 0.0 to 1.0.

population densities in the ascending order in the United States. A bar chart, also called a histogram, groups data into equal intervals and uses bars to show the number or frequency of values falling in each class. A bar chart may have vertical bars, as shown in Figure 9.1, or horizontal bars. A scatterplot uses markings to plot the values of two variables along the x- and y-axis. The scatterplot in Figure 9.1 plots percent income change 1996–1998 against percent unemployment in 1997 by state in the United States.

A cumulative distribution graph is one type of line graph that plots the ordered data values against cumulative distribution values calculated as $(i - 0.5) / n$ for the ith ordered value out of n total values (Figure 9.2).

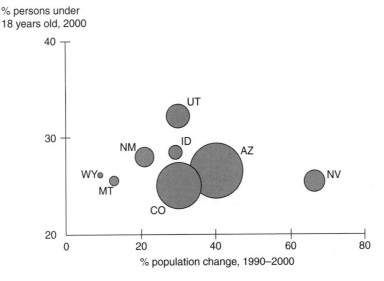

Figure 9.3

This bubbleplot shows percent population change, 1990–2000, along the *x*-axis; percent persons under 18 years old along the *y*-axis; and 2000 population by the bubble size.

Bubble plots are a variation of scatterplots. Instead of using constant symbols as in a scatterplot, a bubble plot has varying-sized bubbles that are made proportional to the value of a third variable. In Figure 9.3, the bubble size shows the state population in 2000, which is a third variable in addition to percent population change 1990–2000 and percent persons under 18 years old in 2000.

Boxplots, also called the "box and whisker" plots, summarize the distribution of five statistics from a data set: the minimum, first quartile (25th percentile), median (the midpoint value), third quartile (75th percentile), and maximum (Figure 9.4). Boxplots may be used in two ways. First, by comparing boxplots from different data sets, one can frequently see the major trends among them. Second, by examining the position of the statistics in a boxplot, one can tell if the distribution of data values is symmetric or skewed and if there are unusual data points (called outliers).

Some graphs are more specialized. Hi-low-close graphs display a range of *y* values at each *x* value. These graphs are often used to show the high, low, opening, and closing of stock values, but

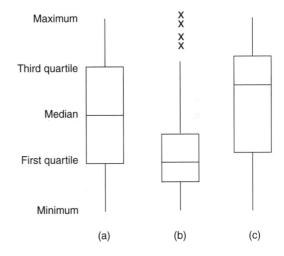

Figure 9.4

Boxplot (a) suggests that the data values follow a normal distribution. Boxplot (b) shows a positively skewed distribution with a higher concentration of data values near the high end. The x's in (b) may represent outliers, which are more than 1.5 box lengths from the end of the box. Boxplot (c) shows a negatively skewed distribution with a higher concentration of data values near the low end.

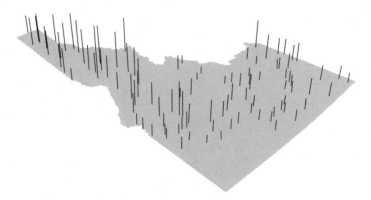

Figure 9.5

This plot shows annual precipitation at 105 weather stations in Idaho. A north to south (along the y-dimension) decreasing trend is apparent in the plot. There is also an increasing trend from east to west (along the x-dimension).

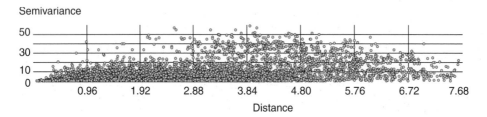

Figure 9.6

This variogram cloud is based on annual precipitation data at 105 weather stations in Idaho. Distances are measured in 10,000 meters, and semivariances are measured in 10 square inches.

they can also be used for other types of data such as air pollutant amounts. Polar graphs display angular data in degrees or radians versus radial distances. The angular data are displayed on a circular angle axis, and the radial distances are extended from the center of the circle. Quantile–quantile plots, also called QQ plots, compare the cumulative distribution of a data set with that of some theoretical distribution such as the normal distribution. The points in a QQ plot fall along a straight line if the data set follows the theoretical distribution.

Some graphs are designed for spatial data. Figure 9.5 shows a plot of spatial data values by raising a bar at each point location so that the height of the bar is proportionate to its value. This kind of plot allows the user to see the general trends among the data values in both the x-dimension (east–west) and y-dimension (north–south).

The variogram cloud (Cressie 1993) is a scatterplot of semivariance values for every pair of data points:

(9.3)

$$\gamma(h) = \frac{1}{2}[z(x_i) - z(x_j)]^2$$

where $\gamma(h)$ is the semivariance for each pair of points, x_i and x_j. The semivariance values are then plotted against the distance between x_i and x_j (Figure 9.6). The variogram is useful for detecting spatial dependency between data points (Chapter 13).

9.2.3 Dynamic Graphics

Dynamic graphics take advantage of multiple windows that are dynamically linked. Therefore, highlighting in one window causes the immediate

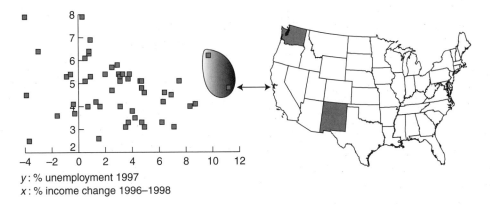

Figure 9.7
The scatterplot on the left is dynamically linked to the map on the right. The "brushing" of two points in the scatterplot highlights the corresponding states (Washington and New Mexico) on the map.

highlighting of corresponding data points in other windows. This kind of interaction and experimentation with the data set helps the user recognize data patterns and data relationships. Common methods of data manipulation in dynamic graphics include selection, deletion, rotation, and transformation of data points.

Brushing is a method for selecting and highlighting a data subset in multiple views (Becker and Cleveland 1987). This method allows a user to select a subset of points from a scatterplot and view related data points that are highlighted in other scatterplots. One can also use brushing in the geographic context by including maps (Monmonier 1989). Figure 9.7 illustrates a brushing example linking a scatterplot and a map.

9.2.4 Data Exploration and GIS

Data exploration in GIS is similar to exploratory data analysis; it allows the user to explore data structure and to focus on a data subset of interest. But the medium for data exploration has expanded to include both vector- and raster-based maps and map features. As maps are dynamically linked to charts, diagrams, and tables, they provide a new interface to the database and offer a spatial view to data analysis. At the same time, the use of maps

has called for specialized methods of exploratory spatial data analysis that consider the special nature of spatial data. Some recent studies have suggested enhancing exploratory data analysis in GIS by using macro programs in a GIS (Batty and Xie 1994) or by linking a GIS to software specifically written for spatial data analysis (Walker and Moore 1988; Haslett et al. 1990; Anselin 1999; Andrienko et al. 2001).

A special nature of data exploration in GIS is the linkage between spatial data and attribute data. For example, a GIS user working with soil conditions would want to know how much of the study area is rated poor, but also where those poor soils are distributed. The linkage must be dynamic and simultaneous: highlighting a subset of attribute data should simultaneously highlight the corresponding map features in the view window and vice versa. Windows-based GIS packages can fulfill this requirement easily by displaying the spatial data and attribute data of a map in separate but dynamically linked windows.

The linkage between spatial and attribute data is an important characteristic of the vector data model. Using the feature attribute table, a vector-based map can be joined or linked with other tables in a relational database for data query. The linkage is not as important for the raster data model for two

---

Q *Box 9.1* **Query Methods in ArcGIS**

Depending on the application, query expressions can vary in ArcGIS. The main difference is between the desktop application of ArcMap, which uses SQL, and ArcInfo Workstation's TABLES module, which uses INFO. RESELECT in TABLES has the same function as "create a new selection" or "select from current selection" in ArcMap in selecting a data subset. ASELECT has the same function as "add to current selection" in adding more records to the subset. And NSELECT has the same function as "switch selection" in switching between the selected subset and the unselected subset.

Query operations in ArcMap itself can also vary. Query operations use SQL if the data set is a geodatabase feature class but a limited version of SQL if the data set is a coverage or a shapefile. Therefore, a geodatabase feature class allows greater flexibility and more query operations than a coverage or a shapefile. Additionally, the syntax can be different. For example, the fields in a query expression have square brackets around them if the fields belong to a geodatabase feature class but are surrounded by double quotes if the fields belong to a coverage or a shapefile.

---

reasons. First, a raster model uses cells with fixed locations. Second, a raster model with floating-point cell values does not have a value attribute table, thus limiting its use in data query. Therefore, this chapter separates vector data from raster data in the discussion of data exploration.

9.3 VECTOR DATA QUERY

9.3.1 Attribute Data Query

Attribute data query retrieves a data subset from a map by working with its attribute data. The selected data subset may be visually inspected or saved for further processing. Attribute data query requires the use of expressions, which must be interpretable by a GIS or a database management system. The structure of these expressions varies from one system to another, although the general concept is the same. ArcGIS Desktop, for example, uses SQL (Structured Query Language) for query expressions (Box 9.1).

9.3.1.1 Query Expressions

Query expressions in GIS follow Boolean algebra (after the English logician George Boole, 1815–1864) and consist of Boolean expressions and connectors. A simple **Boolean expression** con-

tains two operands and a logical operator. For example, "class" = 2 is an expression, in which class and 2 are operands and = is a logical operator. In this example, class is the name of a field, 2 is the field value used in the query, and the expression selects those records that have the class value of 2. Operands may be a field, a number, or a string. Logical operators may be equal to ($=$), greater than ($>$), less than ($<$), greater than or equal to ($>=$), less than or equal to ($<=$), or not equal to ($<>$).

Boolean expressions may contain calculations that involve operands and the arithmetic operators of $+$, $-$, $*$, and $/$. Suppose length is a field measured in feet. One can use the expression, "length" $* 0.3048 > 100$, to find those records that have the length value of greater than 100 meters. Longer calculations, such as "length" $* 0.3048 - 50 > 100$, evaluate the $*$ and $/$ operators first from left to right and then the $+$ and $-$ operators. One can use parentheses to change the order of evaluation. For example, to subtract 50 from length before multiplication by 0.3048, one can use the expression, ("length" $- 50) * 0.3048 > 100$.

Boolean connectors are AND, OR, XOR, and NOT, which are used to connect two or more expressions in a query statement. The connector AND connects two expressions in this example of selecting vegetation stands: "class" = 2 AND "age" > 100. Records selected from the statement must satisfy both "class" = 2 and "age" > 100. If the connector is changed to OR in the example, then records that satisfy either one or both of the expressions are selected. If the connector is changed to XOR, then records that satisfy one and only one of the expressions are selected. The connector NOT negates an expression so that a true expression is changed to false and vice versa. The statement, NOT "class" = 2 AND "age" > 100, for example, selects those records whose class is not equal to 2 and whose age is greater than 100.

Boolean connectors of NOT, AND, and OR are actually keywords used in the operations of COMPLEMENT, INTERSECT, and UNION on sets in probability. The operations are illustrated in Figure 9.8, with A and B representing two subsets of a universal set.

- The COMPLEMENT of A contains elements of the universal set that do NOT belong to A.
- The UNION of A and B is the set of elements that belong to A OR B.
- The INTERSECT of A and B is the set of elements that belong to both A AND B.

9.3.1.2 Type of Operation

Attribute data query begins with a complete data set. A basic query operation is to select a subset and divide the data set into two groups: one containing selected records and the other unselected records. Given a selected data subset, three types of operation can act on it: add more records to the subset, remove records from the subset, and select a smaller subset (Figure 9.9). Operations can also be performed between the selected and unselected subsets. First, one can switch between the selected and the unselected subsets. Second, one can clear the selection by bringing back all records.

These different types of operation allow greater flexibility in data query. For example, instead of using an expression of "class" = 2 OR "age" > 100 OR "density" = 'high,' one can first select "class" = 2 and then, in the next two operations, use the expressions of "age" > 100 and "density" = 'high,' respectively, to add more

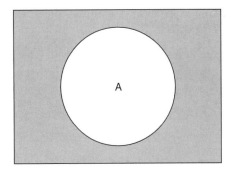

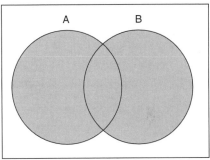

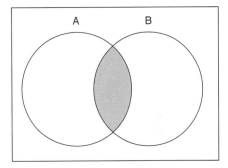

Figure 9.8
The shaded portion represents the complement of data subset A (top), the union of data subsets A and B (middle), and the intersection of A and B (bottom).

records to the subset. (The single quotes indicate that the field value of high is a string.) This alternative allows the user to use shorter expressions, which are easier to set up.

9.3.1.3 Examples of Query Operation

The following examples show different query operations using data from Table 9.1, which has 10 records and 3 fields.

Example 1: Select a Data Subset and Then Add More Records to It

[Create a new selection] "cost" $>=$ 5 AND "soiltype" = 'Ns1'

0 of 10 records selected

[Add to current selection] "soiltype" = 'N3'

3 of 10 records selected

Example 2: Select a Data Subset and Then Switch Selection

[Create a new selection] "cost" $>=$ 5 AND "soiltype" = 'Tn4' AND "area" $>=$ 300

2 of 10 records selected

[Switch Selection]

8 of 10 records selected

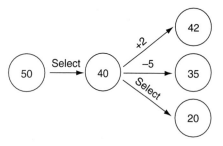

Figure 9.9
Three types of operation may be performed on the subset of 40 records: add more records to the subset (+2), remove records from the subset (−5), or select a smaller subset (20).

Example 3: Select a Data Subset and Then Select a Smaller Subset from It

[Create a new selection] "cost" $>$ 8 OR "area" $>$ 400

4 of 10 records selected

[Select from current selection] "soiltype" = 'Ns1'

2 of 10 records selected

9.3.1.4 Relational Database Query

Relational database query works with a relational database, which may consist of many separate but interrelated tables. A query of a table in a relational database not only selects a data subset in the table but also selects records related to the subset in other tables. This feature is desirable in data exploration because it allows the user to examine related data characteristics from multiple tables.

To use a relational database, one must first be familiar with the overall structure of the database, the designation of keys in relating tables, and a data dictionary listing and describing the fields in each table. For data query involving two or more tables, one can choose to either join or relate the tables. In the first case, attribute data from two or more tables are combined into a single table. In the second case, tables remain separate but are dynamically linked. When a record in one table is selected, the link will automatically select and highlight the corresponding record or records in another table.

TABLE 9.1	A Data Set for Query Operation Examples				
Cost	Soil Type	Area	Cost	Soil Type	Area
1	Ns1	500	6	Tn4	300
2	Ns1	500	7	Tn4	200
3	Ns1	400	8	N3	200
4	Tn4	400	9	N3	100
5	Tn4	300	10	N3	100

As discussed in Chapter 6, the National Map Unit Interpretation Record (MUIR) database is a relational database maintained by the Natural Resources Conservation Service (NRCS). The database contains over 80 estimated soil properties, interpretations, and performance data for each polygon on a soil map. A major challenge for using this database is to sort out where each soil attribute resides and how tables are linked. Suppose we ask the question, What types of plants, in their common names, are found where annual flooding frequency is rated as either frequent or occasional? To answer the question, we need the following three MUIR tables: the map unit components table or comp.dbf, which contains data on annual flood frequency; the woodland native plants table or forest.dbf, which has data on forest types; and a lookup table called plantnm.dbf, which has common plant names. The next step is to find the keys that can link the three tables (Figure 9.10). Finally, we need to link the attribute table of the soil map to comp.dbf by using musym (map unit symbol) as the key.

After the tables are related, we can issue the following query statement to comp.dbf:

"anflood" = 'frequent' OR "anflood" = 'occas'

Evaluation of the query expression selects not only records in comp.dbf that meet the criteria but also the corresponding records in forest.dbf, plantnm.dbf, and the attribute table and the corresponding soil polygons in the map. This dynamic selection is possible because the tables are interrelated in the MUIR database and are dynamically linked to the map. A detailed description of relational database query is included in Task 4 of the Applications section.

9.3.1.5 Use SQL to Query a Database

SQL (Structured Query Language) is a data query language designed for relational databases. IBM developed SQL in the 1970s, and many commercial database management systems such as Oracle, Informix, DB2, Access, and Microsoft SQL Server have adopted the language. To access data in a database that uses SQL, a GIS user must prepare query statements in SQL and follow the structure or the syntax of the language.

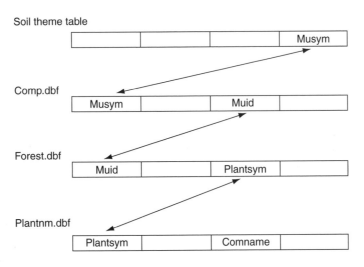

Figure 9.10
The keys relating three dBASE files in the MUIR database and the soil attribute table.

The basic syntax of SQL, with the keywords in bold type, is

select <attribute list>

from <relation>

where <condition>

The **select** keyword selects field(s) from a database, the **from** keyword selects table(s) from a database, and the **where** keyword specifies the condition or criterion for data query.

A simple SQL statement querying the sale date of the parcel, P101, in Figure 9.11 is

select Sale_date

from Parcel

where PIN = 'P101'

The following SQL statement, querying the sale date of the parcel owned by Costello, involves two tables and the AND connector:

select Parcel.Sale_date

from Parcel, Owner

where Parcel.PIN = Owner.PIN AND Owner_name = 'Costello'

The prefixes of Parcel and Owner distinguish the two tables used in the query. Parcel.PIN and Owner.PIN are the keys in connecting the two tables. More examples of SQL statements are included in Box 9.2.

GIS packages such as ArcGIS, MGE, and MapInfo either allow GIS users to use SQL in

Relation 1: Parcel

PIN	Sale_date	Acres	Zone_code	Zoning
P101	1-10-98	1.0	1	Residential
P102	10-6-68	3.0	2	Commercial
P103	3-7-97	2.5	2	Commercial
P104	7-30-78	1.0	1	Residential

Relation 2: Owner

PIN	Owner_name
P101	Wang
P101	Chang
P102	Smith
P102	Jones
P103	Costello
P104	Smith

Figure 9.11
The key PIN relates the parcel and owner tables and allows use of SQL with both tables.

querying a local or external database. Data query in ArcGIS, for example, uses a dialog with the key words of select, from, and where already prepared. The user can just enter the query expression in the dialog box.

9.3.2 Spatial Data Query

Spatial data query refers to the process of retrieving data from a map by working with map features. One may select map features using a cursor, a graphic, or the spatial relationship between map features. Spatial data query is the geographic interface to the database and is therefore useful for tasks that cannot be easily accomplished by attribute data query, such as selecting contiguous areas. As in attribute data query, results of spatial data query may be visually inspected or saved as new maps for further processing.

9.3.2.1 Feature Selection by Cursor

The simplest spatial data query is to select a map feature by pointing at the feature itself. One can also use the cursor to select map features within a box.

9.3.2.2 Feature Selection by Graphic

This query method uses a graphic such as a circle, a box, a line, or a polygon to select map features that fall inside or are intersected by the graphic object (Figure 9.12). One can draw the graphic to be used for selection by using the mouse in real time. Feature selection by graphic is similar to brushing in exploratory data analysis. Like the brush, the graphic to be used for selection can be moved around the computer screen and can be made in any size or shape. Examples of query by graphic include selecting restaurants within a 1-mile radius of a hotel, selecting land parcels that intersect a

---

 Box 9.2   More Examples of SQL Statements

Continuing with the second example in the text, we can use the keyword **like** in the where clause:

  where Parcel.PIN = Owner.PIN AND
  Owner_name **like** 'C%'

The query result will show the sale dates of the parcels owned by Chang and Costello. Next we can use the keyword **in:**

  where Parcel.PIN = Owner.PIN AND
  Owner_name **in** ('Wang', 'Smith','Jones')

The query result will show the sale dates of the parcels owned by Wang, Smith, and Jones.

  A national forest in Idaho uses Oracle to manage a centralized database, which is accessible to its ranger districts through phone lines. Although SQL statements used for querying the national forest database look complicated because of many attributes involved, the syntax structure remains the same. The following shows two query examples. The first code queries a table called stand_activities for records that

have values in the sale_name column by using the keywords of **not** and **null:**

  select sale_name, sale_cntr_no, stand_id,
  sale_unit_no_1, sale_unit_no_2

  from stand_activities

  where sale_name is **not null**

  The second code queries two tables, sa (stand-activities) and sn (sale_names), for those records that must satisfy all three expressions about stand_id, activity_code, and accomp_year by using the AND connector and the keywords of **between** and **null:**

  select sale_name, stand_id, activity_code,
  activity_units, accomp_year

  from sa, sn

  where stand_id **between** '23200000' and
  '29400000' AND activity_code **between** 4431
  and 4432 AND accomp_year is **null**

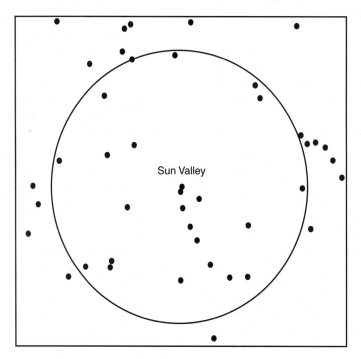

Figure 9.12
A circle with a specified radius is drawn around Sun Valley. The circle is then used as a graphic object to select point features within the circular area.

proposed highway, and finding owners of land parcels within a proposed nature reserve.

9.3.2.3 Feature Selection by Spatial Relationship

This query method selects map features based on their spatial relationships to other features. Map features to be selected may be in the same map as map features used for selection. Or, more commonly, they are in different maps. An example of the first type of query is to find roadside rest areas within a radius of 50 miles of a selected rest area; in this case, map features to be selected and used for selection are in the same map. An example of the second type of query is to find rest areas within each county. Two maps are required for this query: one map showing county boundaries and the other roadside rest areas.

Spatial relationships used in query include the following:

- **Containment**—selects features that fall completely within features used for selection. Examples include finding schools within a selected county, and finding state parks within a selected state.
- **Intersect**—selects features that intersect features used for selection. Examples include selecting land parcels that intersect a proposed road, and finding settlements that intersect an active fault line.
- **Proximity**—selects features that are within a specified distance of features used for selection. Examples include finding state parks within 10 miles of an interstate highway, and finding pet shops within 1 mile of selected streets. If features to be selected and features used for selection share common boundaries and if the specified distance is 0, then proximity becomes **adjacency.** Examples of spatial adjacency include

Box 9.3   Expressions of Spatial Relationship in ArcMap

ArcMap handles feature selection by spatial relationship through the Select by Location dialog. The dialog requires the user to specify one or more layers whose features will be selected, and a layer whose features will be used for selection. Eleven expressions of spatial relationship connect features to be selected and used for selection. These expressions may be subdivided by the relationships of containment, intersect, and proximity/adjacency:

- **Containment:** "are completely within," "completely contain," "have their center in," "contain," and "are contained by."
- **Intersect:** "intersect" and "are crossed by the outline of."

- **Proximity/adjacency:** "are within a distance of," "share a line segment with," "touch the boundary of," and "are identical to."

A complete query expression in the Select by Location dialog may read as follows: "I want to select features from cities that are completely within the features in quake," where cities is a city map and quake is a earthquake-prone map.

The basic type of spatial data query is to select a subset of map features from a layer. After a subset has been selected, three types of operation can act on it: add to the currently selected features, remove from the currently selected features, and select from the currently selected features.

---

selecting land parcels that are adjacent to a flood zone, and finding vacant lots that are adjacent to a new theme park.

Box 9.3 shows the types of spatial relationship that can be used for spatial data query in ArcMap.

9.3.2.4 Combining Attribute and Spatial Data Queries

So far we have approached data exploration through attribute data query or spatial data query. In many cases data exploration requires both types of query. For example, both types of query are needed to find gas stations that are within 1 mile of a freeway exit in southern California and have an annual revenue exceeding $2 million each. Assuming that the maps of gas stations and freeway exits are available, there are at least two ways to solve the question.

1. Locate all freeway exits in the study area, and draw a circle around each exit with a 1-mile radius. Select gas stations within the circles through spatial data query. Then use attribute data query to find gas stations that have annual revenues exceeding $2 million.

2. Locate all gas stations in the study area, and select those stations with annual revenues exceeding $2 million through attribute data query. Next, use spatial data query to narrow the selection of gas stations to those within 1 mile of a freeway exit.

The first option queries spatial data and then attribute data. The process is reversed with the second option. Assuming that there are many more gas stations than freeway exits, the first option may be a better option, especially if the gas station map must be linked to other attribute tables for getting the revenue data. As shown in the above example, the combination of spatial and attribute data queries opens wide the possibilities of data exploration. Some GIS users might even consider this kind of data exploration as data analysis because that is what they need to do to solve most of their tasks.

9.4 RASTER DATA QUERY

Raster data may be queried using the cell value or the graphic method. An integer ESRI grid has a value attribute table, which functions like the feature attribute table of a vector-based map in data management and query. Without a value attribute table, a floating-point grid is more limited in query functions.

9.4.1 Query by Cell Value

A common method to query a grid is to use a logical expression involving cell values. As examples, the expression, [road] = 1, can query an integer road grid and the expression, [elevation] >= 1243.26, can query a floating-point elevation grid. Because a floating-point elevation grid contains continuous values, querying a specific value is not likely to find any cell in the grid. Therefore, using numeric ranges in query expressions is recommended for floating-point grids. A query statement separates cells of a grid that satisfy the statement from cells that do not.

Raster data query can also use the Boolean connectors of AND, OR, and NOT to string together separate expressions. These separate expressions can apply to different attributes of an integer grid. For example, the statement, ([anflood] = 'freq') AND ([wtdepth] < 1), selects cells that have frequent annual flood and water table depth of less than 1 meter from a soil grid. This kind of query is not possible with a floating-point grid.

A compound statement with separate expressions can apply to multiple grids, which may be integer, or floating point, or a mix of both types. For example, the statement, ([grid1] = 13) AND ([grid2] = 3) AND ([grid3] = 4), selects cells that have the value of 13 in grid1, 3 in grid2, and 4 in grid3. Querying multiple grids by cell value is unique with raster data. For vector data, all attributes to be used in a compound expression must be associated with the same map.

9.4.2 Query Using Graphic Method

One can query a grid by using a point or a graphic object such as a circle, a box, or a polygon. In the case of a point, the query returns the cell value that corresponds to the point location. In the case of a graphic object, the query returns the values of those cells that fall within the graphic object and saves the result into an output grid.

9.5 GRAPHS

Graphs are the basic tools for exploratory data analysis. Prepared from tabular data, charts can be displayed in separate windows from maps and tables. Because charts are dynamically linked to maps and tables, a query of attribute data will simultaneously highlight the selected records in the table, the corresponding features in the map, and the positions of the selected records in the chart.

Graphs can also complement a map by summarizing information from the map's attribute data. This is why graphs are usually included in a GIS package. ArcGIS users can make graphs using ArcMap or the Geostatistical Analyst extension. ArcMap has a tool menu for creating graph types of line graphs, bar charts, scatterplots, bubble plots, polar graphs, and hi-low-close graphs. The Geostatistical Analyst extension has menu selections for creating histograms, QQ plots, variograms, 3-D plots of spatial data for trend analysis, and other graphs for data exploration.

Commercial statistical analysis packages such as SAS, SPSS, SYSTAT, S-PLUS (now Insightful), and Excel offer more tools to work with charts than does a GIS package. Therefore, one way to explore data with charts is to export data from a GIS to a statistical analysis package (Scott 1994).

9.6 GEOGRAPHIC VISUALIZATION

Geographic visualization, sometimes called cartographic visualization, refers to the use of maps for setting up a context for processing visual information and for formulating research questions or hypotheses (MacEachren 1995; MacEachren et al. 1998). Geographic visualization therefore has the same objective and the same types of interactivity as exploratory data analysis (Crampton 2002). The techniques of brushing, linking, and query have already been discussed in the previous sections. This

Box 9.4   Data Classification Methods

Data classification methods available in GIS packages may include the following five methods: equal interval, equal frequency, mean and standard deviation, natural breaks, and user-defined. Equal interval uses a constant class interval in classification. Equal frequency, also called quantile, divides the total number of data values by the number of classes and ensures that each class contains the same number of data values. Mean and standard deviation sets the class breaks at units of the standard deviation (0.5, 1.0, etc.) above or below the mean. The natural breaks method, also called the Jenks optimization method, optimizes the grouping of data values in classification (Slocum 1999). Typically, the natural breaks method uses a computing algorithm to minimize differences between data values in the same class and to maximize differences between classes. The user-defined method lets the user choose the appropriate or meaningful class breaks. For example, in mapping rates of population change by state, the user may choose zero or the national average as a class break.

section covers three methods that are map-based: data classification, spatial aggregation, and map comparison.

9.6.1 Data Classification

Data classification is probably the most common method for map manipulation. The method is useful for data visualization as well as for creating new attribute data and new maps.

9.6.1.1 Data Classification for Visualization

Classification involves the use of a classification method and a number of classes for aggregating data and map features (Box 9.4). By changing the classification parameters, the same data can produce different-looking maps, each with its own spatial pattern. Of all map types, choropleth maps are most affected by classification (Chapter 8). Cartographers often make several versions of the choropleth map from the same data and choose one—typically one with a good spatial organization of classes—for final map production.

In the context of data exploration, classification is most useful in separating spatial data by some descriptive statistics. Suppose we want to explore rate of unemployment by state in the United States. To get a preliminary look at the data, we may place rates of unemployment into two classes based on the national average: above and below the national average (Figure 9.13a). Although generalized, the map divides the country into contiguous regions, which may suggest factors that can explain why some states have done better than others.

To further isolate those states that are way above or below the national average, we can classify rates of unemployment using the mean and standard deviation method (Figure 9.13b). We can now focus our attention on states that are, for example, more than 1 standard deviation above the mean.

To explore the topic further, we can link the classified map with tables showing other attributes. For example, we can link the map with a table showing percent change in median household income and find out whether states that have lower unemployment rates tend to have higher rates of income growth, and vice versa.

9.6.1.2 Data Classification for Creating New Data

Chapter 6 has already covered the procedure for creating new attribute data for a vector-based map: create a new field and repeat the process of selecting records that fall within a class and assigning a field value to the selected data subset.

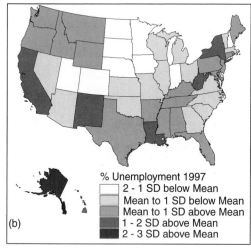

Figure 9.13
The top map shows rate of unemployment in 1997 as either above or below the national average of 4.9%. The bottom map uses the mean and standard deviation (SD) for data classification.

Creating a new raster data set by classification is often referred to as **reclassification,** recoding, or transforming through lookup tables (Tomlin 1990). Two reclassification methods may be used. The first method is a one-to-one change, meaning that a cell value in the input grid is assigned a new value in the output grid. For example, irrigated cropland in a land-use grid is assigned a value of 1 in the output grid. The second method assigns a

Population density (persons per sq. mi.)	Value after reclassification
0–25	1
26–50	2
51–75	3
>=76	4

Figure 9.14
Reclassification assigns a new value to a range of cell values in the input grid such as 1 for 0–25, 2 for 26–50, and so on.

new value to a range of cell values in the input grid. For example, cells with population densities between 0 and 25 persons per square mile in a population density grid are assigned a value of 1 in the output grid and so on (Figure 9.14). An integer grid can be reclassified by either method, but a floating-point grid can only be reclassified by the second method.

Raster data reclassification serves an important function in data exploration. It has the following applications:

- **Data Simplification.** Reclassification can create a simplified grid. For example, instead of having continuous slope values, a grid can have 1 for slopes of 0 to 10%, 2 for 10 to 20%, and so on.
- **Data Isolation.** Reclassification can create a new grid that contains a unique category or value such as irrigated cropland or slopes of 10 to 20%.
- **Data Ranking.** Reclassification can create a new grid that shows the ranking of cell values in the input grid. For example, a reclassified grid can show the ranking of 1 to 5, with 1 being least suitable and 5 being most suitable.

9.6.2 Spatial Aggregation

Spatial aggregation is functionally similar to data classification except that it groups data spatially. Figure 9.15 shows percent population change in the United States by state and by region. Regions in this case are spatial aggregates of states and therefore give a more general view of population growth in the country. Figure 9.15 uses two levels of geographic units defined by the U.S. Census Bureau for data collection. Other geographic levels are county, census tract, block group, and block. Because these levels of geographic units form a hierarchical order, we can aggregate data from one lower level to a higher level. Displaying aggregated data at different levels offers a view of the effect of spatial scaling.

If distance is the primary factor in a study, one can aggregate spatial data by distance measures from points, lines, or areas. An urban area, for example, may be aggregated into distance zones away from the city center or from its streets (Batty and Xie 1994). Unlike the geography of census, these distance zones require the following steps in data processing: establish distance zones by buffering the city center or the streets, overlay distance zones with the base maps of census data, and recompile census data for different distance zones.

Spatial aggregation for raster data means aggregating cells of the input grid to produce a coarser-resolution grid. For example, one can aggregate cells of a grid by a factor of 3. Each cell in the output grid corresponds to a 3-by-3 matrix in the input grid, and the cell value is a computed statistic such as mean, median, minimum, maximum, or sum from the nine input cell values.

9.6.3 Map Comparison

Map comparison can help a GIS user sort out the relationship between different maps. For example, the display of wildlife locations on a vegetation layer may reveal the association between the wildlife species and the distribution of vegetation covers (Figure 9.16).

If the maps to be compared consist of only point or line features, they can be coded in different colors and superimposed on one another in a single view. But this process becomes difficult if they include polygon features or raster data. One option is to use transparency as a visual variable (Chapter 8). A semitransparent map allows another map to show

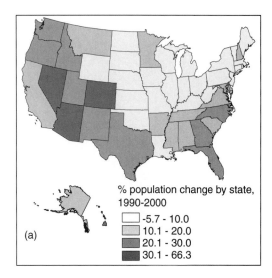

(a)

% population change by state, 1990-2000
- ☐ -5.7 - 10.0
- ☐ 10.1 - 20.0
- ☐ 20.1 - 30.0
- ■ 30.1 - 66.3

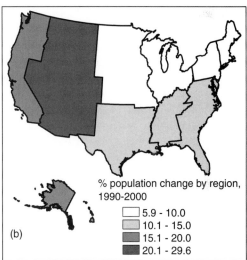

(b)

% population change by region, 1990-2000
- ☐ 5.9 - 10.0
- ☐ 10.1 - 15.0
- ☐ 15.1 - 20.0
- ■ 20.1 - 29.6

Figure 9.15
The top map shows percent population change by state, 1990–2000. The darker the symbol, the higher the percent increase. The bottom map shows percent population change by region.

through. Therefore, if the maps to be compared are both in raster format, one grid can be displayed in a color scheme and the other in semitransparent shades of gray. The gray shades simply darken the color symbols and do not produce confusing color

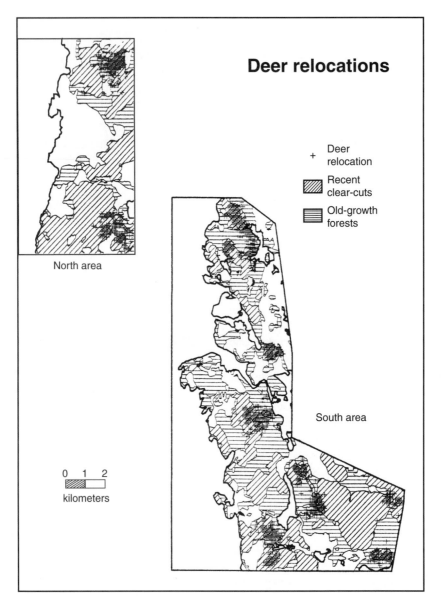

Figure 9.16
An example of using multiple maps in data exploration. In this view of deer relocations in SE Alaska, the focus is on the distribution of deer relocations along the clear-cut/old forest edge.

mixtures. Application of transparency is, however, limited to two raster or polygon layers.

There are three other options for comparing polygon or raster layers. The first option is to place all polygon and raster layers, along with other point and line layers, onto the screen but to turn on and off polygon and raster layers so that only one of them is viewed at a time. Used by many websites for interactive mapping, this option is designed for casual users.

The second option is to use a set of adjacent views, one for each polygon or raster layer, similar to the use of small multiples (Tufte 1983) or a scatterplot matrix (Cleveland 1993). This option is especially useful for observing temporal changes. For example, land cover maps of 1970, 1980, and 1990 for a study area can be placed in separate views so that the user can observe the changes over time. With maps for many points in time, one can run map animation showing continuous changes (DiBiase et al. 1992; Weber and Buttenfield 1993; Peterson 1995).

The third option is to use map symbols that can show data from two polygon maps. One method for this option is to use cartographic point symbols such as bar charts. A bar chart is placed in each polygon, and each bar chart has two or more bars representing, for example, amounts of timber harvest from two or more years for comparison. Another method is the bivariate choropleth map (Meyer et al. 1975). A bivariate choropleth map combines two color schemes, one for each variable. For example, a yellow-to-red color scheme for rate of unemployment may be combined with a yellow-to-blue color scheme for rate of income change to produce a unemployment/income change bivariate map. But if each variable is grouped into four classes, the bivariate map will have 16 color symbols, and the lack of logical progression between the mixed color symbols can be

Figure 9.17
A bivariate map showing the combinations of (1) rate of unemployment in 1997, either above or below the national average, and (2) rate of income change, 1996–1998, either above or below the national average.

a problem (Olson 1981). An obvious improvement to the readability of a bivariate map is to reduce the number of classes for each variable. Figure 9.17, for example, classifies the rate of unemployment and the rate of income change into either > or <= the national average and then shows the combinations of the two variables on a single map.

KEY CONCEPTS AND TERMS

Adjacency: A spatial relationship that can be used to select features that share common boundaries.

Attribute data query: The process of retrieving data from a map by working with its attribute data.

Boolean connector: A keyword such as AND, OR, XOR, or NOT that is used to construct complex logical expressions.

Boolean expression: A combination of a field, a value, and a logical operator, such as "class" = 2, from which an evaluation of True or False is derived.

Brushing: A data exploration technique for selecting and highlighting a data subset in multiple views.

Containment: A spatial relationship that can be used in data query to select features that fall within specified features.

Data exploration: Data-centered query and analysis.

Dynamic graphics: A data exploration method in which the user can directly manipulate data points in charts and diagrams that are displayed in multiple and dynamically linked windows.

Exploratory data analysis: Use of a variety of techniques such as charts, diagrams, and scatterplots to examine data as the first step in statistical analysis.

Geographic visualization: A method that is used to display geographic data in visual representations and to set up a context for visual information processing, leading to formulation of research questions or hypotheses.

Intersect: A spatial relationship that can be used in data query to select features that intersect specified features.

Mean: The average of all values in a data set.

Proximity: A spatial relationship that can be used to select features that are within a distance of specified features.

Range: The difference between the minimum and maximum values in a data set.

Reclassification: The process of reclassifying cell values of an input grid to create a new grid.

Relational database query: Query in a relational database, which not only selects a data subset in a table but also selects records related to the subset in other tables.

Spatial data query: The process of retrieving data from a map by working with map features.

Standard deviation: A measure of data dispersion, which is defined as the square root of the average of the squared deviations of each data value about the mean.

Structured Query Language (SQL): A data query and manipulation language designed for relational databases.

Variance: A measure of data dispersion, which is defined as the average of the squared deviations of each data value about the mean.

APPLICATIONS: DATA EXPLORATION

This applications section covers six tasks. Tasks 1 and 2 present an overview of data exploration in ArcMap: you will perform "select feature by location" in Task 1 and "select feature by graphic" in Task 2. In Task 3, you will query a joint attribute table; examine the query results using a magnifier window; and bookmark the area with the query results. Task 4 covers relational database query. Task 5 combines spatial and attribute data queries. Task 6 deals with raster data query.

Task 1. Select Feature by Location

What you need: *idcities.shp,* with 654 places in Idaho; and *snowsite.shp,* with 206 snow stations in Idaho and the surrounding states.

Task 1 lets you use the select feature by location method to select snow stations within 40 miles of Sun Valley, Idaho, and plot the snow station data in charts.

1. Start ArcCatalog, and make connection to the Chapter 9 database. Launch ArcMap. Add *idcities.shp* and *snowsite.shp* to Layers. Right-click Layers and select Properties. Under the General tab, rename the data frame from Layers to Tasks 1&2 and choose Miles from the Display dropdown list.

2. This step is to select Sun Valley from *idcities.* Choose Select By Attributes from the Selection menu. Select *idcities* from the Layer dropdown list and "Create a new

selection" from the Method list. Then prepare an SQL statement in the expression box as: "CITY_NAME" = 'Sun Valley'. (If Sun Valley does not appear in the values box, click on Complete List.) Click Apply. Sun Valley is now highlighted in the map.

3. Choose Select By Location from the Selection menu. Do the following in the Select By Location dialog: choose "select features from" from the first dropdown list, check *snowsite,* choose "are within a distance of" from the second list, choose *idcities* from the third list, enter 40 (miles) for the distance to buffer, and click Apply. Snow courses that are within 40 miles of Sun Valley are highlighted in the map.

4. Right-click *snowsite* and select Open Attribute Table. Click Selected to show only the 24 snow courses within 40 miles of Sun Valley.

5. Now you will graph the elevation (ELEV) and maximum snow water equivalent (SWE_MAX) of the 24 snow courses. Click the Tools menu, point to Graphs, and select Create. In the first panel of Graph Wizard, click on the Graph type of Scatter and the upper-left Graphic subtype. In the second panel, make sure that *snowsite* is the layer that contains data. Check ELEV, and uncheck all others, for the Y axis field. Check SWE_MAX for the X axis field. In the third panel, enter Elev-SweMax for the title, uncheck Show Legend, and click Finish. A scatter diagram showing a positive relationship between ELEV and SWE_MAX appears.

Task 2. Select Feature by Graphic

What you need: *idcities.shp* and *snowsite.shp,* as in Task 1.

Task 2 lets you use the select feature by graphics method to select snow stations within 40 miles of Sun Valley.

1. Make sure that the Tasks 1&2 data frame still contains *idcities* and *snowsite* in ArcMap.

Choose Clear Selected Features from the Selection menu to clear features selected in Task 1. Choose Select By Attributes from the Selection menu. Select *idcities* from the Layer dropdown list and "Create a new selection" from the Method list. Then prepare an SQL statement in the expression box as: "CITY_NAME" = 'Sun Valley'. (If Sun Valley does not appear in the values box, click on Complete List.) Click Apply. Sun Valley is now highlighted in the map.

2. This step is to draw a circle around Sun Valley. Zoom in the area around Sun Valley. Click the New Rectangle dropdown arrow on the Draw toolbar and select New Circle. Click Sun Valley in the map and drag the mouse pointer to draw a circle. (A radius reading in meters should appear in the lower left of the screen.) Right-click anywhere within the circle and select Properties. Under the Size and Position tab, first click the center Anchor Point and then enter 128720 meters for the Size Width. (A width of 128720 meters is the same as a diameter of 80 miles or a radius of 40 miles.) Click OK. If the circle you have drawn is not right, you can delete it and redraw it.

3. Now you will use the circle to select features in *snowsite*. Click the Selection menu and click Set Selectable Layers. Check *snowsite,* uncheck *idcities,* and close the dialog. Make sure that the circle is active with the handles; if not, click the circle to activate it. Click the Selection menu and click Select by Graphics. Snow courses that are within 40 miles of Sun Valley are highlighted in the map. Open the attribute table of *snowsite*. Click Selected to show only the selected snow courses, which should be exactly the same as those in Task 1.

Task 3. Query Attribute Data from a Joint Table

What you need: *wp.shp,* a vegetation stand shapefile; *wpdata.dbf,* a dBASE file containing stand data for *wp.shp.*

As explained in the text, query or data selection is the most important element of data exploration. Data query can be approached from either attribute data or spatial data. Task 3 focuses on attribute data query.

1. Select Data Frame from the Insert menu in ArcMap. Rename the new data frame Task 3, and add *wp.shp* to Task 3. Right-click *wp* and select Open Attribute Table. The field ID to the far right of the attribute table will be used in joining tables. Close the *wp* attribute table.

2. Click the Add Data button, navigate the Chapter 9 database, and add *wpdata.dbf* to Task 3. Right-click *wpdata* and select Open. *Wpdata* contains attributes such as Origin, As, and Elev for polygons in *wp*. The field ID will be used in joining tables. Close *wpdata.*

3. Right-click *wp,* point to Joins and Relates, and select Join. At the top of the Join Data dialog, opt for joining attributes from a table. Then select ID in the first dropdown list, *wpdata* in the second list, and ID in the third list. Click OK to dismiss the dialog.

4. *Wpdata* is now joined to the *wp* attribute table. Right-click *wp* and select Open Attribute Table. The joined table has two sets of attributes, distinguished by the prefix. Attributes with the prefix of wp such as wp.ID are from the *wp* attribute table, and attributes with the prefix of wpdata such as wpdata.ID are from *wpdata.*

5. To perform query using the joined table, click the Options dropdown arrow and choose Select by Attributes. In the Select by Attributes dialog, make sure that the method is to create a new selection. Then prepare an SQL statement in the expression box that reads: "wpdata.ORIGIN" > 0 AND "wpdata.ORIGIN" <= 1900. Click Apply.

6. Click Selected at the bottom of the table so that only the selected records (175 out of 856) are shown. Polygons of the selected records are highlighted in the *wp* map. To

narrow the selected records, click the Options dropdown arrow and choose Select by Attributes. In the Select by Attributes dialog, make sure that the method is to select from current selection. Then prepare an SQL statement in the expression box that reads: "wpdata.ELEV" <= 30. Click Apply. This selection results in 32 records (polygons) out of a total of 856. To take a closer look at the selected polygons in the map, click the Window menu and select Magnifier. When the magnifier window appears, click the window's title bar and drag the window over the map to see a magnified view.

7. Before moving to the next part of the task, select Clear Selected Features from the Selection menu to clear the selection.

8. Click the Selection menu and click Select by Attributes. Enter the following SQL statement in the expression box: ("wpdata.ORIGIN" > 0 AND "wpdata.ORIGIN" <= 1900) AND "wpdata.ELEV" > 40. [Notice that the first part of the expression about ORIGIN is enclosed in a pair of parentheses. You can either click () first or enter them manually.] Click Apply. Right-click *wp* and select Open Attribute Table. Click Selected at the bottom of the Attributes of *wp* table. Four records are selected. The selected polygons are near the top of the map. Use the Zoom In tool to zoom in the area around the selected polygons. You can bookmark the zoom-in area for future reference. Click the View menu, point to Bookmarks, and select Create. Enter protect for the Bookmark Name. To view the zoom-in area next time, click the View menu, point to Bookmarks, and select protect.

Task 4. Query Attribute Data from a Relational Database

What you need: *mosoils.shp,* a soil shapefile; *comp.dbf, forest.dbf,* and *plantnm.dbf,* three dBASE files from the National Map Unit Interpretation Record (MUIR) database maintained by the Natural Resources Conservation Service (NRCS).

Task 4 lets you work with the MUIR database. By linking the tables in the database properly, you can explore many soil attributes in the database from any table. And, because the tables are linked to the soil map, you can also see where selected records are located.

1. Select Data Frame from the Insert menu in ArcMap. Rename the new data frame Task 4, and add *mosoils.shp, comp.dbf, forest.dbf,* and *plantnm.dbf* to Task 4. *Mosoils* is a soil shapefile. *Comp, forest,* and *plantnm* are dBASE files with soil attributes.

2. First, relate *mosoils* to *comp.* Right-click *mosoils* in the Table of Contents, point to Joins and Relates, and click Relate. In the Relate dialog, select MUSYM in the first dropdown list, comp in the second list, and MUSYM in the third list, and enter soil_comp as the relate name. Click OK to dismiss the dialog.

3. Follow the same procedure as in step 2 to prepare two other relates: forest_comp, relating *forest* to *comp* by using MUID as the common field; and forest_plantnm, relating *forest* to *plantnm* by using PLANTSYM as the common field.

4. The four tables (the mosoils attribute table, comp, forest, and plantnm) are now related in pairs by three relates. Right-click *comp* and select Open. Click the Options dropdown arrow and choose Select by Attributes. In the next dialog, create a new selection by entering the following SQL statement in the expression box: "ANFLOOD" = 'FREQ' OR "ANFLOOD" = 'OCCAS'. Click Apply. Click Selected at the bottom of the table so that only the selected records are shown.

5. To see which records in the *mosoils* attribute table are related to the selected records in *comp*, go through the following steps. Click the Options dropdown arrow in the *comp* table, point to Related Tables, and click soil_comp: mosoils. The Attributes of *mosoils* table appears with the related

records. And the *mosoils* map shows where those selected records are located.

6. You can follow the same procedure as in the previous step to see which records in *forest* are related to those polygons that have frequent or occasional annual flooding.

7. Because no relate exists between *comp* and *plantnm,* you cannot open selected records in *plantnm* directly from *comp.* Instead, you can open selected records in *plantnm* from the Selected Attributes of *forest* table by using the related tables of forest_plantnm: plantnm.

Task 5. Combine Spatial and Attribute Data Queries

What you need: *thermal.shp,* a shapefile with 899 thermal wells and springs; *idroads.shp,* showing major roads in Idaho.

Task 5 assumes that you are asked by a company to locate potential sites for a hot spring resort in Idaho. You are given two criteria for selecting potential sites:

- The hot spring must be within 2 miles of a major road.
- The temperature of the water must be greater than 60°C.

The field type in *thermal.shp* uses *s* to denote springs and *w* to denote wells. The field temp shows the water temperature in °C.

1. Select Data Frame from the Insert menu in ArcMap. Add *thermal.shp* and *idroads.shp* to the new data frame. Right-click the new data frame and select Properties. Under the General tab, rename the data frame Task 5 and choose Miles from the Display dropdown list.

2. First select thermal springs and wells that are within 2 miles of major roads in Idaho. Choose Select By Location from the Selection menu. Do the following in the Select By Location dialog: choose "select features from" from the first dropdown list,

check *thermal,* select "are within a distance of" from the second list, select *idroads* from the third list, and enter 2 (miles) for the distance to buffer. Click Apply. Thermal springs and wells that are within 2 miles of roads are highlighted in the map.

3. Next, narrow the selection of map features by using the second criterion. Choose Select By Attributes from the Selection menu. Select *thermal* from the Layer dropdown list and "Select from current selection" from the Method list. Then prepare an SQL statement in the expression box as: "TYPE" = 's' AND "TEMP" > 60. Click Apply.

4. Right-click *thermal* and open its attribute table. Click Selected at the bottom of the attribute table so that only the selected records are shown. The selected records all have Type of *s* and Temp above 60.

5. Map tips are useful for examining the water temperature of the selected hot springs. Right-click *thermal* in the Table of Contents and select Properties. Under the Display tab, check the box to Show Map Tips (uses primary display field). Under the Fields tab, select TEMP from the Primary display field dropdown list. Click OK to dismiss the Properties dialog. Click the Select Elements button. Move the mouse pointer to a highlighted hot spring, and a map tip will display the water temperature of the spring.

Task 6. Query Raster Data

What you need: *slope_gd,* a slope grid; and *aspect_gd,* an aspect grid.

Task 6 shows you different methods for querying a single grid or multiple grids (two for Task 6).

1. Select Data Frame from the Insert menu in ArcMap. Rename the new data frame Task 6, and add *slope_gd* and *aspect_gd* to Task 6.

2. Select Extension from the Tools menu and make sure that Spatial Analyst is checked. Click the View menu, point to Toolbars, and check Spatial Analyst. The Spatial Analyst toolbar appears. Select Raster Calculator from the Spatial Analyst toolbar. In the Raster Calculator dialog, prepare the following statement in the expression box: [slope_gd] = 2. (The = sign appears as == in the expression box.) Click Evaluate. The layer *Calculation* is added to the Table of Contents. Cells with the value of 1 are areas with slopes between 10 and 20 degrees.

3. Select Raster Calculator from the Spatial Analyst toolbar, and prepare the following statement in the expression box: [slope_gd] = 2 AND [aspect_gd] = 4. (The word "and" is shown as & in the expression box.) Click Evaluate. Cells with the value of 1 in *Calculation2* are areas with slopes between 10 and 20 degrees and the south aspect.

REFERENCES

Andrienko, N., G. Andrienko, A. Savinov, H. Voss, and D. Wettschereck. 2001. Exploratory Analysis of Spatial Data Using Interactive Maps and Data Mining. *Cartography and Geographic Information Science* 28: 151–65.

Anselin, L. 1999. Interactive Techniques and Exploratory Spatial Data Analysis. In P. A. Longley, M. F. Goodchild, D. J. MaGuire, and D. W. Rhind, eds., *Geographical Information Systems,* 2d ed., pp. 253–66, New York: Wiley.

Batty, M., and Y. Xie. 1994. Modelling Inside GIS: Part I. Model Structures, Exploratory Spatial Data Analysis and Aggregation. *International Journal of Geographical Information Systems* 8: 291–307.

Becker, R. A., and W. S. Cleveland. 1987. Brushing Scatterplots. *Technometrics* 29: 127–42.

Cleveland, W. S. 1993. *Visualizing Data.* Summit, NJ: Hobart Press.

Cleveland, W. S., and M. E. McGill, eds. 1988. *Dynamic Graphics for Statistics.* Belmont, CA: Wadsworth.

Crampton, J. W. 2002. Interactivity Types in Geographic Visualization. *Cartography and Geographic Information Science* 29: 85–98.

Cressie, N. A. C. 1993. *Statistics for Spatial Data,* rev. New York: Wiley.

DiBiase, D., A. M. MacEachren, J. B. Krygier, and C. Reeves. 1992. Animation and the Role of Map Design in Scientific Visualization. *Cartography and Geographic Information Systems* 19: 201–14, 265–66.

Haslett, J., G. Wills, and A. Unwin. 1990. SPIDER—an Interactive Statistical Tool for the Analysis of Spatially Distributed Data. *International Journal of Geographical Information Systems* 3: 285–96.

MacEachren, A. M. 1995. *How Maps Work: Representation, Visualization, and Design.* New York: Guilford Press.

MacEachren, A. M., F. P. Boscoe, D. Haug, and L. W. Pickle. 1998. Geographic Visualization: Designing Manipulable Maps for Exploring Temporally Varying Georeferenced Statistics. *Proceedings of the IEEE Symposium on Information Visualization,* pp. 87–94.

Meyer, M. A., F. R. Broome, and R. H. J. Schweitzer. 1975. Color Statistical Mapping by the U.S. Bureau of the Census. *American Cartographer* 2: 100–17.

Monmonier, M. 1989. Geographic Brushing: Enhancing Exploratory Analysis of the Scatterplot Matrix. *Geographical Analysis* 21: 81–84.

Olson, J. 1981. Spectrally Encoded Two-Variable Maps. *Annals of the Association of American Geographers* 71: 259–76.

Peterson, M. P. 1995. *Interactive and Animated Cartography.* Englewood Cliffs, NJ: Prentice Hall.

Scott, L. M. 1994. Identification of GIS Attribute Error Using Exploratory Data Analysis. *The Professional Geographer* 46: 378–86.

Slocum, T. A. 1999. *Thematic Cartography and Visualization.* Upper Saddle River, NJ: Prentice Hall.

Tomlin, C. D. 1990. *Geographic Information Systems and Cartographic Modeling.* Englewood Cliffs, NJ: Prentice Hall.

Tufte, E. R. 1983. *The Visual Display of Quantitative Information.* Cheshire, CT: Graphics Press.

Tukey, J. W. 1977. *Exploratory Data Analysis.* Reading, MA: Addison-Wesley.

Walker, P. A., and D. M. Moore. 1988. SIMPLE—an Inductive Modeling and Mapping Tool for Spatially-Oriented Data. *International Journal of Geographical Information Systems* 2: 347–63.

Weber, C. R., and B. P. Buttenfield. 1993. A Cartographic Animation of Average Yearly Surface Temperatures for the 48 Contiguous United States; 1897–1986. *Cartography and Geographic Information Systems* 20: 141–50.

10

VECTOR DATA ANALYSIS

10.1 INTRODUCTION

The scope of GIS analysis varies among disciplines that use GIS. GIS users in hydrology will likely emphasize the importance of terrain analysis and hydrologic modeling, whereas GIS users in wildlife management will be more interested in analytical functions dealing with wildlife point locations and their relationship to the environment. This is why GIS companies have taken two general approaches in packaging their products. One is to prepare a set of analytical tools used by most GIS users, and the other is to prepare modules or extensions designed for specific applications such as hydrologic modeling. This chapter, the first of several chapters on data analysis, covers basic analytical tools for vector data analysis.

The vector data model uses points and their x-, y-coordinates to construct spatial features of points, lines, and polygons. Therefore, vector data analysis is based on the geometric objects of point, line, and polygon and the accuracy of analysis results depends on the accuracy of these objects in terms of location and shape. Because vector data may be topology-based or nontopological, topology can also be a factor for some vector data analysis.

This chapter is grouped into the following four sections. Section 10.2 covers buffering and its applications. Section 10.3 discusses map overlay, types of overlay, problems with overlay, and applications of overlay. Section 10.4 covers tools for measuring distances between points and between points and lines. Section 10.5 includes tools for map manipulation.

10.2 BUFFERING

Based on the concept of proximity, **buffering** separates a map into two areas: one area that is within a specified distance of selected map features and the other area that is beyond. The area that is within the specified distance is called the buffer zone.

Selected map features for buffering may be points, lines, or areas (Figure 10.1). Buffering around points creates circular buffer zones. Buffering around lines creates a series of elongated buffer zones. And buffering around polygons creates buffer zones, extending outward from the polygon boundaries.

There are several variations from Figure 10.1 in buffering. The buffer distance or buffer size does not have to be constant; it can vary according to the values of a given field (Figure 10.2). For example, stream buffer sizes may vary depending on the intensity of adjacent land use. A map feature may have more than one buffer zone. For example, a nuclear power plant may be buffered with distances of 5, 10, 15, and 20 miles, thus forming multiple rings around the plant (Figure 10.3). These buffer zones, although spaced equally from

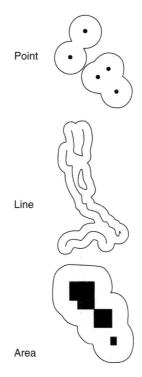

Figure 10.1
Buffering around points, lines, and areas.

the plant, are not equal in area. The second ring from the plant in fact covers an area about three times larger than the first ring. One must consider this area difference if the buffer zones are part of an evacuation plan. Buffering around line features does not have to be on both sides of the lines. Given that line topology is present, buffering can be on either the left side or the right side of the line feature. Boundaries of buffer zones may remain intact so that each buffer zone is a separate polygon. Or these boundaries may be dissolved so that there are no overlapped areas between buffer zones (Figure 10.4).

Regardless of its variations, buffering uses distance measurements from map features to create the buffer zones. The GIS user must therefore know the measurement unit of map features (e.g., meters or feet) and, if necessary, input that

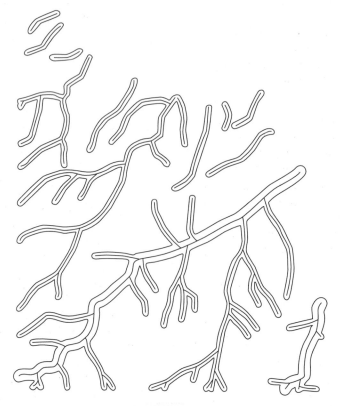

Figure 10.2
Buffering with different buffer distances.

information prior to buffering. ArcGIS, for example, can use the map unit as the default for the distance unit or allow the user to select a different distance unit such as miles instead of feet. Because buffering uses distance measurements from map features, the positional accuracy of map features determines the accuracy of buffer zones.

10.2.1 Applications of Buffering

Buffering creates a buffer zone map, which sets the buffering operation apart from spatial data query. Spatial data query can select map features that are located within a certain distance of other features but cannot create a buffer zone map.

A buffer zone is often treated as a protection zone and is used for planning or regulatory purposes. Many examples can be cited as follows:

- A city ordinance may stipulate that no liquor stores or pornographic shops shall be within 1000 feet of a school or a church.
- Government regulations may stipulate that logging operations must be at least 2 miles away from any stream to minimize the sedimentation problem and set the 2-mile buffer zones of streams as the exclusion zones.
- A national forest may restrict oil and gas well drilling within 500 feet of roads or highways; within 200 feet of trails; within 500 feet of streams, lakes, ponds, or reservoirs; or within 400 feet of springs.
- An urban planning agency may set aside land along the edges of streams to reduce the effects of nutrient, sediment, and pesticide

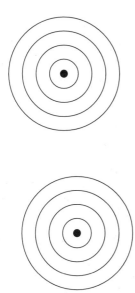

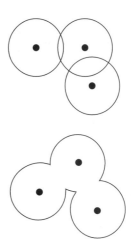

Figure 10.4
Buffer zones not dissolved (top) or dissolved (bottom).

Figure 10.3
Buffering with four rings.

runoff; to maintain shade to prevent the rise of stream temperature; and to provide shelter for wildlife and aquatic life (Thibault 1997).

- A resource agency may establish stream buffers or vegetated filter strips to protect aquatic resources from adjacent agricultural land-use practices (Castelle et al. 1994; Daniels and Gilliam 1996).

A buffer zone may be treated as a neutral zone and as a tool for conflict resolution. In controlling the protesting mass, police may require protesters to be at least 300 feet from a building. Perhaps the best-known neutral zone is the demilitarized zone separating North Korea from South Korea along the 38° N parallel.

Sometimes buffer zones may represent the inclusion zones in GIS applications. For example, the siting criteria for an industrial park may stipulate that a potential site must be within 1 mile of a heavy-duty road. In this case, the 1-mile buffer zones of all heavy-duty roads become the inclusion zones.

Rather than serving as a screening device, buffer zones themselves may become the object

for analysis. A forest management plan may define areas that are within 300 meters of streams as riparian zones and evaluate riparian habitat for wildlife (Iverson et al. 2001). Another example comes from urban planning in developing countries, where urban expansion typically occurs near existing urban areas and major roads. Management of future urban growth should therefore concentrate on the buffer zones of existing urban areas and major roads.

Finally, buffering with multiple rings can be useful as a sampling method. Schutt and colleagues (1999), for example, buffered stream networks in 10-meter increments to a distance of 300 meters so that they could analyze the composition and pattern of woody vegetation as a function of distance from the stream network. One can also apply incremental banding to other studies such as land-use change around urban areas.

10.3 MAP OVERLAY

Map overlay combines the geometries and attributes of two feature maps to create the output (Figure 10.5). One of the two maps is called the input map and the other the overlay map. The geometry,

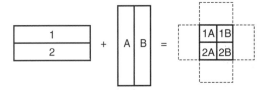

Figure 10.5
Map overlay combines the geometry and attribute data from two maps into a single map. INTERSECT is the map overlay method in this illustration. The output from the intersect includes areas that are common to both input maps. The dashed lines are not included in the output.

or the spatial data, of the output represents the geometric intersection of map features from the input and overlay maps. Therefore, the number of map features on the output map is not the sum of map features on the input and overlay maps but is usually much larger than the sum. Each map feature on the output contains a combination of attributes from the input and overlay maps, and this combination differs from its neighbors.

Feature maps to be overlaid must be spatially registered and based on the same coordinate system (Chapter 2). In the case of the UTM (Universal Transverse Mercator) coordinate system or the SPC (State Plane Coordinate) system, the maps must also be in the same zone. Because GIS users are still migrating from NAD27 to NAD83, they must also check to ensure that the datum is the same between the maps.

Map overlay can work with only two polygon maps at a time. For instance, if three polygon maps are to be overlaid, the operation begins with the first two and then uses the output from them with the third. The process, especially the tracking of the intermediate outputs, can be tedious. The problem of using only two maps in an overlay operation disappears with the regions data model, which allows overlaying more than two layers in a single operation (Chapter 15).

10.3.1 Feature Type and Map Overlay

In practice, the first consideration of map overlay is feature type. The input map may contain points,

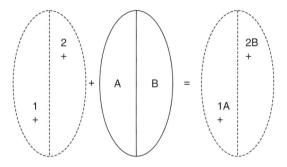

Figure 10.6
Point-in-polygon overlay. The input is a point map (the dashed lines are for illustration only and are not part of the point map). The output is also a point map, which has attribute data from the overlay polygon map.

lines, or polygons; the overlay map must be a polygon map; and the output has the same feature type as the input. Map overlay can therefore be grouped by feature type into point-in-polygon, line-in-polygon, and polygon-on-polygon.

In a **point-in-polygon** operation, the same point features in the input are included in the output but each point is assigned with attributes of the polygon within which it falls (Figure 10.6). An example of the point-in-polygon overlay is to find the association between wildlife locations and vegetation types.

In a **line-in-polygon** operation, the output contains the same line features as in the input but each of them is dissected by the polygon boundaries on the overlay map (Figure 10.7). Thus the output has more line segments than does the input. Each segment on the output combines attributes from the line map and the polygon within which it falls. An example of the line-in-polygon overlay is to find soil data for a proposed road. The input map includes the proposed road. The overlay map is a soil map. And the output shows a dissected proposed road, each road segment having a different set of soil data from its adjacent segments.

The most common overlay operation is **polygon-on-polygon,** involving two polygon maps. The output combines the polygon boundaries from the input and overlay maps to create a new set of polygons (Figure 10.8). Each new

Box **10.1** | **Methods of Map Overlay**

Different GIS packages offer different map overlay methods, although the methods are all based on Boolean operations. ArcView offers UNION and INTERSECT. ArcInfo provides UNION, INTERSECT, and IDENTITY through ArcToolbox and ArcInfo Workstation. MGE has the overlay methods of UNION, INTERSECT, MINUS, and DIFFERENCE. MINUS is functionally similar to IDENTITY, and DIFFERENCE uses the Boolean operation of XOR.

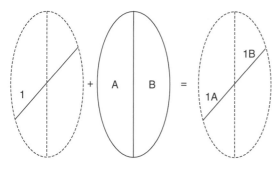

Figure 10.7
Line-in-polygon overlay. The input is a line map (the dashed lines are for illustration only and are not part of the line map). The output is also a line map. But the output differs from the input in two aspects: the line is broken into two segments, and the line segments have attribute data from the overlay polygon map.

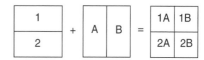

Figure 10.8
Polygon-on-polygon overlay. In the illustration, the two maps to be overlaid have the same area extent. The output combines the geometry and attribute data from the two maps into a single polygon map.

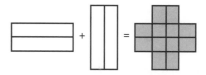

Figure 10.9
The UNION method keeps all areas of the two input maps in the output.

polygon carries attributes from both maps, and these attributes differ from those of adjacent polygons. An example of the polygon-on-polygon overlay is to analyze the association between elevation zones and vegetation types.

10.3.2 Map Overlay Methods

If the input map has the same area extent as the overlay map, then that area extent also applies to the output. But, if the input map has a different area extent than the overlay map, then the area extent of the output may vary depending on the overlay method used. The three common map overlay methods are called UNION, INTERSECT, and IDENTITY (Box 10.1).

UNION preserves all map features from the input and overlay maps by combining the area extents from both maps (Figure 10.9). It is the Boolean operation that uses the keyword OR: (input map) OR (overlay map). The output corresponds to the area extent of the input map, or the overlay map, or both. UNION requires that both the input and overlay maps be polygon maps.

INTERSECT preserves only those features that fall within the area extent common to both the input and overlay maps (Figure 10.10). INTERSECT is the Boolean operation that uses the

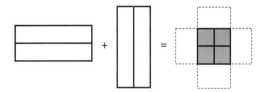

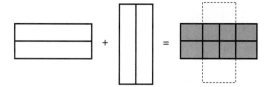

Figure 10.10
The INTERSECT method preserves only the area common to the two input maps in the output. The dashed lines are for illustration only; they are not part of the output.

Figure 10.11
The IDENTITY method produces an output that has the same extent as the input map. But the output includes the geometry and attribute data from the identity map.

keyword AND: (input map) AND (overlay map). The output must correspond to the area extent of both the input and overlay maps. INTERSECT is often a preferred method of overlay because any map feature on its output has attribute data from both of its inputs. For example, a forest management plan may call for an inventory of vegetation types within riparian zones. INTERSECT will be a more efficient overlay method than UNION in this case because the output contains only riparian zones and vegetation types within the zones. The input map for INTERSECT may contain points, lines, or polygons.

IDENTITY preserves only map features that fall within the area extent defined by the input map (Figure 10.11). Expressed in a Boolean expression, IDENTITY represents the operation: [(input map) AND (overlay map)] OR (input map). Features of the overlay map that fall outside the area extent of the input map are left out of the output. The input map for IDENTITY may contain points, lines, or polygons.

The offering of the three overlay methods may differ among GIS packages. For example, the GeoProcessing Wizard in ArcGIS offers UNION and INTERSECT and limits the input map for INTERSECT to either a line or a polygon map (Box 10.2). To overlay a point map with a polygon map, one can use the spatial join method (an option for joining attribute data), which can join data from the polygon map to the point map based on the spatial correspondence between features (Figure 10.12). For example, if a point corresponds to a polygon with an attribute value of agriculture,

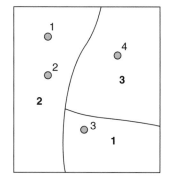

Point-ID	Polygon-ID of joined attribute data
1	2
2	2
3	1
4	3

Figure 10.12
The spatial join method involves two maps: the map to assign data from, and the map to assign data to. Each map can be a point, line, or polygon map. Data assignment is based on the spatial relationship of nearest, part of, or intersects. In this illustration, each point is assigned attribute data of its bounding polygon.

then agriculture will be joined with the point's record.

10.3.3 Slivers

A common error from overlaying two polygon maps is **slivers,** very small polygons along correlated or shared boundary lines of the two input

Box 10.2 **Map Overlay Operations Using Shapefiles**

The GeoProcessing Wizard in ArcGIS does not automatically update the area and perimeter values of polygons in an output shapefile created by UNION or INTERSECT. This shortcoming can be frustrating to shapefile users who need to know the total area of some selected polygons from the output. There are two ways to update the area and perimeter values of polygons in an output shapefile. First, the shapefile can be imported as a layer into a geodatabase. The conversion creates two new fields called shape length and shape area, which contain the updated perimeter and area values, respectively. Second, one can use a Visual Basic script in ArcMap to calculate feature geometry values, including area and perimeter values. The Applications section covers both options.

maps (Figure 10.13). The existence of slivers often results from digitizing errors. Because of the high precision of manual digitizing or scanning, the shared boundaries on the input maps are not right on top of one another. When the two maps are overlaid, the digitized boundaries intersect to form slivers. Other causes of slivers include errors in the source map or errors in interpretation. Polygon boundaries on soil and vegetation maps are usually interpreted from field survey data, aerial photographs, and satellite images. A wrong interpretation can create erroneous polygon boundaries.

Most GIS packages incorporate fuzzy tolerance in map overlay operations to remove slivers. The **fuzzy tolerance** forces points that make up the lines to be snapped together if the points fall within the specified distance (Figure 10.14). The fuzzy tolerance is either defined by the user or based on the default value in the GIS package. Slivers that remain on the output of a map overlay operation are those that were not removed by the built-in fuzzy tolerance.

One option to reduce the sliver problem is to increase the fuzzy tolerance value in map overlay. A larger fuzzy tolerance will result in a smaller number of slivers on the output. But because the fuzzy tolerance applies to the entire map, a large tolerance value will snap shared boundaries as well as lines that are not shared on the input maps, thus creating distorted map features on the overlay output. Another option is to first integrate features to

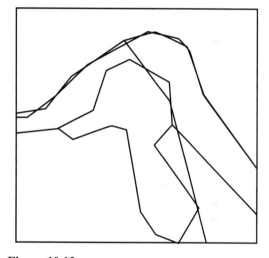

Figure 10.13
The top boundary in this illustration represents a coastline, which has a series of slivers. These slivers are formed between the coastlines from two maps used in an overlay operation. If the coastlines register perfectly between the two maps, then slivers will not be present.

be overlaid by using cluster tolerance (Chapter 5). If feature boundaries fall within the specified cluster tolerance, they are considered to be identical and integrated. Integrated features will not produce slivers. But like the fuzzy tolerance, the cluster tolerance must be used cautiously because it can alter the geometries of other map features.

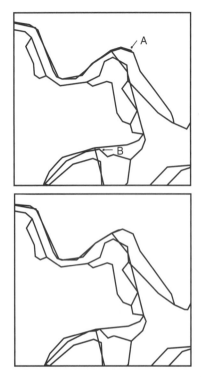

Figure 10.14
Points (and lines) are snapped together if they fall within a specified fuzzy tolerance. Many slivers along the top boundary (A) are removed by use of a fuzzy tolerance. The fuzzy tolerance can also snap arcs that are not slivers (B).

Perhaps a better option for removing slivers is to apply the concept of minimum mapping unit. The **minimum mapping unit** represents the smallest area unit that will be managed by a government agency or an organization. For example, if a national forest adopts 5 acres as its minimum mapping unit, then one can remove any slivers smaller than 5 acres by combining them with adjacent polygons. This map feature manipulation is covered in Section 10.5.

10.3.4 Error Propagation in Map Overlay

Slivers are examples of errors in digital maps that can propagate to the output of a map overlay operation. **Error propagation** refers to the generation

of errors that are due to inaccuracies of the input maps. Error propagation usually involves two types of error: positional and identification (Mac-Dougall 1975; Chrisman 1987). Slivers represent positional errors in map overlay if they result from the inaccuracies of boundaries that are due to digitizing or interpretation errors. Error propagation can also be caused by the inaccuracies of attribute data, or identification errors, such as the incorrect coding of polygon values. Every map overlay product will have some combinations of positional and identification errors.

How serious can error propagation in map overlay be? The answer depends on the number of input maps and the spatial distribution of errors in the input maps. The accuracy of a composite map tends to decrease as (1) the number of input maps increases and (2) the spatial correspondence or coincidence of errors in the input maps decreases.

An error propagation model proposed by Newcomer and Szajgin (1984) considers the spatial coincidence of errors by computing a conditional probability. Assuming that two input maps have square polygons, the model calculates the probability of the event that both maps are correct on the composite map, or $Pr(E_1 \cap E_2)$, using the equation

(10.1)

$$Pr(E_1 \cap E_2) = Pr(E_1)Pr(E_2|E_1)$$

where $Pr(E_1)$ is the probability of the event that the first input map is correct, and $Pr(E_2|E_1)$ is the probability of the event that the second input map is correct given that the first input map is correct. This model suggests that the highest accuracy that can be expected of a composite map is equal to the accuracy of the least accurate individual input map, and the lowest accuracy is equal to

(10.2)

$$1 - \sum_{i=1}^{n} Pr(E_i')$$

where n is the number of input maps in map overlay and $Pr(E_i')$ is the probability that the input map i is incorrect. In other words, the lowest accuracy of the composite map is 1 minus the sum of percent inaccuracies from all input maps.

Suppose a map overlay operation is conducted with three input maps and the accuracy levels of these maps are known to be 0.9, 0.8, and 0.7, respectively. According to Newcomer and Szajgin's model, the accuracy of the composite map is between 0.4 (1 − 0.6) and 0.7.

Newcomer and Szajgin's model clearly illustrates the severity of error propagation in map overlay. But it is a simple model, quite different from the real-world operations. First, the model is based on square polygons, a much simpler geometry than real maps used in map overlay. Second, the model deals only with the Boolean operation of AND, that is, input map 1 is correct AND input map 2 is correct. The Boolean operation of OR in map overlay is different because it requires only one of the input maps to be correct, that is, input map 1 is correct OR input map 2 is correct. Therefore, the probability of the event that the composite map is correct actually increases as more input maps are overlaid (Veregin 1995). Third, Newcomer and Szajgin's model applies only to binary data, meaning that an input map is either correct or incorrect. The model does not work with interval or ratio data and cannot measure the magnitude of errors. Modeling error propagation with numeric data is more difficult than binary data (Arbia et al. 1998; Heuvelink 1998).

10.3.5 Applications of Map Overlay

A map overlay operation computes the geometric intersections of two feature maps and creates an output with combined features and attributes from the two input maps. This output map can then be used for query and modeling purposes. For example, an investment company can use map overlay to find a land parcel that is zoned commercial, not subject to flooding, and not more than 1 mile from a heavy-duty road. After the 1-mile buffer zones from heavy-duty roads are created, the buffer zone map can be overlaid with the zoning and flood plain maps. A query of the output map will show the land parcels that satisfy the investment com-

pany's selection criteria. A site analysis example using both buffering and map overlay is included in the Applications section, and other examples using map overlay operations in modeling are included in Chapter 14.

A more specific application of map overlay is to help solve the areal interpolation problem (Goodchild and Lam 1980). **Areal interpolation** involves transferring known data from one set of polygons (source polygons) to another (target polygons). For example, census tracts may represent source polygons with known population in each tract from the U.S. Census Bureau and school districts may represent target polygons with unknown population data. A common method for estimating the population in each school district is as follows (Figure 10.15):

- Overlay the maps of census tracts and school districts.
- Query the areal proportion of each census tract that is within each school district.
- Apportion the population for each census tract to school districts according to the areal proportion.
- Sum the apportioned population from each census tract for each school district.

The above method for areal interpolation, however, assumes a uniform population distribution

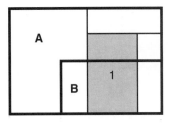

Figure 10.15

In this illustration, census tracts are shown in thick lines and school districts in thin lines. Census tract A has a known population of 4000 and B has 2000. Queried from the result of map overlay, the areal proportion of census tract A in school district 1 is found to be ⅛ and the areal proportion of census tract B, ½. Therefore, the population in school district 1 can be estimated to be 1500, or [(4000 × ⅛) + (2000 × ½)].

within each census tract, which may not be realistic. More recent studies have proposed different methods for improving areal interpolation. One of them uses statistical methods and incorporates additional information for transferring data (Flowerdew and Green 1989; Flowerdew and Green 1995). Another method first partitions the population of source polygons by street segment and then aggregates street segments and their population data for each target polygon (Xie 1995).

10.4 DISTANCE MEASUREMENT

Distance measurement refers to measuring straight-line (Euclidean) distances between points, or between points and their corresponding nearest points or lines. Euclidean distance is one type of distance measure used in GIS. Non-Euclidean distance measures based on cost or time are covered in Chapters 11 and 16. Box 10.3 describes distance measurement capabilities in ArcGIS.

Distance measures can be used directly in data analysis. For example, Chang et al. (1995) used distance measures to test whether deer relocation points were closer to old-growth/clear-cut edges than random points located within the deer's relocation area. Distance measures can also be used as input to data analysis. For example, the Gravity Model, a spatial interaction model commonly used in migration studies and business applications, uses distance measures between points as the input.

10.5 MAP MANIPULATION

Many GIS packages provide tools for manipulating and managing maps in a database. Like buffering and map overlay, these tools are considered basic GIS tools often needed for data preprocessing and data analysis. Map manipulation is easy to follow graphically, even though terms describing the various tools may differ between GIS packages. Box 10.4 summarizes map manipulation tools in ArcGIS.

Most tools covered in this section involve two maps and are sometimes classified as map overlay methods. This chapter, however, limits map overlay to only those methods that can combine spatial and attribute data from two maps into a single map.

Dissolve aggregates map features that have the same value of a selected attribute (Figure 10.16). For example, one can aggregate roads by highway number or counties by state. An important application of dissolve is to simplify a classified polygon map. Classification groups values of a selected attribute into classes and makes obsolete boundaries of adjacent polygons, which have different values initially but are now grouped into the

---

 Box 10.3 **Distance Measurement Using ArcGIS**

ArcView covers distance measurement under the spatial join method. The map to assign data to must be a point map. The map to assign data from may be a point or line map. And the spatial relationship between the two maps is nearest. The spatial join method joins attributes of the closest point or line to the appropriate record of the point attribute table. In addition, a new field called distance, showing the distance measurement, is added to the table.

ArcInfo has NEAR and POINTDISTANCE that measure distances. These two commands are accessible in ArcToolbox and ArcInfo Workstation. NEAR calculates distances between points on a point coverage and their nearest points or lines on another coverage. POINTDISTANCE measures distances for each point on a point coverage to all points on another coverage. POINTDISTANCE does not use the nearest relationship.

same class. Dissolve can remove these unnecessary boundaries and creates a new, simpler map with the classification results as its attribute values. Another application of dissolve is to aggregate both spatial and attribute data of the input map. For instance, to dissolve a county map, one can choose state name as the attribute to dissolve and county population to aggregate. The output map is a state map with an attribute showing the state population (the sum of county populations).

Clip creates a new map that includes only those features of the input map that falls within the area extent of the clip map (Figure 10.17). Clip is a useful tool, for example, for cutting a map acquired elsewhere to fit a study area. The input may be a point, line, or polygon map, but the clip map must be a polygon map.

Merge creates a new map by piecing together two or more maps (Figure 10.18). For example, Merge can put together a map from four input maps, each corresponding to the area extent of a USGS 7.5-minute quadrangle. The output can then

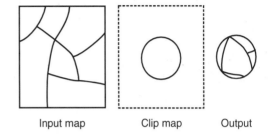

Input map   Clip map   Output

Figure 10.17
Clip creates an output that contains only those features of the input map that fall within the area extent of the clip map. The dashed lines are for illustration only; they are not part of the clip map.

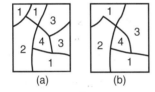

(a)   (b)

Figure 10.16
Dissolve removes boundaries of polygons that have the same attribute value in (a) and creates a simplified map (b).

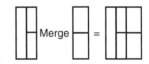

Figure 10.18
Merge pieces together two adjacent maps into a single map. Merge does not remove the shared boundary between the maps.

Box 10.4 Map Manipulation Using ArcGIS

The GeoProcessing Wizard in ArcGIS offers the map manipulation tools of dissolve, clip, and merge. The wizard saves the output either as a shapefile or as a geodatabase feature class. Besides those geoprocessing tools, ArcInfo offers the additional commands of RESELECT (called SELECT in ArcToolbox), ELIMINATE, UPDATE, ERASE, and SPLIT. These commands are accessible in both ArcToolbox and ArcInfo Workstation and are designed to work with coverages.

The output from map manipulation may vary depending on whether the output is a shapefile or a coverage. The Geoprocessing Wizard does not automatically update the area and perimeter values of an output shapefile. One can use the same methods described in Box 10.2 to update the values. A shapefile allows a polygon to have spatially disjoint components. Therefore, a record in the feature table of an output shapefile may include multiple components.

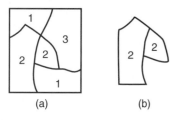

(a)   (b)

Figure 10.19
Select creates a new map (b) with selected map
features from the input map (a).

be used as a single map for data query or display.
But the boundaries separating the input maps still
remain on the output and divide a feature into sep-
arate features if the feature crosses the boundary.
Merge is therefore different from edgematching
(Chapter 5), which can create a seamless coverage.

Select, also called *reselect,* creates a new map
that contains map features selected from a user-
defined logical expression (Figure 10.19). For example, one can create a map showing high-canopy
closure by selecting stands that have 60 to 80 percent closure from a vegetation map.

Eliminate creates a new map by removing
map features that meet a user-defined logical expression from the input map (Figure 10.20). For
example, eliminate can implement the minimum
mapping unit concept by removing polygons that
are smaller than the defined unit in a map.

Update uses a "cut and paste" operation to replace the input map by the update map and its map
features (Figure 10.21). As the name suggests, update is useful for updating an existing map with
new map features in limited areas of the map. It
saves the GIS user from redigitizing the map.

Erase removes from the input map those map
features that fall within the area extent of a map
named as the erase map (Figure 10.22). Suppose a
suitability analysis stipulates potential sites must
be at least 300 meters away from any stream.
A stream buffer map can be used in this case as
the erase map to remove itself from further
consideration.

Split divides the input map into two or more
maps (Figure 10.23). A split map, which shows
area subunits, is used as the template for divid-

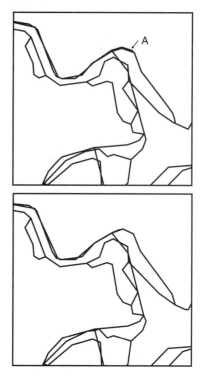

Figure 10.20
Eliminate can remove polygons that are smaller than a
specified size. Some slivers along the top boundary (A)
are therefore eliminated.

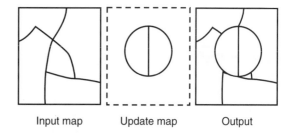

Input map   Update map   Output

Figure 10.21
Update replaces the input map with the update map and
its map features.

ing the input map. For example, a national forest
can use split to divide a vegetation stand map by
district so that each district office can have its
own map.

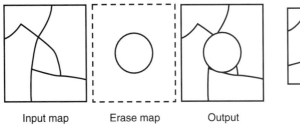

Input map   Erase map   Output

Figure 10.22
Erase removes features from the input map that fall within the area extent of the erase map. The dashed lines are for illustration only; they are not part of the erase map.

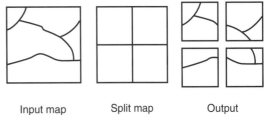

Input map   Split map   Output

Figure 10.23
Split uses the geometry of the split map to divide the input map into four separate maps.

KEY CONCEPTS AND TERMS

Areal interpolation: A process of transferring known data from one set of polygons to another.

Buffering: A GIS operation, in which areas that are within a specified distance of selected map features are separated from areas that are beyond.

Clip: A GIS operation that creates a new map, which includes only those features of the input map that fall within the area extent of the clip map.

Dissolve: A GIS operation that removes boundaries between polygons, which have the same value of a selected attribute.

Eliminate: A GIS operation that creates a new map by removing map features that meet a user-defined logical expression from the input map.

Erase: A GIS operation that removes from the input map those map features that fall within the area extent of the erase map.

Error propagation: The generation of errors in the map overlay output that are due to inaccuracies of the input maps.

Fuzzy tolerance: A distance tolerance used in a GIS package to force points and lines to be snapped together if they fall within the specified distance.

IDENTITY: An overlay method that preserves only features that fall within the area extent defined by the input map.

INTERSECT: An overlay method that preserves only those features that fall within the area extent common to the input and overlay maps.

Line-in-polygon overlay: A GIS operation, in which a line map is dissected by the polygon boundaries on the overlay map, and each arc on the output combines attributes from the line map and the polygon within which it falls.

Map overlay: A GIS operation that combines the geometry and attributes of two digital maps to create the output.

Merge: A GIS operation that creates a new map by piecing together two or more maps.

Minimum mapping unit: The smallest area unit that is managed by a government agency or an organization.

Point-in-polygon overlay: A GIS operation, in which each point of a point map is assigned attribute data of the polygon within which it falls.

Polygon-on-polygon overlay: A GIS operation, in which the output combines the polygon boundaries from the input and overlay maps to create a new set of polygons and each new polygon carries attributes from both maps.

Select: A GIS operation that uses a logical expression to select map features from the input map for the output map. Also called *reselect*.

Slivers: Very small polygons found along the shared boundary of the two input maps in map overlay. Slivers often result from digitizing errors.

Split: A GIS operation that divides the input map into two or more maps.

UNION: A polygon-on-polygon overlay method that preserves all features from the input and overlay maps.

Update: A GIS operation that replaces the input map by the update map and its map features.

APPLICATIONS: VECTOR DATA ANALYSIS

This applications section has three tasks. Task 1 covers the basic functions of vector data analysis such as buffering, map overlay, attribute data manipulation, and dissolve. Because the GeoProcessing Wizard does not automatically update the area and perimeter values of a shapefile, Task 1 converts the output shapefile to a geodatabase layer to update the area and perimeter values. Task 2, on the other hand, uses a Visual Basic script to update the area and perimeter values of the output shapefile. Task 3 covers distance measurement.

Task 1: Perform Buffering and Map Overlay

What you need: shapefiles of *landuse*, *soils*, and *sewers*.

Task 1 simulates GIS analysis for a real-world project and introduces common vector-based analyses such as buffer, map overlay, dissolve, and tabular data manipulation. The task is to find a suitable site for a new university aquaculture lab by using the following selection criteria:

- Preferred land use is brushland (i.e., lucode = 300 in *landuse.shp*).
- Choose soil types suitable for development (i.e., suit >= 2 in *soils.shp*).
- Site must be within 300 meters of sewer lines.

1. Start ArcCatalog, and make connection to the Chapter 10 database. Launch ArcMap. Add

sewers.shp, *soils.shp*, and *landuse.shp* to Layers, and rename Layers Tasks 1&2. All three shapefiles are measured in meters.

2. This step is to buffer *sewers*. Select Buffer Wizard from the Tools menu. Select features of *sewers* to buffer. In the second panel, choose Meters as the distance units and enter 300 as the specified distance. In the third panel, opt to dissolve barriers between buffers, name the output shapefile *sewerbuf.shp*, and click Finish. *Sewerbuf* is added to Tasks 1&2. Right-click *sewerbuf* and open its attribute table. The table has only one record. The BufferDist field has a value of 300 (meters).

3. The next operation is to overlay *soils* and *landuse*. Select GeoProcessing Wizard from the Tools menu. Check Union two layers in the first panel. In the second panel, use the dropdown lists to choose *soils* as the input layer and *landuse* as the polygon overlay layer. Then specify *landsoil.shp* as the output shapefile, and click Finish. *Landsoil* is added to Tasks 1&2. Right-click *landsoil* and open its attribute table. The table includes fields from *soils* and *landuse*.

4. Next overlay *landsoil* and *sewerbuf*. Select GeoProcessing Wizard from the Tools menu. Check Intersect two layers in the first panel. In the second panel, select *sewerbuf* as the

input layer and *landsoil* as the polygon overlay layer. Specify the output shapefile as *finalcov.shp*, and click Finish. *Finalcov* is added to Tasks 1&2.

5. Right-click *finalcov* and open its attribute table. Click the Options dropdown arrow and choose Select by Attributes. Make sure the method is to create a new selection. Enter the following expression in the expression box: "SUIT" >= 2 AND "LUCODE" = 300. Click Apply.

6. Click Selected at the bottom of *finalcov*'s attribute table. The polygons corresponding to the selected records are highlighted in the map. To save the selected features to a separate layer, you can right-click *finalcov*, point to Selection, and choose Create Layer From Selected Features. *Finalcov selection* is added to Tasks 1&2.

7. *Finalcov selection* can remain as a layer for display. It can also be exported to a shapefile. Right-click *finalcov selection*, point to Data, and select Export Data. In the Export Data dialog, choose to export all features, opt to use the same coordinate system as the layer's source data, and save the output shapefile as *sites.shp*. Click OK to dismiss the dialog. Add s*ites.shp* to Tasks 1&2.

8. Right-click *sites.shp* and open its attribute table. Notice that some area values are the same because they have not been updated. You can update the area and perimeter values by importing *sites.shp* as a layer to a personal geodatabase. Right-click the Chapter 10 database in ArcCatalog, point to New, and select Personal Geodatabase. You can rename the new geodatabase as *trial.mdb*. Right-click *trial.mdb*, point to Import, and select Shapefile to Geodatabase. In the next dialog, select *sites.shp* for the input shapefile and click OK. A layer called *sites* is added to *trial.mdb*. Go to ArcMap, add the new layer of *sites*, and open its attribute table. Two new fields in the table, called Shape-Length and

Shape-Area, contain the updated perimeter and area values, respectively.

Task 2: Use a Visual Basic Script to Update Area and Perimeter of a Shapefile

What you need: *sites.shp* from Task 1.

You have updated the perimeter and area values of the output in Task 1 by converting *sites.shp* to a geodatabase layer. An alternative is to use a Visual Basic script in ArcMap to perform the update. This option requires you to add two new fields to *sites.shp* in ArcCatalog and then use the Field Calculator in ArcMap to calculate the perimeter and area values.

1. Remove *sites.shp* from Tasks 1&2 in ArcMap temporarily so that you can add two new fields to *sites.shp* in ArcCatalog.

2. Right-click *sites.shp* in the Catalog tree and select Properties. Under the Fields tab, add shape_area in the first empty cell under Field Name and select Double for the Data Type. Add shape_leng under shape_area and select Double for its Data Type. Click Apply and OK to dismiss the dialog.

3. Add *sites.shp* back to ArcMap. Right-click *sites.shp* to open its attribute table. The new fields of shape_area and shape_leng should appear to the right of the table.

4. Right-click shape_area and select Calculate Values. Click Yes to dismiss the next dialog. When the Field Calculator dialog appears, check Advanced.

5. You will now prepare a Visual Basic script. The first line of the script declares the variable dblArea to store the updated area values. The second line declares the variable pArea, which points to the Iarea interface. The Iarea interface provides access to the area of a geometric object (a polygon in this case). The third line sets the pArea value to be that of shape. And the fourth line assigns the area property of pArea to dblArea. Enter

the following VBA statements in the first text box under "Pre-Logic VBA Script Code":

```
Dim dblArea as Double
Dim pArea as Iarea
Set pArea = [shape]
dblArea = pArea.area
```

6. This step is to assign dblArea to the field shape_area. Type dblArea in the second text box under "shape_area =." Click OK to dismiss the Field Calculator dialog. The calculated area values should now appear in the attribute table of *sites.shp*, and the values should be the same as those of the Geodatabase layer *sites*.

7. Now you are ready to calculate the perimeter values. Right-click shape_leng and select Calculate Values. Click Yes to dismiss the next dialog. When the Field Calculator dialog appears, click Advanced.

8. The Visual Basic script you will use to update the perimeter values has the same structure as the previous script for updating the area values. Enter the following VBA statements in the first text box:

```
Dim dblPerimeter as Double
Dim pCurve as Icurve
Set pCurve = [shape]
dblPerimeter = pCurve.Length
```

9. Type dblPerimeter in the second text box under "shape_leng =." Click OK to dismiss the Field Calculator dialog. The attribute table of *sites.shp* should have the calculated perimeter values.

Task 3: Join Data by Location

What you need: *deer.shp* and *edge.shp*.

Task 3 asks you to use the Join Data by Location method to measure each deer location in *deer.shp* to its closest old-growth/clear-cut edge in *edge.shp*.

1. Select Data Frame from the Insert menu in ArcMap. Rename the new data frame Task 3, and add *deer.shp* and *edge.shp* to Task 3.

2. Right-click *deer*, point to Joins and Relates, and select Join. Click the first dropdown arrow in the Join Data dialog, and select to join data from another layer based on spatial location. Make sure that *edge* is the layer to join to *deer*. Click the radio button stating that each point will be given all the attributes of the line that is closest to it (*deer*), and a distance field showing how close that line is (in map units). Specify *deer_edge.shp* for the output shapefile. Click OK to dismiss the dialog.

3. Right-click *deer_edge* and open its attribute table. The field to the far right of the table is Distance, which lists for each deer location the distance to its closest edge. Click the Options dropdown arrow and choose Select by Attributes. Enter the following SQL statement in the expression box: "Distance" <= 50. Click Apply. Those deer locations that are within 50 meters of their closest edge are highlighted in the table as well as in the map.

REFERENCES

Arbia, G., D. A. Griffith, and R. P. Haining. 1998. Error Propagation Modeling in Raster GIS: Overlay Operations. *International Journal of*

Geographical Information Science 12: 145–67.

Castelle, A. J., A. W. Johnson, and C. Conolly. 1994. Wetland and Stream Buffer Requirements: A

Review. *Journal of Environmental Quality* 23: 878–82.

Chang, K., D. L. Verbyla, and J. J. Yeo. 1995. Spatial Analysis of

Habitat Selection by Sitka Black-Tailed Deer in Southeast Alaska, USA. *Environmental Management* 19: 579–89.

Chrisman, N. R. 1987. The Accuracy of Map Overlays: A Reassessment. *Landscape and Urban Planning* 14: 427–39.

Daniels, R. B., and J. W. Gilliam. 1996. Sediment and Chemical Load Reduction by Grass and Riparian Filters. *Soil Science Society of America Journal* 60: 246–51.

Flowerdew, R., and M. Green. 1989. Statistical Methods for Inference between Incompatible Zonal Systems. In M. F. Goodchild and S. Gopal, eds., *Accuracy of Spatial Databases,* pp. 21–34. London: Taylor and Francis.

Flowerdew, R., and M. Green. 1995. Areal Interpolation and Types of Data. In S. Fotheringham and P. Rogerson, eds., *Spatial Analysis*

and GIS, pp. 121–45. London: Taylor and Francis.

Goodchild, M. F., and N. S. Lam. 1980. Areal Interpolation: A Variant of the Traditional Spatial Problem. *Geoprocessing* 1: 293–312.

Heuvelink, G. B. M. 1998. *Error Propagation in Environmental Modeling with GIS*. London: Taylor and Francis.

Iverson, L. R., D. L. Szafoni, S. E. Baum, and E. A. Cook. 2001. A Riparian Wildlife Habitat Evaluation Scheme Developed Using GIS. *Environmental Management* 28: 639–54.

MacDougall, E. B. 1975. The Accuracy of Map Overlays. *Landscape Planning* 2: 23–30.

Newcomer, J. A., and J. Szajgin. 1984. Accumulation of Thematic Map Errors in Digital Overlay Analysis. *The American Cartographer* 11: 58–62.

Schutt, M. J., T. J. Moser, P. J. Wigington, Jr., D. L. Stevens, Jr., L. S. McAllister, S. S. Chapman, and T. L. Ernst. 1999. Development of Landscape Metrics for Characterizing Riparian-Stream Networks. *Photogrammetric Survey and Remote Sensing* 65: 1157–67.

Thibault, P. A. 1997. Ground Cover Patterns Near Streams for Urban Land Use Categories. *Landscape and Urban Planning* 39: 37–45.

Veregin, H. 1995. Developing and Testing of an Error Propagation Model for GIS Overlay Operations. *International Journal of Geographical Information Systems* 9: 595–619.

Xie, Y. 1995. The Overlaid Network Algorithms for Areal Interpolation Problem. *Computer, Environment and Urban Systems* 19: 287–306.

CHAPTER

11

RASTER DATA ANALYSIS

11.1 INTRODUCTION

The raster data model uses a regular grid to cover the space and the value in each grid cell to correspond to the characteristic of a spatial phenomenon at the cell location. This simple data structure of a grid with fixed cell locations not only is computationally efficient, but also facilitates a large variety of data analyses.

In contrast to vector data analysis, which is based on the geometric objects of point, line, and polygon, raster data analysis is based on cells and grids. Raster data analysis can be performed at the level of individual cells, or groups of cells, or cells within an entire grid. Some raster data operations use a single grid; others use two or more grids. An important consideration in raster data analysis is the type of cell value. Statistics such as mean and standard deviation are designed for numeric values, whereas others such as majority (the most frequent cell value) are designed for both numeric and categorical values.

Various types of data are stored in raster format. Raster data analysis, however, operates only on software-specific raster data such as ESRI grids. Therefore, to use digital elevation models (DEMs), satellite images, and other raster data in data analysis, they must first be processed and imported to software-specific raster data.

This chapter covers the basic tools for raster data analysis in the following six sections. Section 11.2 describes the raster data analysis environment including the parameters of area extent and cell size. Sections 11.3 through 11.6 cover four common types of raster data analysis: local operations, neighborhood operations, zonal operations, and distance measures. Section 11.7 focuses on spatial autocorrelation, a spatial statistic well suited for raster data.

11.2 DATA ANALYSIS ENVIRONMENT

The analysis environment refers to the area for analysis and the output cell size. The area extent for analysis may correspond to a specific grid, or an area defined by its minimum and maximum x-, y-coordinates, or a combination of grids. Given a combination of grids with different area extents, the area extent for analysis can be based on the union or intersect of the grids. The union option uses an area extent that encompasses all input grids, whereas the intersect option uses an area extent that is common to all input grids.

A **mask grid** can also determine the area extent for analysis. A mask grid limits analysis to cells that do not carry the cell value of no data. No data differs from zero. Zero is a valid cell value, whereas no data means the absence of data. For example, an elevation grid converted from a DEM often contains no-data cells along its border (Chapter 7). But in many other cases the user enters no-data cells intentionally to limit the area extent for analysis. For example, one option to limit analysis of soil erosion to only private lands is to code public lands with no data. One can create a mask grid by using data query and reclassification (Box 11.1).

 Box **11.1** **How to Make a Mask Grid**

A mask grid has cell values in areas to be analyzed and no data in areas to be excluded from analysis. One can create a study area mask grid by going through the following steps. First, prepare an outline map for the study area. Make sure that the study area has a different ID value than the outside area. Second, convert the map to a grid and specify the ID value for the cell value. Third, use reclassify, a local operation covered in Chapter 9, and assign no data to the outside area. The output grid from reclassify is a mask grid. A separate mask grid is required for some raster data operations but not for local operations. For local operations, cells with no data will automatically be excluded from analysis.

The GIS user can define the output cell size at any scale deemed suitable. Typically, the output cell size is set to be equal to, or larger than, the largest cell size among the input grids. This follows the rationale that the accuracy of the output should correspond to that of the least accurate input grid. For instance, if the input grids have cell sizes of 10 and 30 meters, the cell size for output should be 30 meters or larger.

11.3 LOCAL OPERATIONS

Constituting the core of raster data analysis, **local operations** are cell-by-cell operations. A local operation creates a new grid from either a single input grid or multiple input grids, and the cell values of the new grid are computed by a function relating the input to the output.

11.3.1 Local Operations with a Single Grid

Given a single grid as the input, a local operation computes each cell value in the output grid as a mathematical function of the cell value in the input grid. Many GIS packages offer arithmetic, logarithmic, trigonometric, and power functions for local operations (Figure 11.1).

Converting a floating-point grid to an integer grid, for example, is a simple local operation,

which uses the integer function to truncate the cell value at the decimal point on a cell-by-cell basis. Converting a slope grid measured in percent to one measured in degrees is also a local operation but requires a more complex mathematical expression. In Figure 11.2, the expression, [slope_d] = 57.296 * arctan ([slope_p]/100), can convert slope_p measured in percent to slope_d measured in degrees. Because computer packages typically use radian instead of degree in trigonometric functions, the constant 57.296 ($360/2\pi$, π = 3.1416) changes the angle measure to degrees.

Reclassification, covered in Chapter 9, is also a local operation. Using either a one-to-one or one-to-many change, reclassification (also called reclassify) assigns a new cell value to each cell in the output grid.

11.3.2 Local Operations with Multiple Grids

Local operations with multiple grids are also referred to as compositing, overlaying, or superimposing maps (Tomlin 1990). These local operations are similar to vector-based map overlay in combin-

(a)

15.2	16.0	18.5
17.8	18.3	19.6
18.0	19.1	20.2

(b)

8.64	9.09	10.48
10.09	10.37	11.09
10.20	10.81	11.42

Figure 11.2
A slope grid may be measured in percent (a) or in degrees (b). A local operation can be used to convert between the two measurement systems.

Arithmetic	+, -, /, *, absolute, integer, floating-point
Logarithmic	exponentials, logarithms
Trigonometric	sin, cos, tan, arcsin, arccos, arctan
Power	square, square root, power

Figure 11.1
The arithmetic, logarithmic, trigonometric, and power functions for local operations.

ing spatial and attribute data, but are much more efficient. Because cells are the same in size and location among the grids in a local operation, there is no need to work on the geometry of the output grid. In contrast, the computation of intersections between map features is a necessary part of map overlay with vector data.

A greater variety of local operations have multiple input grids than have a single input grid. Besides mathematical functions that can be used on individual grids, other measures that are based on the cell values or their frequencies in the input grids can also be derived and stored on the output grid. Some of these measures are limited to grids with numeric data.

Summary statistics, including maximum, minimum, range, sum, mean, median, and standard deviation, are measures that apply to grids with numeric data. For example, a local operation using the mean statistic can calculate a mean annual precipitation grid from 20 input grids, each of which has annual precipitation data as its cell values.

Other measures that are suitable for grids with numeric or categorical data are statistics such as majority, minority, and number of unique values.

For each cell, a majority output grid tabulates the most frequent cell value among the input grids, a minority grid tabulates the least frequent cell value, and a variety grid tabulates the number of different cell values.

Some local operations do not involve statistics or computation. A local operation called combine assigns a unique output value to each unique combination of input values. Suppose a slope grid has three cell values (0 to 20%, 20 to 40%, and greater than 40% slope), and an aspect grid has four cell values (north, east, south, and west aspects). The combine operation creates an output grid with a unique value for each combination of slope and aspect, such as 1 for greater than 40% slope and the south aspect, 2 for (20 to 40%) slope and the south aspect, and so on (Figure 11.3).

Box 11.2 describes local operations available in ArcGIS as well as the way they are designed for use.

11.3.3 Applications of Local Operations

As the core of raster data analysis, local operations have many applications. As an example, a change

---

 ***Box* 11.2 Local Operations in ArcGIS**

The Spatial Analyst extension to ArcGIS is designed for raster data analysis. Its menu includes Reclassify, Cell Statistics, and Raster Calculator for local operations. Both Reclassify and Cell Statistics are menu-driven and easy to use. But the proper use of Raster Calculator requires some understanding of functions. The syntax for use of a function is: function ([grid]). For example, Int([grid1]) uses the integer (int) function to convert the floating-point grid [grid1] to an integer grid. Another example, ATan([grid1] / 100) * 57.296, converts the measurement unit of a slope grid from percent to degrees.

The Raster Calculator dialog incorporates a large assortment of operators and functions in its menu choices: arithmetic operators, logical operators, Boolean connectors, and mathematical functions (arithmetic, logarithmic, trigonometric, and power functions). ArcGIS users can prepare a local operation statement by selecting from the Raster Calculator's menus or by typing directly in the expression box.

Besides the functions listed in the Raster Calculator dialog, additional functions can also be used. Combine is an example. The statement to combine grid1 and grid2 is as follows: combine([grid1], [grid2]). In fact, functions available in the Grid module of ArcInfo Workstation can now be used in the same way in the Raster Calculator. There is really no difference between ArcGIS Desktop and ArcInfo Workstation in raster data analysis.

	3	2	1
(a)	2	1	2
	1	2	3

	3	2	4
(b)	3	2	4
	2	4	1

	1	3	6
(c)	2	4	5
	4	5	7

(d) Combine code	1	2	3	4	5	6	7
(slope, aspect)	(3,3)	(2,3)	(2,2)	(1,2)	(2,4)	(1,4)	(3,1)

Figure 11.3
The combine operation sorts out the unique combinations of cell values in (a) and (b) and assigns a unique value for each combination in (c). The combination codes and their representations are shown in (d).

detection study can use the unique combinations produced by the combine operation to trace the change of cell values, such as change of vegetation covers. Local operations are perhaps most useful for GIS models that require mathematical computation on a cell-by-cell basis.

The Revised Universal Soil Loss Equation (RUSLE) (Wischmeier and Smith 1978; Renard et al. 1997) uses six environmental factors in the equation

(11.1)

$$A = RKLSCP$$

where A is the average soil loss, R is the rainfall–runoff erosivity factor, K is the soil erodibility factor, L is the slope length factor, S is the slope steepness factor, C is the crop management factor, and P is the support practice factor. With each factor prepared as an input grid, one can multiply the grids in a local operation to produce the output grid of average soil loss.

A study by Mladenoff et al. (1995) uses the logistic regression model for predicting favorable wolf habitat,

(11.2)

$$\text{logit } (p) = -6.5988 + 14.6189 \, R,$$

and

$$p = 1 \, / \, [1 + e^{\text{logit}(p)}]$$

where p is the probability of occurrence of a wolf pack, R is road density, and e is the natural exponent. Logit (p) can be calculated in a local operation using a road density grid as the input. Likewise, p can be calculated in another local operation using logit (p) as the input. Chapter 14 discusses in more detail both the wolf habitat predictive model and RUSLE.

Because grids are superimposed in local operations, error propagation can be an issue in interpreting the output. Unlike vector data, raster data do not directly involve digitizing errors. Instead, the main source of errors is the quality of the cell values, which in turn can be traced to other data sources. For example, if raster data are converted from satellite images, statistics for assessing the classification accuracy of satellite images can be used to assess the quality of raster data (Congalton 1991; Veregin 1995). But these statistics are based on binary data (i.e., correctly or incorrectly classified), similar to those used in Newcomer and Szajgin's model (1984) in Chapter 10. It is much more difficult to model error propagation with interval and ratio data (Heuvelink 1998).

11.4 NEIGHBORHOOD OPERATIONS

A **neighborhood operation** involves a focal cell and a set of its surrounding cells. The surrounding cells are chosen for their distance and/or directional relationship to the focal cell. Two common neighborhoods are a focal cell and its four immediate neighbors, and a focal cell and its eight adjacent neighbors in a 3-by-3 window (Figure 11.4). Other types of neighborhood use circles, annuluses, and wedges. A circle neighborhood extends from the focal cell with a specified radius. An annulus or doughnut-shaped neighborhood consists of the ring area between a smaller circle and a

−1,−1	0,−1	1,−1
−1,0	0,0	1,0
−1,1	0,1	1,1

Figure 11.4
The spatial relationship between the focal cell and its neighbors can be based on column and row. The focal cell has the column and row readings of 0 and 0. The cells to the left of the focal cell have the column reading of −1, while the cells to the right have the column reading of 1. The cells above the focal cell have the row reading of −1, while the cells below have the row reading of 1.

larger circle centered at the focal cell. A wedge neighborhood consists of a piece of a circle centered at the focal cell.

A neighborhood operation typically uses the cell values within the neighborhood, with or without the focal cell value, in a computation, and then assigns the computed value to the focal cell. Although a neighborhood operation works on a single grid, its process is similar to that of a local operation with multiple grids. Instead of using cell values from different input grids, a neighborhood operation uses the cell values from a defined neighborhood.

The output from a neighborhood operation can show summary statistics including maximum, minimum, range, sum, mean, median, and standard deviation, as well as tabulation of measures such as majority, minority, and variety. These statistics and measures are the same as those from local operations with multiple grids. To complete a neighborhood operation on a grid, the focal cell is moved from one cell to another until all cells are visited.

A **block operation** is a neighborhood operation that uses a rectangular neighborhood (block) and assigns the calculated value to all block cells in the output grid. A block operation therefore does not move from cell to cell, but from block to block.

Box 11.3 describes neighborhood operations available in ArcGIS and their intended use.

Box 11.3 | **Neighborhood Operations in ArcGIS**

Spatial Analyst has a menu selection called Neighborhood Statistics, which is designed for neighborhood operations using a grid or a point feature map as the input. (A point feature map is converted to a point grid on the fly.) If the input is a point feature map, the value of a field in its attribute table is used for computation. The Neighborhood Statistics dialog requires two inputs: statistic and neighborhood. The statistic input includes the options of minimum, maximum, mean, median, sum, range, standard deviation, major-

ity, minority, and variety. The neighborhood input has the options of rectangle, circle, doughnut, and wedge. Depending on which neighborhood is selected, the dialog displays the input fields for the neighborhood dimension, such as the width and height for a rectangular neighborhood.

To run block operations, one can use block functions such as blockmajority directly in the Raster Calculator. Blockmajority assigns the majority within a block to all block cells.

11.4.1 Applications of Neighborhood Operations

An important application of neighborhood operations is data simplification. The moving average method, for instance, reduces the level of cell value fluctuation in the input grid (Figure 11.5). The method typically uses a 3 × 3 or a 5 × 5 rectangle as the neighborhood. As the neighborhood is moved from one focal cell to another, the average of cell values within the neighborhood is computed and assigned to the focal cell. The output grid of moving averages represents a generalization of the original cell values. Another example is a neighborhood operation that uses variety as a measure, tabulates how many different cell values in the neighborhood, and assigns the number to the focal cell. One can use this method to show, for example, the variety of vegetation types or wildlife species in an output grid.

Neighborhood operations are common in image processing. These operations are variously called filtering, convolution, or moving window operations for spatial feature manipulation (Lillesand and Kiefer 2000). Edge enhancement, for example, can use a range filter, essentially a neighborhood operation using the range statistic (Figure 11.6). The range measures the difference between the maximum and minimum cell values within the

(a)

1	2	2	2	2
1	2	2	2	3
1	2	1	3	3
2	2	2	3	3
1	2	2	2	3

(b)

1.56	2.00	2.22
1.67	2.11	2.44
1.67	2.11	2.44

Figure 11.5
The moving average method calculates the mean of cell values within a moving window and assigns the mean to the focal cell. In this illustration, the cell values in (b) are the moving averages of the shaded cells in (a) using a 3 × 3 moving window. For example, 1.56 in the output grid is calculated from (1 +2 +2 +1 +2 +2 +1 +2 +1) / 9.

defined neighborhood. A high range value therefore indicates the existence of an edge within the neighborhood. The opposite to edge enhancement is a smoothing operation that is based on the ma-

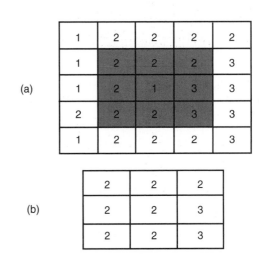

200	200	110	210	210
200	200	110	210	210
150	150	100	170	170
140	140	130	160	160
140	140	130	160	160

(a)

100	110	110
100	110	110
50	70	70

(b)

1	2	2	2	2
1	2	2	2	3
1	2	1	3	3
2	2	2	3	3
1	2	2	2	3

(a)

2	2	2
2	2	3
2	2	3

(b)

Figure 11.6
In this illustration, the cell values in (b) are derived for the shaded cells in (a) from a 3 × 3 neighborhood operation using the range statistic. For example, the upper-left cell in the output grid has a cell value of 100, which is calculated from (200 − 100).

Figure 11.7
In this illustration, the cell values in (b) are derived for the shaded cells in (a) from a 3 × 3 neighborhood operation using the majority statistic. For example, the upper-left cell in the output grid has a cell value of 2 because there are five 2s and four 1s in its neighborhood.

jority measure (Figure 11.7). The majority operation assigns the most frequent cell value to every cell within the neighborhood, thus creating a smoother grid than the original grid.

Another area of study that heavily depends on neighborhood operations is terrain analysis (Chapter 12). Slope, aspect, and surface curvature measures of a cell are all derived from neighborhood operations using elevation values of its adjacent neighbors.

Neighborhood operation can also be important to studies that need to select cells by their neighborhood characteristics. For example, installing a gravity sprinkler irrigation system requires information about elevation drop within a circular neighborhood of a cell. Suppose that a system requires an elevation drop of 130 feet within a distance of 0.5 mile to make it financially feasible. A neighborhood operation on an elevation grid can answer the question by using a circle with a radius of 0.5 mile as the neighborhood and (elevation) range as the statistic. A query of the output grid can show which cells meet the criterion.

Because of its ability to summarize statistics within a defined area, a neighborhood operation can be used to select sites that meet a study's specific criteria. An example is a study by Crow et al. (1999), which used neighborhood operations to select a stratified random sample of 16 plots that represented two ownerships located within two regional ecosystems.

11.5 ZONAL OPERATIONS

A **zonal operation** works with groups of cells of same values or like features. These groups are called zones. Zones may be contiguous or noncontiguous. A contiguous zone includes cells that are spatially connected, whereas a noncontiguous zone includes separate regions of cells.

A zonal operation may work with a single grid or two grids. Given a single input grid, zonal operations describe the geometry of zones, such as area, perimeter, thickness, and centroid (Figure 11.8). The area is the sum of the cells that fall within the

Zone	Area, in sq. m.	Perimeter, in m.	Thickness, in m.
1	36,224,000,000	1,708,000	77,554
2	48,268,001,280	1,464,000	77,414

Figure 11.8

This illustration shows two large watersheds and their zonal geometric measures. The cell resolution is 2000 meters; therefore, the areas are measured in square meters, the perimeters in meters, and the thickness measures in meters. The centroid of each zone is shown by an x.

zone times the cell size. The perimeter of a contiguous zone is the length of its boundary, and the perimeter of a noncontiguous zone is the sum of the length of each region. The thickness calculates the radius (in cells) of the largest circle that can be drawn within each zone. The centroid is the geometric center of a zone located at the intersection of the major axis and the minor axis of an ellipse that best approximates the zone. These geometric measures of zones are particularly useful for studies of landscape ecology (Forman and Godron 1986).

Given two grids in a zonal operation, one input grid and one zonal grid, a zonal operation produces an output grid, which summarizes cell values in the input grid for each zone in the zonal grid. The summary statistics and measures include area, minimum, maximum, sum, range, mean, standard deviation, median, majority, minority, and variety. (The last four measures are not available if the input grid is a floating-point grid.) Figure 11.9 shows a zonal operation of computing the mean by zone. Figure 11.9a is the zonal grid with three

zones, Figure 11.9b is the input grid, and Figure 11.9c is the output grid.

Box 11.4 describes zonal operations available in ArcGIS and their intended use.

11.5.1 Applications of Zonal Operations

Area, perimeter, thickness, and centroid are part of a large group of geometric measures used in landscape ecology (McGarigal and Marks 1994). Other geometric measures can also be derived from area and perimeter. For example, a shape index called the areal roundness is defined as 354 times the square root of a zone's area divided by its perimeter (Tomlin 1990). Essentially, the areal roundness compares the shape of a contiguous zone to a circle. The area roundness value approaches 100 for a circular shape and 0 for a highly distorted shape. The expression for computing areal roundness is: 354 * Sqrt(zonalarea([grid1])) / zonalperimeter ([grid1]), where grid1 is the input grid and Sqrt is the square root function.

Box 11.4 Zonal Operations in ArcGIS

One can use a zonal function in the Raster Calculator to calculate the zonal geometry of an input grid. The zonal functions include the geometric measures of area, perimeter, thickness, and centroid. For example, the statement, zonalarea([grid1]), computes the area of each zone in grid1.

Zonal Statistics in the Spatial Analyst menu can perform zonal operations with two grids. Given a zonal grid and an input grid, Zonal Statistics calculates the summary statistics of area, minimum, maximum, range, mean, standard deviation, sum, variety, majority, minority, and median and save the statistics into a table. Additionally, the user can select a summary statistic to make a histogram with zones and the statistic along the x- and y-axis.

Figure 11.9
The zonal operation in this illustration uses the zones of 1, 2, and 3 in (a) and the cell values in (b) to calculate the zonal means of 2.17, 2.25, and 4.17. The output grid is shown in (c).

Zonal operations with two grids are useful in generating descriptive statistics for comparison purposes. For example, to compare topographic characteristics of different soil textures, we can use a soil grid that contains the categories of sand, loam, and clay as the zonal grid and slope, aspect, and elevation as the input grids. By running a series of zonal operations, we can summarize the slope, aspect, and elevation characteristics associated with the three soil textures.

11.6 DISTANCE MEASURE OPERATIONS

Distance measure operations calculate distances away from cells designated as the source cells. The cells, for which distances are measured, and the source cells are in the same grid. An example of this type of operation is to calculate for each cell the distance to the closest stream in a stream grid. Distance measure operations are also called extended neighborhood operations (Tomlin 1990) or global operations because they cover the entire grid.

Distance measures in a grid follow the node–link relationship (Figure 11.10). A node represents the center of a cell, and a link—either a lateral link or a diagonal link—connects the node to its adjacent cells. Distances are calculated along the links in cells: a lateral link is 1.0 cell, and a diagonal link is 1.4142 cells.

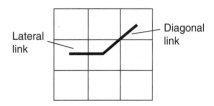

Figure 11.10
Distance measures in a grid follow the links, which connect the cells at their centers. A lateral link connects two direct neighbors, whereas a diagonal link connects two diagonal neighbors.

6	7	8
5	0	1
4	3	2

Figure 11.11
Direction measures in a direction grid are numerically coded. The focal cell has a code of 0. The numeric codes 1 to 8 represent the direction measures of 90°, 135°, 180°, 225°, 270°, 315°, 360°, and 45° in a clockwise direction.

Distances in raster data analysis may be expressed as physical (Euclidean) distance or cost distance. The **physical distance** measures the distance by summing the links between two cells and multiplying the sum by the cell size, whereas the **cost distance** measures the cost for traversing the physical distance. A truck driver, for example, is more interested in the time or the fuel cost for covering a route than its physical distance. The cost distance in this case is determined not only by the physical distance, but also by the speed limit and road condition.

Some distance measure operations to be discussed later also involve direction. Direction measures in a grid follow the eight principal directions from a focal cell to its eight neighboring cells. Expressed in degrees and in a clockwise direction, they are 0° or 360°, 45°, 90°, 135°, 180°, 225°, 270°, and 315°. When direction measures are included in a grid, they are coded in numbers from 0 to 8 rather than in degrees (Figure 11.11).

11.6.1 Physical Distance Measure Operations

A **physical distance measure operation** uses cells as units in distance measurement. There are two types of physical distance measure operations. The first is to buffer the source cells with continuous distances, thus creating a series of wavelike distance zones over the entire grid (Figure 11.12). By using the reclassify function, one can convert this continuous distance grid to a grid containing dis-

crete distance buffers of the source cells. The second operation (called allocation) determines for each cell in a grid its closest source cell by the physical distance measure (Figure 11.13).

11.6.2 Cost Distance Measure Operations

A **cost distance measure operation** uses the cost or impedance to traverse each cell as a distance unit. Cost distance measure operations are much more complex than physical distance measure operations. To begin with, a cost distance operation requires another grid defining the cost or impedance to move through each cell. The cost for each cell in the cost grid is often the sum of different costs. As an example, the cost for constructing a pipeline may include construction and operational costs as well as the potential costs of environmental impacts (Box 11.5). Given a cost grid, the cost distance of a lateral link is the average of the costs in the linked cells. The cost distance of a diagonal link is the average cost times 1.4142 (Figure 11.14).

The objective of a cost distance measure operation is no longer to calculate the distance from each cell to the closest source cell but to find the path with the least accumulative cost. The inclusion of cost creates many paths connecting a cell and the source cell, each with a different accumulative cost. The algorithm for finding the least accumulative cost follows an iterative process, which begins by activating cells adjacent to the source

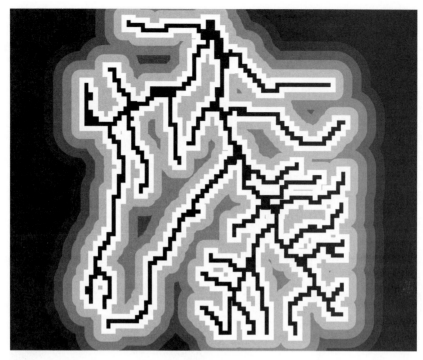

Figure 11.12
An example of continuous distance measures from a stream network.

Box 11.5   Cost Grid for a Site Analysis of Pipelines

A site analysis of a pipeline project must consider the construction and operational costs. Some of the variables that can influence the costs include the following:

- Distance from source to destination
- Topography, such as slope and grading
- Geology, such as rock and soils
- Number of stream, road, and railroad crossings
- Right-of-way costs
- Proximity to population centers

In addition, the site analysis should consider the potential costs of environmental impacts during construction and liability costs that may result from accidents after the project has been completed. Environmental impacts of a proposed pipeline project may involve the following:

- Cultural resources
- Land use, recreation, and aesthetics
- Vegetation and wildlife
- Water use and quality
- Wetlands

Each of the above variables must be evaluated and measured in actual or, more likely, relative cost values. Relative costs are ranked values. For example, costs may be ranked from 1 to 5, with 5 being the highest rank value. A grid is made for each cost variable after it has been evaluated. The final step is to make the (total) cost grid by summing the individual cost grids.

cell and by computing costs to the cells. The cell with the lowest cost distance is chosen from the active cell list, and its value is assigned to the out-put grid. Next, cells adjacent to the chosen cell are activated and added to the active cell list. Again, the lowest cost cell is chosen from the list and its

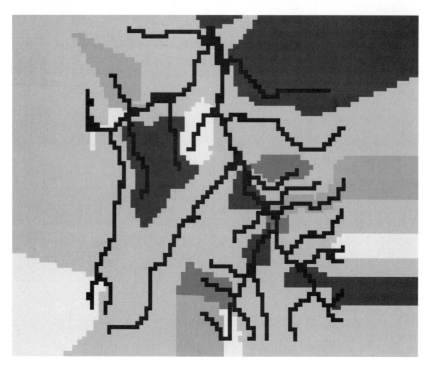

Figure 11.13
An example of allocating each cell to its closest stream.

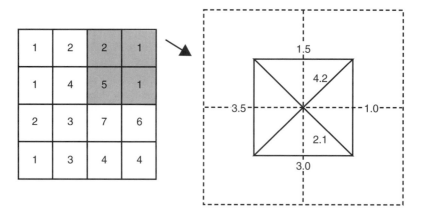

Figure 11.14
The cost distance of a lateral link is the average of the costs in the linked cells, for example, $(1 + 2) / 2 = 1.5$. The cost distance of a diagonal link is the average cost times 1.4142, for example, $1.4142 \times [(1 + 5) / 2] = 4.2$.

neighboring cells are activated. Each time a cell is reactivated, meaning that the cell is accessible to the source cell through a different path, its accumulative cost must be recomputed. The lowest accumulative cost is then assigned to the reactivated cell. This process continues until all cells in the output grid are assigned with their least accumulative costs to the source cell.

Figure 11.15 illustrates the cost distance measure operation. Figure 11.15a shows a grid with the source cells at the opposite corners. Figure 11.15b represents a cost grid. To simplify the computation, both grids are set to have a cell size of 1. Figure 11.15c shows the cost of each lateral link and the cost of each diagonal link. Figure 11.15d shows for each cell the least accumulative cost. Box 11.6 explains how Figure 11.15d is derived.

A cost distance measure operation can result in different types of output. The first is a least accumulative cost grid as shown in Figure 11.15d.

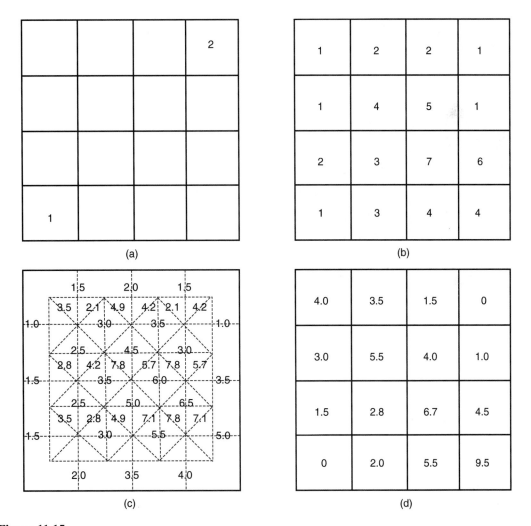

Figure 11.15
Using the grid with two source cells (a) and the cost grid (b), this illustration shows the cost distance for each link (c), and the least accumulative cost distance from each cell to a source cell (d).

The second is a direction grid, showing the direction of the least cost path from each cell to a source. The third is an allocation grid, showing the assignment of each cell to a source on the basis of cost distance measures. Using the same data as in Figure 11.15, Figure 11.16a shows the least cost paths of two cells and Figure 11.16b shows the assignment of each cell to a source cell. The darkest cell in Figure 11.16b can be assigned to either one of the two sources. The fourth type is a shortest

Box 11.6 Derivation of the Least Accumulative Cost Grid

Step 1. Activate cells adjacent to the source cells, place the cells in the active list, and compute cost values for the cells. The cost values for the active cells are as follows: 1.0, 1.5, 1.5, 2.0, 2.8, and 4.2.

		1.5	0
		4.2	1.0
1.5	2.8		
0	2.0		

Step 2. The active cell with the lowest value is assigned to the output grid, and its adjacent cells are activated. The cell at row 2, column 3, which is already on the active list, must be reevaluated, because a new path has become available. As it turns out, the new path with a lateral link from the chosen cell yields a lower accumulative cost of 4.0 than the previous cost of 4.2. The cost values for the active cells are as follows: 1.5, 1.5, 2.0, 2.8, 4.0, 4.5, and 6.7.

		1.5	0
		4.0	1.0
1.5	2.8	6.7	4.5
0	2.0		

Step 3. The two cells with the cost value of 1.5 are chosen, and their adjacent cells are placed in the active list. The cost values for the active cells are as follows: 2.0, 2.8, 3.0, 3.5, 4.0, 4.5, 5.7, and 6.7.

	3.5	1.5	0
3.0	5.7	4.0	1.0
1.5	2.8	6.7	4.5
0	2.0		

Step 4. The cell with the cost value of 2.0 is chosen and its adjacent cells are activated. Of the three adjacent cells activated, two have the accumulative cost values of 2.8 and 6.7. The values remain the same because the alternative paths from the chosen cell yield

higher cost values (5 and 9.1, respectively). The cost values for the active cells are as follows: 2.8, 3.0, 3.5, 4.0, 4.5, 5.5, 5.7, and 6.7.

	3.5	1.5	0
3.0	5.7	4.0	1.0
1.5	2.8	6.7	4.5
0	2.0	5.5	

Step 5. The cell with the cost value of 2.8 is chosen. Its adjacent cells all have accumulative cost values assigned from the previous steps. These values remain unchanged because they are all lower than the values computed from the new paths.

	3.5	1.5	0
3.0	5.7	4.0	1.0
1.5	2.8	6.7	4.5
0	2.0	5.5	

Step 6. The cell with the cost value of 3.0 is chosen. The cell to its right has an assigned cost value of 5.7, which is higher than the cost of 5.5 via a lateral link from the chosen cell.

4.0	3.5	1.5	0
3.0	5.5	4.0	1.0
1.5	2.8	6.7	4.5
0	2.0	5.5	

Step 7. All cells have been assigned with the least accumulative cost values, except the cell at row 4, column 4. The least accumulative cost for the cell is 9.5 from either source.

4.0	3.5	1.5	0
3.0	5.5	4.0	1.0
1.5	2.8	6.7	4.5
0	2.0	5.5	9.5

path grid, which shows the least cost path from each cell to a source.

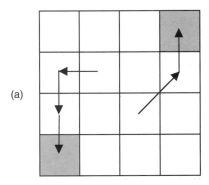

(a)

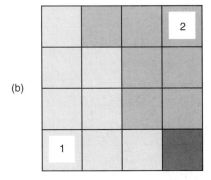

(b)

Figure 11.16
Using the same data as in Figure 11.15, this illustration shows the least cost path (a) and the allocation grid (b).

Box 11.7 describes distance measure operations available in ArcGIS as well as the way these operations are organized.

11.6.3 Applications of Distance Measure Operations

Like buffering around vector-based features, physical distance measure operations have many applications. An example is a raster model of the potential nesting habitat of greater sandhill cranes in northwestern Minnesota (Herr and Queen 1993). The study used a cell resolution of 30 meters and compiled a vegetation map from Landsat Thematic Mapper data. Using continuous distance zones measured from undisturbed vegetation, roads, buildings, and agricultural land, the study categorized potentially suitable nesting vegetation as optimal, suboptimal, marginal, or unsuitable. Distance measure operations provided the tools to implement the model.

Cost distance measure operations are commonly used for site analysis of roads and pipelines to find the least accumulative cost path. But they can also be applied to other types of movement. For example, Hepner and Finco (1995) built a model of gas dispersion using an impedance grid that represented the product of surface distance, slope impedance, and wind impedance. The accumulative impedance in their study represented the

---

 Box **11.7** **Distance Measure Operations in ArcGIS**

The Distance menu of Spatial Analyst provides four selections: Straight Line Distance, Allocation, Cost Weighted, and Shortest Path. The first two use physical distance measures, and the last two use cost distance measures. Straight Line Distance creates an output grid containing continuous distance measures away from the source cells in a grid. Allocation creates a grid, in which each cell is assigned the value of its closest source cell. Cost Weighted calculates for each cell the least accumulative cost, over a cost grid,

to its closest source cell. To define the threshold for the least accumulative cost, one can enter a maximum distance for the operation. The Cost Weighted operation can also produce a direction grid, showing the direction of the least cost path, or an allocation grid, showing the assignment of each cell to a source. Shortest Path uses the distance and direction grids created by the cost distance measure operation to generate the least cost path from any cell or zone.

difficulty of gas travel between the release point (source cell) and any given cell in the study area. And equal steps of the accumulative impedance values supposedly defined the progression over time of a dense gas cloud over the terrain.

Because distance measure operations are based on cells, they are not as precise as buffering around vector-based features. The same is true with directions because direction measures in these operations follow the eight principal directions. Therefore, a least accumulative cost path often displays a zigzag pattern.

11.7 SPATIAL AUTOCORRELATION

Spatial autocorrelation measures the relationship among values of a variable according to the spatial arrangement of the values (Cliff and Ord 1973). The relationship may be described as highly correlated if like values are spatially close to each other and independent or random if no pattern can be discerned from the arrangement of values. The absence of significant spatial autocorrelation can validate the use of standard statistical tests of hypotheses. On the other hand, the presence of significant spatial autocorrelation should encourage the researcher to incorporate spatial dependency in the analysis (Legendre 1993). Spatial autocorrelation is well suited to raster data analysis because cells in a grid follow a well-defined spatial arrangement.

Spatial autocorrelation does not belong to any of the above raster data operations. It is probably best classified as a global operation because the computation of spatial autocorrelation uses every cell in a grid and produces a statistic that applies to the entire grid.

One popular spatial autocorrelation measure is **Moran's I,** which can be computed by

(11.3)

$$\frac{\sum\limits_{i=1}^{n} \sum\limits_{j=1}^{m} w_{ij}(x_i - x_m)(x_j - x_m) / \sum\limits_{i=1}^{n} \sum\limits_{j=1}^{m} w_{ij}}{\sum\limits_{i=1}^{n} (x_i - x_m)^2 / n}$$

where x_i is the value of cell i, x_j is the value of cell i's neighbor j, x_m is the mean cell value of the grid, w_{ij} is a coefficient, and n is the total number of cells in the grid. The coefficient w_{ij} has a value of 1 if j is one of the four cells directly adjacent to i and a value of 0 for other cells or cells with no data. Moran's I is positive when nearby areas have similar attribute values, negative when they have dissimilar values, and close to zero when attribute values are arranged randomly. The values Moran's I takes on tend to range between −1 and 1, but are not restricted to the range.

Another popular spatial autocorrelation measure is **Geary's c,** which can be computed by

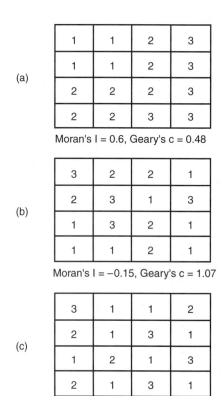

Figure 11.17
Three sets of spatial autocorrelation statistics: (a) a positively correlated pattern, (b) a random pattern, and (c) a negatively correlated pattern.

(11.4)

$$\frac{\sum\limits_{i=1}^{n}\sum\limits_{j=1}^{m} w_{ij}(x_i - x_j)^2 / \sum\limits_{i=1}^{n}\sum\limits_{j=1}^{m} w_{ij}}{\sum\limits_{i=1}^{n}(x_i - x_m)^2 / (n-1)}$$

The notations in the equation are the same as those for Moran's I. Whereas Moran's I uses the covariance $(x_i - x_m)(x_j - x_m)$ in the computation, Geary's c uses the variance $(x_i - x_j)^2$. Geary's c has a value of 1 for a random pattern, less than 1 for a positively correlated pattern, and greater than 1 for a nega-tively correlated pattern. Figure 11.17 shows three patterns and the associated values of Moran's I and Geary's c for comparison.

ArcInfo Workstation, IDRISI, GRASS, and ILWIS are all capable of computing Moran's I and Geary's c. Recent developments in spatial statistics have included measures of spatial autocorrelation in multiple distances and local autocorrelation statistics (Getis and Ord 1996; Lam et al. 1996; Lee and Wong 2001; Rogerson 2001). These new statistics are not yet available in commercial GIS packages.

KEY CONCEPTS AND TERMS

Block operation: A neighborhood operation that uses a rectangular neighborhood (block) and assigns the calculated value to all block cells in the output grid.

Cost distance: Distance measured by the cost or impedance of moving between cells.

Cost distance measure operation: A distance measure operation that uses the cost or impedance to move through each cell as distance unit.

Distance measure operation: A raster data operation that calculates distances away from cells designated as the source cells.

Geary's c: A spatial autocorrelation statistic that uses the variance in computations.

Local operation: A cell-by-cell operation in raster data analysis.

Mask grid: A grid that limits raster data analysis to cells that do not carry the cell value of no data.

Moran's I: A spatial autocorrelation statistic that uses the covariance in computations.

Neighborhood operation: A raster data analysis operation that involves a focal cell and a set of its surrounding cells.

Physical distance: Distance measured by summing the links between cells and multiplying the sum by the cell size.

Physical distance measure operation: A distance measure operation that uses cells as units.

Spatial autocorrelation: A spatial statistic that measures the relationship among values of a variable according to the spatial arrangement of the values.

Zonal operation: A raster data analysis operation that involves groups of cells of same values or like features.

APPLICATIONS: RASTER DATA ANALYSIS

This applications section covers the basic operations of raster data analysis. Task 1 covers a local operation. Task 2 uses a neighborhood operation. Task 3 uses a zonal operation. Task 4 includes the physical distance measure operation in data query. Task 5 solves a least accumulative cost distance problem.

Task 1: Perform a Local Operation

What you need: *emidalat*, an elevation grid with a cell resolution of 30 meters.

Task 1 lets you use a local operation to convert the elevation values of *emidalat* from meters to feet.

1. Start ArcCatalog, and make connection to the Chapter 11 database. Select Properties from the context menu of *emidalat* in the Catalog tree. The General tab shows that *emidalat* has 214 rows, 186 columns, and a cell size of 30 (meters). Also, *emidalat* is a floating-point grid.

2. Launch ArcMap. Add *emidalat* to Layers, and rename Layers Tasks 1&2. Select Extensions from the Tools menu and check the box for Spatial Analyst. Click the View menu, point to Toolbars, and make sure that the box for Spatial Analyst is checked.

3. Select Raster Calculator from the Spatial Analyst dropdown list. Enter the following expression in the Raster Calculator's expression box: [emidalat] * 3.28. Click Evaluate. *Calculation* shows *emidalat* in feet.

4. *Calculation* is a temporary grid. To make it a permanent grid, right-click *Calculation* and select Make Permanent. Enter a name for the permanent grid and click Save.

Task 2: Perform a Neighborhood Operation

What you need: *emidalat*, as in Task 1.

Task 2 asks you to use a neighborhood operation to generalize the elevation values of *emidalat*.

1. Select Neighborhood Statistics from the dropdown menu of Spatial Analyst. Click the downward arrow and specify *emidalat* for the Input data. The default setup in the Neighborhood Statistics dialog is to calculate a neighborhood mean of *emidalat* by using a 3-by-3 rectangle and to create a temporary output raster with a cell size of 30 (meters). Take the default values and click OK. *NbrMean of emidalat* shows the neighborhood mean of *emidalat*.

2. *NbrMean of emidalat* is a temporary grid. To make it a permanent grid, right-click *NbrMean of emidalat* and select Make Permanent. Enter a name for the permanent grid and click Save.

Task 3: Perform a Zonal Operation

What you need: *precipgd*, a grid showing the average annual precipitation in Idaho; *hucgd*, a watershed grid.

Task 3 asks you to derive annual precipitation statistics by watershed in Idaho. Both *precipgd* and *hucgd* are projected onto the Idaho Transverse Mercator coordinate system and are measured in meters. The precipitation measurement unit for *precipgd* is 1/100 of an inch; for example, the cell value of 675 means 6.75 inches.

1. Select Data Frame from the Insert menu in ArcMap. Rename the new data frame Task 3, and add *precipgd* and *hucgd* to Task 3.

2. Make sure that the Spatial Analyst toolbar is available. Select Zonal Statistics from the Spatial Analyst dropdown list. Select *hucgd* for the Zone dataset, Value for the Zone field, and *precipgd* for the Value raster. Check the boxes to Ignore NoData in calculations, to Join output table to zone layer, and to Chart statistic with Mean. Save the output table as *zstats.dbf*. Click OK to dismiss the dialog.

3. The chart shows the mean of *precipgd* within each zone of *hucgd*. The table shows the statistics of *precipgd*, including minimum, maximum, range, mean, standard deviation, sum, variety, majority, minority, and median within each zone of *hucgd*.

Task 4: Measure Physical Distances

What you need: *strmgd*, a grid showing streams; *elevgd*, a grid showing elevation zones.

Task 4 asks you to locate the potential habitats of a plant species. The cell values in *strmgd* are the ID values of streams. The cell values in *elevgd* are elevation zones 1, 2, and 3. Both grids have the cell resolution of 100 meters. The potential habitats of the plant species must meet the following criteria:

- Elevation zone 2
- Within 200 meters of streams

1. Select Data Frame from the Insert menu in ArcMap. Rename the new data frame Task 4, and add *strmgd* and *elevgd* to Task 4.

2. Make sure that the Spatial Analyst toolbar is available. Click the Spatial Analyst downward arrow, point to Distance, and select Straight Line. In the Straight Line dialog, select *strmgd* for Distance to, enter 100 (meters) for the Output cell size, and opt for a Temporary Output raster. Click OK to dismiss the dialog.

3. *Distance to strmgd* shows continuous distance zones away from streams in *strmgd*.

4. This step is to create a new grid that shows area within 200 meters of streams. Select Reclassify from the Spatial Analyst dropdown list. In the Reclassify dialog, select *Distance to strmgd* for the Input raster and click Classify. In the Classification dialog, first select Equal Interval for the Method and 2 for the Classes. (The selection of Equal Interval is merely to activate the Classes button. You are not using the Equal Interval classification method.) Then click the first value under Break Values in the lower right and enter 200. Click the empty space in the Break Values frame to unselect the second cell. Click OK to dismiss the Classification dialog. Opt for a Temporary Output raster in the Reclassify dialog, and click OK. *Reclass of Distance to strmgd* grid separates areas that are within 200 meters of streams from areas that are beyond.

5. Select Raster Calculator from the Spatial Analyst dropdown list. Enter the following expression in the Raster Calculator's expression box: [Reclass of Distance to strmgd] = 1 AND [elevgd] = 2. Click Evaluate. The *Calculation* layer shows areas that meet the criteria with the value of 1.

6. You can also complete Task 4 without reclassifying *Distance to strmgd*. In that case,

the expression in the Raster Calculator dialog should be changed to: [Distance to strmgd] <= 200 AND [elevgd] = 2.

Task 5: Compute the Least Accumulative Cost Distance

What you need: *sourcegrid* and *costgrid*, the same grids as in Figure 11.15; *pathgrid*, a grid to be used with the shortest path function.

1. Select Data Frame from the Insert menu in ArcMap. Rename the new data frame Task 5, and add *sourcegrid, costgrid,* and *pathgrid* to Task 5.

2. Make sure that the Spatial Analyst toolbar is available. Click the Spatial Analyst downward arrow, point to Distance, and select Cost Weighted. In the Cost Weighted dialog, do the following: select *sourcegrid* for Distance to, select *costgrid* for the Cost raster, check Create direction, check Create allocation, and opt for Temporary grids including the Output raster. Click OK to dismiss the dialog.

3. *CostDistance to sourcegrid* shows the least accumulative cost distance from each cell to a source cell. You can use the Identify tool to click a cell and find its accumulative cost.

4. *CostDirection to sourcegrid* shows the least cost path from each cell to a source cell. The cell value in the raster indicates which neighboring cell to traverse to reach a source cell.

5. *CostAllocation to sourcegrid* shows the allocation of cells to each source cell.

6. Click the Spatial Analyst downward arrow, point to distance, and select Shortest Path. In the Shortest Path dialog, specify *pathgrid* for Path to, *costgrid* for the Cost distance raster, *CostDirection to sourcegrid* for the Cost direction raster, and specify the path type to be for each cell. Click OK and dismiss the Shortest Path dialog. The output shapefile shows the path from each cell in *pathgrid* to its closest source.

References

Cliff, A. D., and J. K. Ord. 1973. *Spatial Autocorrelation*. New York: Methuen.

Congalton, R. G. 1991. A Review of Assessing the Accuracy of Classification of Remotely Sensed Data. *Photogrammetric Engineering & Remote Sensing* 37: 35–46.

Crow, T. R., G. E. Host, and D. J. Mladenoff. 1999. Ownership and Ecosystem as Sources of Spatial Heterogeneity in a Forested Landscape, Wisconsin, USA. *Landscape Ecology* 14: 449–63.

Forman, R. T. T., and M. Godron. 1986. *Landscape Ecology*. New York: Wiley.

Getis, A., and J. K. Ord. 1996. Local Spatial Statistics: An Overview. In P. Longley and M. Batty, eds., *Spatial Analysis: Modelling in a GIS Environment,* pp. 261–77. Cambridge, England: GeoInformation International.

Hepner, G. H., and M. V. Finco. 1995. Modeling Dense Gas Contaminant Pathways over Complex Terrain Using a Geographic Information System. *Journal of Hazardous Materials* 42: 187–99.

Herr, A. M., and L. P. Queen. 1993. Crane Habitat Evaluation Using GIS and Remote Sensing. *Photogrammetric Engineering & Remote Sensing* 59: 1531–38.

Heuvelink, G. B. M. 1998. *Error Propagation in Environmental Modelling with GIS*. London: Taylor and Francis.

Lam, N. S., M. Fan, and K. Liu. 1996. Spatial-Temporal Spread of the AIDS Epidemic, 1982–1990: A Correlogram Analysis of Four Regions of the United States. *Geographical Analysis* 28: 93–107.

Lee, J., and D. W. S. Wong. 2001. *Statistical Analysis with ArcView GIS*. New York: Wiley.

Legendre, P. 1993. Spatial Autocorrelation: Trouble or New Paradigm? *Ecology* 74: 1659–73.

Lillesand, T. M., and R. W. Kiefer. 2000. *Remote Sensing and Image Interpretation*, 4th ed. New York: Wiley.

McGarigal, K., and B. J. Marks. 1994. *Fragstats: Spatial Pattern Analysis Program for Quantifying Landscape Structure*. Forest Science Department, Oregon State University.

Mladenoff, D. J., T. A. Sickley, R. G. Haight, and A. P. Wydeven. 1995. A Regional Landscape Analysis and Prediction of Favorable Gray Wolf Habitat in the Northern Great Lakes Regions. *Conservation Biology* 9: 279–94.

Newcomer, J. A., and J. Szajgin. 1984. Accumulation of Thematic Map Errors in Digital Overlay Analysis. *The American Cartographer* 11: 58–62.

Renard, K. G., G. R. Foster, G. A. Weesies, D. K. McCool, and D. C. Yoder (coordinators). 1997. Predicting Soil Erosion by Water: A Guide to Conservation Planning with the Revised Universal Soil Loss Equation (RUSLE). *Agricultural Handbook 703*. Washington, DC: U.S. Department of Agriculture.

Rogerson, P. A. 2001. *Statistical Methods for Geography*. London, England: Sage.

Tomlin, C. D. 1990. *Geographic Information Systems and Cartographic Modeling*. Englewood Cliffs, NJ: Prentice Hall.

Veregin, H. 1995. Developing and Testing of an Error Propagation Model for GIS Overlay Operations. *International Journal of Geographical Information Systems* 9: 595–619.

Wischmeier, W. H., and D. D. Smith. 1978. Predicting Rainfall Erosion Losses: A Guide to Conservation Planning. *Agricultural Handbook 537*. Washington, DC: U.S. Department of Agriculture.

TERRAIN MAPPING AND ANALYSIS

12.1 INTRODUCTION

The terrain with its undulating, continuous land surface is a familiar phenomenon to GIS users. The land surface has been the object of mapping and analysis for hundreds of years. Mapmakers have devised various techniques for terrain mapping such as contouring, hill shading, hypsometric tinting, and perspective views. Geomorphologists have developed measures of the land surface including slope, aspect, and surface curvature. And more recently terrain mapping and analysis has also included viewshed analysis and watershed analysis.

Terrain mapping and analysis techniques are no longer tools for specialists. GIS has made it relatively easy to incorporate them into a variety of applications. Slope and aspect play a regular role in hydrologic modeling, snow cover evaluation, soil mapping, landslide delineation, soil erosion, and predictive mapping of vegetation communities (Lane et al. 1998; Wilson and Gallant 2000). Hill shading and perspective views are common features in presentation and reports.

Most GIS packages treat elevation data (z values) as attribute data at point or cell locations rather than as an additional coordinate to x- and y-coordinates as in a true 3-D model. In raster format, the z values correspond to cell values. In vector format, the z values are stored in a field of a feature attribute table. Terrain mapping and analysis can use raster data, vector data, or both as the input. This is perhaps why GIS vendors typically group terrain mapping and analysis functions into a module or an extension, separate from the basic GIS tools (Box 12.1).

This chapter is organized into the following four sections. Section 12.2 covers two common data sources for terrain mapping and analysis: DEM (digital elevation model) and TIN (triangulated irregular network). Section 12.3 describes different methods for terrain mapping. Section 12.4 discusses terrain analysis including slope, aspect, surface curvature, viewshed analysis, and watershed analysis. Section 12.5 compares DEM and TIN for terrain mapping and analysis.

12.2 DATA FOR TERRAIN MAPPING AND ANALYSIS

12.2.1 DEM

A DEM represents a regular array of elevation points. Most GIS users in the United States use DEMs from the U.S. Geological Survey (USGS). Alternative sources for DEMs come from satellite images, radar data, and LIDAR (light detection and ranging) data (Chapter 7). Regardless of its origin, a point-based DEM must be converted to software-specific raster data (e.g., ESRI grid) before it can be used for terrain mapping and analysis. This conversion simply places each elevation point in a DEM at the center of a cell in an elevation grid. DEM and elevation grid are often used interchangeably in the literature as well as in this chapter.

The quality of a DEM can influence the accuracy of terrain measures including slope and aspect. The USGS classifies the quality of 7.5-minute DEMs into three levels, with level 1 having the poorest quality. Using known sources such as benchmarks (vertical control points) and spot elevations as test points, the USGS calculates the root mean square error (RMSE) of the DEM data. Level-1 accuracy has an RMSE target of 7 meters and a maximum RMSE of 15 meters. Level-2 accuracy has a maximum RMSE of one-half the contour interval. And level-3 accuracy has a maximum RMSE of one-third contour interval—not to exceed 7 meters. Most USGS DEMs in use are either level 1 or level 2.

Errors in USGS DEMs may be classified as either global or relative (Carter 1989). Global errors are systematic errors caused by displacements of the DEM, as evidenced by mismatching elevations along the boundaries of adjacent DEMs. One can usually correct global errors by applying geometric transformations, including translation, rotation, and scaling (Chapter 7). Relative errors are local but significant errors relative to the neighboring elevations. Examples of relative errors are artificial peaks and blocks of elevations extending above the surrounding lands, especially along

---

 Box 12.1 | **A Survey of Terrain Analysis Functions among GIS Packages**

Terrain mapping and analysis is an important component of GIS packages, especially those that are raster-based. The following lists some of the terrain analysis functions covered in this chapter and the GIS packages that carry them:

Slope, aspect: ArcGIS, IDRISI, SPANS, GRASS, ILWIS, PAMAP

Surface curvature: ArcGIS, IDRISI, GRASS

Viewshed analysis: ArcGIS, GRASS, IDRISI, SPANS, PAMAP, MFworks

Watershed analysis: ArcGIS, GRASS, IDRISI, PAMAP, MFworks

---

ridgelines. Relative errors can be corrected only by editing the DEM data.

Although not yet widely used for terrain mapping and analysis, LIDAR data have the advantage over USGS DEMs in providing high-quality DEMs with a spatial resolution of 0.5 to 2 meters and a vertical accuracy of about 15 centimeters (Flood 2001). LIDAR data are therefore ideal for studies that require detailed topographic data such as floodplain mapping, telecommunications, transportation, and other applications (Hill et al. 2000).

12.2.2 TIN

A TIN approximates the land surface with a series of nonoverlapping triangles. Elevation values (z values) along with x-, y-coordinates are stored at nodes that make up the triangles. In contrast to a DEM, a TIN is based on an irregular distribution of elevation points.

GIS users typically compile TINs using DEMs as the primary data source in a process sometimes referred to as conversion of DEM to TIN. But a TIN can use other data sources. Additional point data may include surveyed elevation points, GPS (global positioning system) data, and LIDAR data. Line data may include contour lines and breaklines. **Breaklines** are line features that represent changes of the land surface such as streams, shorelines, ridges, and roads. And area data may include lakes and reservoirs.

Because triangles in a TIN can vary in size by the complexity of topography, not every point in a DEM needs to be used to create a TIN. Instead, the process is to select points that are more important in representing the terrain. Several algorithms for selecting significant points from a DEM have been proposed in GIS (Lee 1991; Kumler 1994). Here we examine two algorithms: **VIP** (very important points) used by ArcInfo Workstation, and **maximum z-tolerance** used by 3D Analyst and ArcInfo Workstation.

To select points from a DEM, VIP first converts a DEM to a grid and then evaluates the importance of each point (i.e., each cell in the elevation grid) by measuring how well its value can be estimated from the neighboring point values (Chen and Guevara 1987). The conversion of a point-based DEM to an elevation grid is necessary for the selection process, which is essentially a neighborhood operation (Chapter 11).

Figure 12.1a shows a 3 × 3 moving window in an elevation grid. To assess the importance of P, VIP first estimates the elevation at P by using four pairs of its neighbors: up and down (B–F), left and right (H–D), upper right and lower left (C–G), and upper left and lower right (A–E). Figure 12.1b shows a vertical profile for the case with C–G. P_e is the estimated elevation at P from the elevations at G and C (Z_G and Z_C), and P_h is the actual elevation at P. The difference between P_h and P_e, d, therefore represents the elevation offset at P. But

VIP uses s, which measures the distance of the perpendicular line from P_h to line $Z_G Z_C$, for the offset measure at P. The offset s is a better measure than d in flat or steep slope areas according to Chen and Guevara (1987). After calculating four offset values, one for each pair of neighbors around P, VIP uses their average as an indicator of the significance of P.

Using the above procedure, VIP computes the significance value for each cell in an elevation grid. A frequency distribution of the significance values from a grid usually looks like a normal distribution, with more cells having lower significance values (Chen and Guevara 1987). The selection of points by VIP can be based on either a desired number or a specified significance level.

The maximum z-tolerance algorithm selects points from an elevation grid to construct a TIN such that, for every point in the elevation grid, the difference between the original elevation and the estimated elevation from the TIN is within the specified maximum z-tolerance. The algorithm uses an iterative process. The process begins by constructing a candidate TIN. Then, for each triangle in the TIN, the algorithm computes the elevation difference from each point in the grid to the enclosing triangular facet. The algorithm determines the point with the largest difference. If the difference is greater than a specified z-tolerance, the algorithm flags the point for addition to the TIN. After every triangle in the current TIN is checked, a new triangulation is recomputed with the selected additional points. This process continues until all points in the grid are within the specified maximum z-tolerance.

Elevation points selected by the VIP algorithm or the maximum z-tolerance algorithm, plus additional elevation points from contour lines, survey data, GPS data, or LIDAR data are connected to form a series of nonoverlapping triangles in an initial TIN. A common algorithm for connecting points is called the **Delaunay triangulation** (Tsai 1993). Triangles formed by the Delaunay triangulation have the following characteristics: all nodes (points) are connected to the nearest neighbors to

(a)

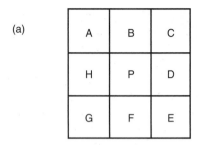

(b)

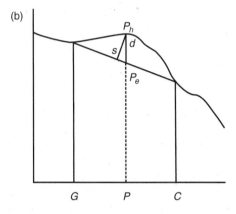

Figure 12.1

VIP evaluates the significance of an elevation point by measuring how well its value can be estimated from the neighboring point values. Figure 12.1b shows the case of using elevations at G and C to estimate the elevation at P. P_e is the estimated elevation, P_h is the actual elevation, and d represents the offset between P_e and P_h. Rather than using d as the measure of significance, VIP uses s.

form triangles; and triangles are as equiangular, or compact, as possible.

Depending on the software design, breaklines may be included in the initial TIN or used to modify the initial TIN. Breaklines provide the physical structure in the form of triangle edges to show changes of the land surface (Figure 12.2). One often notices that triangles along the border of a TIN are stretched and elongated, thus distorting the landform features derived from those triangles. The cause of this irregularity stems from the sudden drop of elevation along the edge. One way to

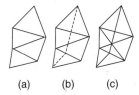

Figure 12.2
The dashed line in (b) represents a breakline, which subdivides the triangles in (a) into a series of smaller triangles in (c).

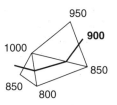

Figure 12.3
The contour line of 900 connects points that are interpolated to have the value of 900 along the triangle edges.

solve this problem is to include elevation points beyond the border of a study area for processing and then to clip the study area from the larger coverage. The process of building a TIN is certainly more complex than preparing an elevation grid from a DEM.

The discussion has so far dealt with the conversion of a DEM to a TIN. One can also convert a TIN to a DEM. The process requires each elevation point of the DEM to be estimated (interpolated) from its neighboring nodes that make up the TIN. Each of these nodes has its x-, y-coordinates as well as a z (elevation) value. The default method used by ArcGIS to convert a TIN to a DEM is local first-order polynomial interpolation. Local polynomial interpolation is one of the spatial interpolation methods covered in Chapter 13.

12.3 TERRAIN MAPPING

12.3.1 Contouring

Contouring is the most common method for terrain mapping. **Contour lines** connect points of equal elevation, and the **contour interval** represents the vertical distance between contour lines. The arrangement and pattern of contour lines reflect the topography. For example, contour lines are closely spaced in steep terrain and are curved in the upstream direction along a stream. With some training and experience, we can visualize the terrain by simply studying contour lines. Contour lines can also be used for manually measuring

slope and aspect, although the practice is becoming rare with the use of GIS.

Automated contouring follows two basic steps: (1) detecting a contour line intersecting a grid cell or a triangle, and (2) drawing the contour line through the grid cell or triangle (Jones et al. 1986). A TIN is a good example for illustrating automated contouring because it has elevation readings for all nodes from triangulation. Given a contour line, every triangle edge is examined to determine if the line should pass through the edge. If it does, linear interpolation, which assumes a constant gradient between the end nodes of the edge, can determine the contour line's position along the edge. After all the positions are calculated, they are connected to form the contour line (Figure 12.3). The initial contour line consists of straight-line segments, which can be smoothed by fitting a mathematical function such as splines to points that make up the line. Another way of producing smooth contour lines is to divide a triangle into a series of smaller triangles and to use these smaller triangles for contouring.

Contour lines do not intersect one another or stop in the middle of a map, although they can be close together in cases of cliffs or form closed lines in cases of depressions or isolated hills. Contour maps created from a GIS sometimes contain irregularities or even errors (Figure 12.4). Irregularities are often caused by use of large cells, whereas errors are caused by use of incorrect parameter values in the smoothing algorithm (Clarke 1995).

12.3.2 Vertical Profiling

A **vertical profile** shows changes in elevation along a line, such as a hiking trail, a road, or a stream (Figure 12.5). The manual method usually involves the following steps:

1. Draw a profile line on a contour map.
2. Mark each intersection between a contour and the profile line and record its elevation.
3. Raise each intersection point to a height proportional to its elevation.
4. Plot the vertical profile by connecting the elevated points.

Automated profiling follows the same procedure but substitutes the contour map with an elevation grid or a TIN.

Figure 12.4
Automated contouring may produce contour lines that are highly irregular.

12.3.3 Hill Shading

Also known as shaded relief or simply shading, **hill shading** simulates how the terrain looks with the interaction between sunlight and surface features (Figure 12.6). A mountain slope directly facing incoming light will be very bright; a slope opposite to the light will be dark. Hill shading helps viewers recognize the shape of landform features. Hill shading can be mapped alone, such as the well-known example of Thelin and Pike's (1991) digital shaded-relief map of the United States(**http://www.usgs.gov/reports/misc/Misc._ Investigations_Series_Maps_(I_Series)/I_2206/ usa_dem.gif**). But often hill shading is the background for terrain or thematic mapping.

Hill shading used to be produced by talented artists. But the computer can now generate high-quality shaded maps. Four factors control the visual effect of hill shading. The sun's azimuth is the direction of the incoming light, ranging from 0° (due north) to 360° in a clockwise direction. Typically, the default for the sun's azimuth is 315°. With the light source located above the upper-left corner of the hill-shaded map, the shadows appear to fall toward the viewer, thus avoiding the pseudoscopic effect (Box 12.2). The sun's altitude is the angle of the incoming light measured above the horizon between 0° and 90°. The other two factors are the surface's slope and aspect: slope ranges from 0° to 90° and aspect from 0° to 360° (Section 12.4.1). Using the above four factors, the follow-

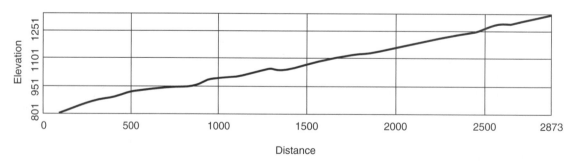

Figure 12.5
A vertical profile showing changes in elevation along a stream tributary. The profile has a vertical exaggeration factor of 1.0 (i.e., no vertical exaggeration).

ing equation can compute the relative radiance value for every cell in an elevation grid or for every triangle in a TIN (Eyton 1991):

(12.1)

$$R_f = \cos(A_f - A_s)\sin H_f \cos H_s + \cos H_f \sin H_s$$

Figure 12.6
An example of hill shading, with the sun's azimuth at 315° (NW) and the sun's altitude at 45°.

where R_f is the relative radiance value of a facet (a grid cell or a triangle), A_f is the facet's aspect, A_s is the sun's azimuth, H_f is the facet's slope, and H_s is the sun's altitude. R_f ranges in value from 0 to 1. If multiplied by the constant 255, R_f can be converted to the illumination value (I_f) for display. An I_f value of 255 would result in white and an I_f value of 0 would result in black on a shaded map. GIS packages such as ArcGIS use I_f for hill shading (Box 12.3).

Relative radiance is similar to another measure called the incidence value (Franklin 1987):

(12.2)

$$\cos(H_f) + \cos(A_f - A_s)\sin(H_f)\cot(H_s)$$

The notations in Eq. (12.2) are the same as Eq. (12.1). One can also derive the incidence value by multiplying the relative radiance value by $\sin(H_s)$. Besides producing hill shading, both relative radiance and incidence can be used in image processing as variables representing the interaction between the incoming radiation and local topography.

12.3.4 Hypsometric Tinting

Hypsometry depicts the distribution of the Earth's mass with elevation. **Hypsometric tinting,** also known as layer tinting, applies color symbols to different elevation zones. The use of well-chosen color symbols can help viewers see the progression in elevation, especially on a small-scale map. One can also use hypsometric tinting to highlight a particular elevation zone, which may be important, for instance, in a wildlife habitat study.

***Box* 12.2   The Pseudoscopic Effect**

A hill-shaded map looks right when the shadows appear to fall toward the viewer. If the shadows appear to fall away from the viewer, such as using 135° as the sun's azimuth, the hills on the map look like depressions and the depressions look like hills in an optical illusion called the pseudoscopic effect (Campbell 1984). Of course, the sun's azimuth at 315° is totally unrealistic for most parts of the Earth's surface.

Box 12.3 **A Worked Example of Computing Relative Radiance**

Suppose a cell in an elevation grid has a slope value of 10° and an aspect value of 297° (W to NW), the sun's altitude is 65°, and the sun's azimuth is 315° (NW). The relative radiance value of the cell can be computed by

$$R_f = \cos(297-315)\sin(10)\cos(65) + \cos(10)\sin(65) = 0.9623$$

The cell will appear bright with an R_f value of 0.9623.

If the sun's altitude is lowered to 25° and the sun's azimuth remains at 315°, then the cell's relative radiance value becomes

$$R_f = \cos(297-315)\sin(10)\cos(25) + \cos(10)\sin(25) = 0.5658$$

The cell will appear in medium gray with an R_f value of 0.5658.

12.3.5 Perspective View

Perspective views are 3-D views of the terrain: the terrain has the same appearance as viewed with an angle from an airplane (Figure 12.7). Four parameters can control the appearance of a 3-D view (Figure 12.8):

- **Viewing azimuth** is the direction from the observer to the surface, ranging from 0° to 360° in a clockwise direction.
- **Viewing angle** is the angle measured from the horizon to the altitude of the observer. A viewing angle is always between 0° and 90°. An angle of 90° means viewing the surface from directly above, whereas an angle of 0° means viewing the surface directly ahead. Therefore, the 3-D effect reaches its maximum as the angle approaches 0° and its minimum as the angle approaches 90°.
- **Viewing distance** is the distance between the viewer and the surface. Adjustment of the viewing distance allows the surface to be viewed up close or from a distance.
- **z-scale** is the ratio between the vertical scale and the horizontal scale. Also called the *vertical exaggeration factor,* z-scale is useful for highlighting minor landform features.

Because of its visual appeal, 3-D perspective view is a display tool in many GIS packages. The 3D Analyst extension to ArcGIS, for example, pro-

vides the graphical interfaces for manipulating the viewing parameters. GIS users can rotate the surface, navigate the surface, or take a close-up view of the surface. To make perspective views even more realistic, one can superimpose these views with thematic layers such as land cover, vegetation, and roads in a process called **3-D draping** (Figure 12.9). One can also add clouds and change the color of the sky. 3-D perspective view is an exciting method for portraying the land surface.

12.4 TERRAIN ANALYSIS

12.4.1 Slope and Aspect

Slope measures the rate of change of elevation at a surface location, and **aspect** is the directional measure of slope. If we define the elevation (z) of a point on the land surface as a function of the point's position (x and y), then we can define slope (S) at the point as a function of the first-order derivatives of the surface in the x and y directions:

(12.3)

$$S = \sqrt{(\partial_z/\partial_x)^2 + (\partial_z/\partial_y)^2}$$

And we can define the slope's directional angle as

(12.4)

$$A = \arctan\left((\partial_z/\partial_y)/(\partial_z/\partial_x)\right)$$

Figure 12.7
A 3-D perspective view.

Slope may be expressed as percent slope or degree slope. Percent slope is 100 times the ratio of rise (vertical distance) over run (horizontal distance), whereas degree slope is the arc tangent of the ratio of rise over run (Figure 12.10). Aspect (*A*) is a directional measure in degrees. Aspect starts with 0° at the north, moves clockwise, and ends with 360° also at the north.

Aspect is a circular measure. An aspect of 10° is closer to 360° than to 30°. GIS users often have to manipulate aspect measures before using them in data analysis. A common method is to classify aspects into the four principal directions (north, east, south, and west) or eight principal directions (north, northeast, east, southeast, south, southwest, west, and northwest) and to treat aspects as categorical data (Figure 12.11). Rather than converting aspects to categorical data, Chang and Li (2000) have proposed a method for capturing the principal direction while retaining the numeric measure. For

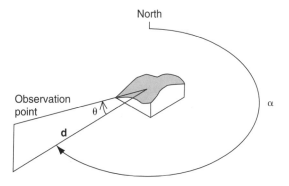

Figure 12.8
The diagram shows three parameters controlling the appearance of a 3-D view. α is the viewing azimuth, measured clockwise from the north. θ is the viewing angle, measured from the horizon. *d* is the viewing distance, measured between the observation point and the 3-D surface.

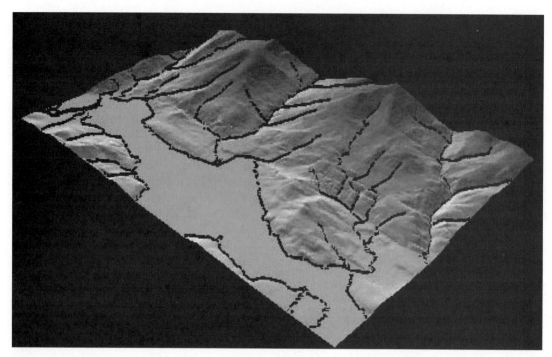

Figure 12.9
An example of 3-D draping. In this case, streams and shorelines are draped on a 3-D surface.

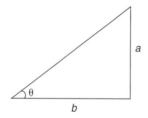

Figure 12.10
Percent slope in the diagram is 100 x (*a/b*). *a* is the vertical distance or rise, and *b* is the horizontal distance or run. Degree slope can be calculated by arctan (*a/b*).

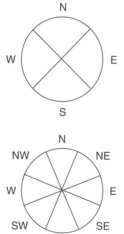

Figure 12.11
Aspect is a directional measure in degrees. Aspect measures are often grouped into the four principal directions (top) or eight principal directions (bottom).

instance, to capture the N-S principal direction, one can set 0° at north, 180° at south, and 90° at both west and east (Figure 12.12). A common method for converting aspect measures to linear measures is to use their sine or cosine values, which range from −1 to 1 (Zar 1984).

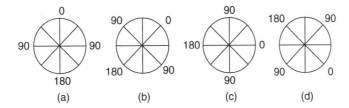

Figure 12.12
Transformation methods to capture the N–S direction (a), the NE–SW direction (b), the E–W direction (c), and the NW–SE direction (d).

As basic elements for analyzing and visualizing landform characteristics, slope and aspect are important in studies of watershed units, landscape units, and morphometric measures (Moore et al. 1991). When used with other variables, slope and aspect can assist in solving problems in forest inventory estimates, soil erosion, wildlife habitat suitability, site analysis, and many other fields.

12.4.1.1 Computing Algorithms for Slope and Aspect Using Grid

When an elevation grid is used as the data source, slope and aspect are computed for each cell in the grid. One can measure the slope and aspect for a cell by the quantity and direction of tilt of the cell's normal vector—a directed line perpendicular to the cell (Figure 12.13). Given a normal vector (n_x, n_y, n_z), the formula for computing the cell's slope is

(12.5)

$$\sqrt{n_x^2 + n_y^2} / n_z$$

And the formula for computing the cell's aspect is

(12.6)

$$\arctan (n_y / n_x)$$

Different approximation methods have been proposed for estimating slope and aspect. Here we will examine three common methods. All three methods use a 3 × 3 moving window to estimate the slope and aspect of the center cell, but they differ in the number of neighboring cells used in the estimation and the weight applying to each cell.

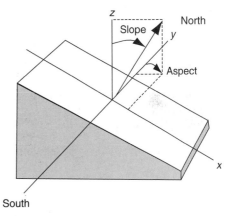

Figure 12.13
The normal vector to the cell is the directed line perpendicular to the cell. The quantity and direction of tilt of the normal vector determine the slope and aspect of the cell. (Redrawn from Hodgson, 1998, *CaGIS* 25, (3): pp. 173–185; reprinted with the permission of the American Congress on Surveying and Mapping.)

The first method, which is attributed to Fleming and Hoffer (1979) and Ritter (1987), uses the four immediate neighbors of the center cell. The slope (S) at C_0 in Figure 12.14 can be computed by

(12.7)

$$S = \sqrt{(e_1 - e_3)^2 + (e_4 - e_2)^2} / 2d$$

where e_i are the neighboring cell values, and d is the cell size. The n_x component of the normal vector to C_0 is $(e_1 - e_3)$, or the elevation difference in the x dimension. The n_y component is $(e_4 - e_2)$, or the elevation difference in the y dimension. To compute the percent slope at C_0, one can multiply S by 100.

S's directional angle D can be computed by

(12.8)

$$D = \arctan((e_4 - e_2) / (e_1 - e_3))$$

D is measured in radians and is with respect to the x-axis. Aspect, on the other hand, is measured in degrees and from a north base of 0°. Box 12.4 shows an algorithm for converting D to aspect (Ritter 1987; Hodgson 1998).

The second method for computing slope and aspect is called Horn's algorithm (1981), an algorithm used by ArcGIS. Horn's algorithm uses eight neighboring cells and applies a weight of 2 to the four immediate neighbors and a weight of 1 to the four corner cells. Horn's algorithm computes slope at C_0 in Figure 12.15 by

Figure 12.14
Ritter's algorithm for computing slope and aspect at C_0 uses the four immediate neighbors of C_0.

(12.9)

$$S = \sqrt{[(e_1 + 2e_4 + e_6) - (e_3 + 2e_5 + e_8)]^2 + [(e_6 + 2e_7 + e_8) - (e_1 + 2e_2 + e_3)]^2} / 8d$$

And the D value at C_0 is computed by

(12.10)

$$D = \arctan([(e_6 + 2e_7 + e_8) - (e_1 + 2e_2 + e_3)] / [(e_1 + 2e_4 + e_6) - (e_3 + 2e_5 + e_8)])$$

D can be converted to aspect by using the same algorithm for the first method except that $n_x = (e_1 + 2e_4 + e_6)$ and $n_y = (e_3 + 2e_5 + e_8)$ (Box 12.5).

The third method called Sharpnack and Akin's algorithm (1969) also uses eight neighboring cells

Figure 12.15
Horn's algorithm for computing slope and aspect at C_0 uses the eight neighboring cells of C_0. The algorithm also applies a weight of 2 to e_2, e_4, e_5, and e_7, and a weight of 1 to e_1, e_6, e_3, and e_8.

Box 12.4   Conversion of D to Aspect

The notations used here are the same as in Eq. (12.5) to (12.8). In the following, the text after an apostrophe is an explanatory note.

If $S <> 0$ then
  $T = D \times 57.296$
  If $n_x = 0$
    If $n_y < 0$ then
      Aspect $= 180$
    Else

      Aspect $= 360$
  ElseIf $n_x > 0$ then
    Aspect $= 90 - T$
  Else '$n_x < 0$
    Aspect $= 270 - T$
Else '$S = 0$
  Aspect $= -1$ 'undefined aspect for flat surface
End If

but applies the same weight to every cell. The formula for computing S is

(12.11)

$$S = \sqrt{\frac{[(e_1 + e_4 + e_6) - (e_3 + e_5 + e_8)]^2 + [(e_6 + e_7 + e_8) - (e_1 + e_2 + e_3)]^2}{6d}}$$

And the formula for computing D is

(12.12)

$$D = \arctan([(e_6 + e_7 + e_8) - (e_1 + e_2 + e_3)] / [(e_1 + e_4 + e_6) - (e_3 + e_5 + e_8)])$$

12.4.1.2 Computing Algorithms for Slope and Aspect Using TIN

The algorithms for computing slope and aspect for a triangle in a TIN also use the bidirectional normal vector, the vector perpendicular to the triangular surface. Suppose a triangle is made of the following three nodes: A (x_1, y_1, z_1), B (x_2, y_2, z_2), and C (x_3, y_3, z_3) (Figure 12.16). The normal vector is a cross product of vector AB, $[(x_2 - x_1), (y_2 - y_1), (z_2 - z_1)]$, and vector AC, $[(x_3 - x_1), (y_3 -$

$y_1), (z_3 - z_1)]$. And the three components of the normal vector are:

(12.13)

$$n_x = (y_2 - y_1)(z_3 - z_1) - (y_3 - y_1)(z_2 - z_1)$$
$$n_y = (z_2 - z_1)(x_3 - x_1) - (z_3 - z_1)(x_2 - x_1)$$
$$n_z = (x_2 - x_1)(y_3 - y_1) - (x_3 - x_1)(y_2 - y_1)$$

The S and D values of the triangle can be derived from Eq. (12.5) and (12.6), and the D value can then be converted to the aspect measured in degrees and from a north base of $0°$ (Box 12.6).

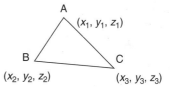

Figure 12.16
The algorithm for computing slope and aspect of a triangle in a TIN uses the x, y, and z values at the three nodes of the triangle.

 Box 12.5 **A Worked Example of Computing Slope and Aspect Using Grid**

The following diagram shows a 3×3 window of an elevation grid. Elevation readings are measured in meters, and the cell size is 30 meters.

1006	1012	1017
1010	1015	1019
1012	1017	1020

This example computes the slope and aspect of the center cell using Horn's algorithm:

$n_x = (1006 + 2 \times 1010 + 1012) - (1017 + 2 \times 1019 + 1020) = -37$

$n_y = (1012 + 2 \times 1017 + 1020) - (1006 + 2 \times 1012 + 1017) = 19$

$S = \sqrt{(-37)^2 + (19)^2} / (8 \times 30) = 0.1733$

$S_p = 100 \times 0.1733 = 17.33$

$D = \arctan(n_y/n_x) = \arctan(19/-37) = -0.4744$

$T = -0.4744 \times 57.296 = -27.181$

Because $S <> 0$ and $n_x < 0$.

Aspect $= 270 - (-27.181) = 297.181$.

For comparison, S_p has a value of 17.16 using the Fleming and Hoffer's algorithm and a value of 17.39 using the Sharpnack and Akin's algorithm. Aspect has a value of 299.06 using the Fleming and Hoffer's algorithm and a value of 296.56 using the Sharpnack and Akin's algorithm.

Box 12.6 **A Worked Example of Computing Slope and Aspect Using TIN**

Suppose a triangle in a TIN is made of the following nodes with their x, y, and z values measured in meters.

Node 1: $x_1 = 532260$, $y_1 = 5216909$, $z_1 = 952$
Node 2: $x_2 = 531754$, $y_2 = 5216390$, $z_2 = 869$
Node 3: $x_3 = 532260$, $y_3 = 5216309$, $z_3 = 938$

The following shows the steps for computing the slope and aspect of the triangle.

$n_x = (5216390 - 5216909)(938 - 952) - (5216309 - 5216909)(869 - 952) = -42534$

$n_y = (869 - 952)(532260 - 532260) - (938 - 952)(531754 - 532260) = -7084$

$n_z = (531754 - 532260)(5216309 - 5216909) - (532260 - 532260)(5216390 - 5216909) = 303600$

$S_p = 100 \times [\sqrt{(-42534)^2 + (-7084)^2} / 303600]$
$= 14.20$

$D = \arctan(-7084 / -42534) = 0.165$

$T = 0.165 \times 57.296 = 9.454$

Because $S <> 0$ and $n_x < 0$, aspect $= 270 - T = 260.546$.

12.4.1.3 Factors Influencing Slope and Aspect Measures

The accuracy of slope and aspect measures can influence the performance of models that use slope and aspect as the input (Srinivasan and Engel 1991). Therefore, it is important to examine several factors that can influence slope and aspect measures.

Slope and aspect measures may vary by the computing algorithm. Using a digitized contour map of moderate topography in Australia, Skidmore (1989) compared six algorithms, including the three described in the previous section. He reported that Horn's algorithm and Sharpnack and Akin's algorithm, both involving eight neighboring cells, were among the best for estimating both slope and aspect. Comparing five algorithms using a synthetic surface and a study area in Tennessee, Hodgson (1998) found Ritter's algorithm involving four immediate neighbors to be consistently more accurate in estimating slope than the methods using eight neighboring cells. Jones (1998) compared eight algorithms using a synthetic surface and a DEM in Scotland and reported that Ritter's algorithm was the best in estimating slope

and aspect, followed by Horn's algorithm and Sharpnack and Akin's algorithm.

The resolution of DEM probably has a greater degree of influence on the accuracy of slope and aspect measures than the computing algorithm. Isaacson and Ripple (1990) reported greater amounts of detail in the slope and aspect maps created from the 7.5-minute DEM than from the 1-degree DEM. The 7.5-minute DEM has a sampling interval of 30 meters, whereas the 1-degree DEM has a sampling interval of about 90 meters. Chang and Tsai (1991) found that the accuracy of the estimated slope and aspect decreased with a decreasing DEM resolution from 20 to 80 meters. Gao (1998) reported that the reliability of both slope and aspect measures decreased with an increasing sampling interval of the DEM from 20 to 60 meters, and the inverse relationship between the reliability of slope measure and the DEM sampling interval was statistically significant.

The quality of DEM can also influence slope and aspect measures. Bolstad and Stowe (1994) compared a USGS 7.5-minute DEM and a DEM produced from a SPOT panchromatic stereopair for the same study area. They reported that slope and

Box 12.7 | **A Worked Example of Computing Surface Curvature**

1017	1010	1017
1012	1006	1019
1015	1012	1020

e_1	e_2	e_3
e_4	e_0	e_5
e_6	e_7	e_8

The diagram above represents a 3×3 window of an elevation grid, with a cell size of 30 meters. This example shows how to compute the profile curvature, plan curvature, and surface curvature at the center cell. The first step is to estimate the coefficients D–H of the quadratic polynomial equation that fits the 3×3 window:

$$D = [(e_4 + e_5) / 2 - e_0] / L^2$$
$$E = [(e_2 + e_7) / 2 - e_0] / L^2$$
$$F = (-e_1 + e_3 + e_6 - e_8) / 4L^2$$
$$G = (-e_4 + e_5) / 2L$$
$$H = (e_2 - e_7) / 2L$$

where e_0 to e_8 are elevation values within the 3×3 window according to the diagram below, and L is the cell size.

Profile curvature $= -2 [(DG^2 + EH^2 + FGH) / (G^2 + H^2)] = -0.0211$

Plan curvature $= 2 [(DH^2 + EG^2 - FGH) / (G^2 + H^2)] = 0.0111$

Curvature $= -2 (D + E) = -0.0322$

All three measures are based on 1/100 (z units). The negative curvature value means the surface at the center cell is upwardly concave. The elevation grid above shows the center cell is like a shallow basin surrounded by higher elevations in the neighboring cells.

aspect errors for the SPOT DEM were statistically significant, whereas slope and aspect errors for the USGS DEM were not significantly different from zero.

Finally, local topography can be a factor in estimating slope and aspect. Chang and Tsai (1991) reported that errors in slope estimates were greater in areas of higher slopes, but errors in aspect estimates were greater in areas of lower relief. Carter (1992) attributed errors in the calculation of aspect, especially in areas of low slope, to the problem of data precision such as the rounding of elevations to the nearest whole number. Bolstad and Stowe (1994) found larger slope errors on steeper slopes and speculated that the correlation was likely due in part to difficulties in stereocorrelation in forested terrain. Florinsky (1998) reported higher aspect and slope errors within flat areas because of the data precision problem.

12.4.2 Surface Curvature

GIS applications in hydrological studies often require computation of surface curvature to determine if the surface at a cell location is upwardly convex or concave (Gallant and Wilson 2000). A common algorithm for computing surface curvature is to fit a 3×3 window with a quadratic polynomial equation (Zevenbergen and Thorne 1987; Moore et al. 1991):

(12.14)

$$z = Ax^2y^2 + Bx^2y + Cxy^2 + Dx^2 + Ey^2 + Fxy + Gx + Hy + I$$

The coefficients A–I can be estimated by using the elevation values in the 3×3 window and the grid cell size (Box 12.7). Three curvature measures can then be computed from the coefficients:

(12.15)

$$\text{Profile curvature} = -2\,[(DG^2 + EH^2 + FGH)\,/\,(G^2 + H^2)]$$

(12.16)

$$\text{Plan curvature} = 2\,[(DH^2 + EG^2 - FGH)\,/\,(G^2 + H^2)]$$

(12.17)

$$\text{Curvature} = -2\,(D + E)$$

Profile curvature is estimated along the direction of maximum slope. Plan curvature is estimated across the direction of maximum slope. And curvature measures the difference between the two: profile curvature − plan curvature. A positive curvature value at a cell means that the surface is upwardly convex at the cell. A negative curvature value means that the surface is upwardly concave. And a 0 value means that the surface is flat.

12.4.3 Viewshed Analysis

A **viewshed** refers to areas of the land surface that are visible from an observation point or observation points. The basis for viewshed analysis is the line-of-sight operation, the operation of determining whether a given target is visible from an observation point (Figure 12.17). Viewshed analysis expands the operation to cover every possible cell in an elevation grid or every possible facet in a TIN as the target. For more than one observation point, the operation is repeated for each point. Depending on the size of the grid or TIN and the number of observation points, viewshed analysis can be computationally intensive.

The output of viewshed analysis is a binary map showing visible and not visible areas. Several parameters can influence the result of a viewshed analysis. Perhaps the most important parameter is the observation point, which should be located in a high elevation with open views to gain maximum visibility. The elevation of the observation point is typically interpolated from the input DEM. Therefore, raster data query by cell value or graphics can

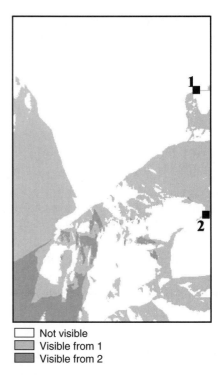

Not visible
Visible from 1
Visible from 2

Figure 12.17a
Viewshed analysis divides the study area into (1) not visible from observation points, (2) visible from observation point 1, and (3) visible from observation point 2.

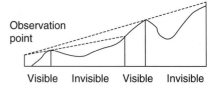

Figure 12.17b
Viewshed analysis is based on the line-of-sight operation.

help GIS users locate areas ideal for the observation point. The elevation of the observation point should be increased by the height of the observer and, in some cases, the height of the physical structure to make it higher than its immediate surroundings. For instance, a forest lookout station is usually 15 to 20 meters high. In ArcGIS, one can add the height of the observation station as an offset value in an attribute field.

Other parameters in viewshed analysis may include radius, viewing azimuth, and tree height. Radius limits the search distance (e.g., within 2 miles of a trail) for viewable areas. The default radius is infinity. The viewing azimuth specifies the angle of the scan (e.g., 120°) so that the observer scans only a portion of the landscape at any one moment. The default azimuth is 360°, which means that the observer constantly scans the surrounding lands in a full 360-degree sweep. Tree height can be an important factor if a viewshed analysis is conducted within a forested land from a road or trail. In that case, estimated tree height can be added to ground elevation to create forest elevation for viewshed analysis (Wing and Johnson 2001).

The accuracy of viewshed analysis depends on the accuracy of DEM data and the data model (i.e., TIN versus elevation grid) (Fisher 1991). Even the definition of the target (e.g., center of a cell or one of the cell's corner points) can influence the analysis outcome (Fisher 1993; Maloy and Dean 2001). A recent study reported that the average level of agreement between GIS-predicted viewsheds and field-surveyed viewsheds was only slightly higher than 50 percent (Maloy and Dean 2001). This finding lends support to the recommendation of defining the viewshed in probabilistic terms, rather than as a binary phenomenon (Fisher 1996).

Viewshed analysis is useful for site selection of forest lookout stations, housing and resort area developments, cell phone receiver stations, and microwave towers. Viewshed analysis is also useful for evaluating the visual landscape such as evaluating the scenic quality along a highway or the visual impact of clear-cuts and logging activities. Given interpreted land cover data over time, viewshed analysis can also monitor and assess the impacts of land cover change on the visual landscape of a protected area such as a national park (Miller 2001).

12.4.4 Watershed Analysis

A **watershed** is an area that drains water and other substances to a common outlet. A watershed is also called a basin or catchment. **Watershed analysis** uses elevation grids and raster data operations to derive topographic features such as watersheds and stream networks, which are important in characterizing the hydrologic process (Jenson 1991; Moore 1996).

Three basic grids in watershed analysis are a filled elevation grid, a flow direction grid, and a flow accumulation grid. A **filled elevation grid** is void of depressions. A depression is a cell or cells in an elevation grid that are surrounded by higher-elevation values, and thus represents an area of internal drainage. Although some depressions are real, such as quarries or glaciated potholes, many are imperfections in the DEM. Therefore depressions must be removed from an elevation grid. One method for removing a depression is to increase its cell value to the lowest cell value surrounding the depression (Jenson and Domingue 1988).

A **flow direction grid** shows the direction water will flow out of each cell of a filled elevation grid. The most common method for determining a cell's flow direction is to find the steepest distance-weighted gradient to one of its eight surrounding cells (Figure 12.18) (O'Callaghan and Mark 1984). Used by ArcGIS, this method does not allow flow to be distributed to multiple cells and tends to produce flow in parallel lines along principal directions (Moore 1996). Other algorithms have been proposed to introduce a degree of randomness into the flow direction computations and to allow flow divergence (Gallant and Wilson 2000).

A **flow accumulation grid** tabulates for each cell the number of cells that will flow to it. In other words, a flow accumulation grid shows how many upstream cells will contribute drainage to each cell (Figure 12.19). Cells having high accumulation values generally correspond to stream channels, whereas cells having an accumulation value of zero generally correspond to ridge lines (Figure 12.20a).

1014	1011	1004
1019	1015	1007
1025	1021	1012

+1	+4	+11
−4		+8
−10	−6	+3

(a)   (b)   (c)

Figure 12.18

The flow direction of the center cell in (a) is determined by first calculating the distance-weighted gradient to each of its eight neighbors. For the four immediate neighbors, the gradient is calculated by dividing the elevation difference between the center cell and the neighbor by 1. For the four corner neighbors, the gradient is calculated by dividing the elevation difference by 1.414. The results in (b) show that the steepest gradient, and therefore the flow direction, is from the center cell to the right cell (+8).

Therefore, using some threshold accumulation value, one can derive from a flow accumulation grid a fully connected drainage network. A threshold value of 1000, for example, means that each cell of the drainage network has a minimum of 1000 contributing cells.

Watersheds can be delineated for an entire grid or for selected points. There are two common methods for areawide watershed delineation. The first method uses the drainage network (Figure 12.20b) derived from a flow accumulation grid (Figure 12.20a) and the flow direction grid for watershed delineation. The output grid has a watershed for each section in the drainage network (Figure 12.20c). As to be expected, the size of the watershed is directly related to the threshold value used in deriving the drainage network: a smaller threshold value will result in smaller watersheds, and vice versa. The second method uses a stream network, rather than a derived drainage network, as an input to watershed delineation (Saunders 1999). This avoids errors that may occur in using a DEM to define a drainage network, especially in areas of low relief. Digital representations of stream networks nationwide are available from the USGS' National Hydrography Dataset (NHD) (**http://nhd.usgs.gov/**).

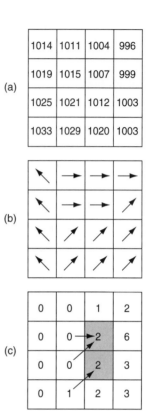

(a)

(b)

(c)

Figure 12.19

The illustration shows a filled elevation grid (a), a flow direction grid (b), and a flow accumulation grid (c). Both shaded cells in (c) have the same flow accumulation value of 2. The top cell receives its flow from its left and lower-left cells. The bottom cell receives its flow from its lower-left cell, which already has a flow accumulation value of 1.

Selected points for watershed delineation may represent stream gauging stations or dams. These points are called pour points in watershed analysis. A watershed for a pour point represents the ascending (upstream) flow paths from the point (Figure 12.21).

12.5 GRID VERSUS TIN

GIS users can often choose either elevation grids or TINs for terrain mapping and analysis. GIS packages, such as ArcGIS, allow use of elevation

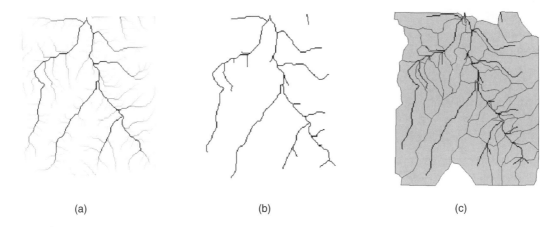

(a)   (b)   (c)

Figure 12.20
A flow accumulation grid is shown in (a). The darkness of the symbol corresponds to the flow accumulation value.
A drainage network derived from the flow accumulation grid by using a threshold value of 500 cells is shown in (b).
The delineation of watersheds is shown in (c).

grids or TINs and can convert a grid to a TIN or a TIN to a grid (Box 12.8). Given the options, GIS users might ask which data model to use. There is no easy answer to the question. Essentially, grids and TINs differ in data flexibility and computational efficiency.

A main advantage of using a TIN lies in the flexibility with input data sources. One can construct a TIN using inputs from DEM, contour lines, GPS data, LIDAR data, and survey data. The user can add elevation points to a TIN at their precise locations and add breaklines, such as streams, roads, ridgelines, and shorelines, to define surface discontinuities. A DEM or an elevation grid cannot indicate a stream in a hilly area and the accompanying topographic characteristics if the stream width is smaller than the DEM resolution. But a TIN will have no problem of depicting the stream as a breakline. Because the GIS user put together a TIN, the quality of the TIN reflects the user's time and effort.

An elevation grid is fixed with a given cell size. One cannot add new sample points to an elevation grid to increase its surface accuracy. Assuming the production method is the same, the only way to improve the accuracy of a grid is to increase its resolution, for example, from 30 meters

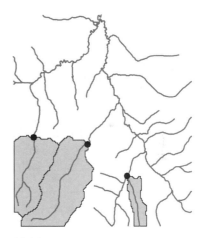

Figure 12.21
A watershed (shaded area) is derived for each of the three pour points.

to 10 meters. Researchers, especially those working with small watersheds, have in fact advocated DEMs with a 10-meter resolution (Zhang and Montgomery 1994) and even higher (Gertner et al. 2002). But increasing DEM resolution is a costly operation because it requires the recompiling of elevation data and more computer memory.

Box 12.8 | **Terrain Mapping and Analysis Using ArcGIS**

Arcinfo Workstation has commands for terrain mapping and analysis operations covered in this chapter. Both the 3D Analyst and Spatial Analyst extensions to ArcGIS have a surface analysis menu with selections for contour, slope, aspect, hillshade, and viewshed. The input to these analysis functions can be either an elevation grid or a TIN, and the output is in grid format except for the contour, which is in shapefile format. 3D Analyst has additional functionalities for working with TINs. 3D Analyst also has a 3-D viewing application called ArcScene, which can be used to prepare and manipulate perspective views, 3-D draping, and 3-D animation.

Besides data flexibility, TIN is also an excellent data model for terrain mapping and 3-D display. The triangular facets of a TIN better define the land surface than an elevation grid and create a sharper image. Most GIS users seem to prefer the look of a map based on a TIN rather than an elevation grid (Kumler 1994).

Computational efficiency is the main advantage of using grids for terrain analysis. The simple data structure makes it relatively easy to perform neighborhood operations on an elevation grid. Therefore, computations of slope, aspect, surface curvature, relative radiance, and other topographic variables using an elevation grid are fast and efficient. In contrast, the computational load using a TIN can increase significantly as the number of triangles increases. For some terrain analysis operations, TINs are in fact converted to elevation grids prior to data analysis.

Finally, which data model is more accurate in measuring elevation, slope, aspect, and other land surface parameters? Kumler (1994) made a series of comparisons between two types of DEM, a TIN derived from the VIP algorithm, a TIN derived from the maximum z-tolerance algorithm, and three other TINs derived from contour lines. He concluded that TINs made of points sampled from DEMs were inferior to the full DEM, and the contour-based TINs were not as efficient as DEMs in terrain modeling. But Kumler's comparisons were limited to elevation estimates and did not cover slope, aspect, or other topographic parameters. A TIN made of sampled points from a DEM obviously will not perform as well as the full DEM in estimating elevations. This is also true with contour-based TINs because they contain much smaller numbers of elevation points than contour-based DEMs.

KEY CONCEPTS AND TERMS

3-D drape: The method of superimposing thematic layers such as vegetation and roads on 3-D perspective views.

Aspect: The directional measure of slope.

Breaklines: Line features that represent changes of the land surface such as streams, shorelines, ridges, and roads.

Contour interval: The vertical distance between contour lines.

Contour lines: Lines connect points of equal elevation.

Delaunay triangulation: An algorithm for connecting points to form triangles such that all points are connected to their nearest neighbors and triangles are as compact as possible.

Filled elevation grid: An elevation grid that is void of depressions.

Flow accumulation grid: A grid that shows for each cell the number of cells that will flow to it.

Flow direction grid: A grid that shows the direction water will flow out of each cell of a filled elevation grid.

Hill shading: A graphic method, which simulates how the land surface looks with the interaction between sunlight and landform features. The method is also known as *shaded relief* or *shading*.

Hypsometric tint: A mapping method, which applies color symbols to different elevation zones. The method is also known as *layer tinting*.

Maximum z-tolerance: A TIN construction algorithm, which ensures that, for each elevation point selected, the difference between the original elevation and the estimated elevation from the TIN is within the specified tolerance.

Perspective view: A graphic method that produces 3-D views of the land surface.

Slope: The rate of change of elevation at a surface location, measured as an angle in degrees or as a percentage.

Vertical profile: A chart showing changes in elevation along a line such as a hiking trail, a road, or a stream.

Viewing angle: A parameter for creating a perspective view, which represents the angle measured from the horizon to the altitude of the observer.

Viewing azimuth: A parameter for creating a perspective view, which represents the direction from the observer to the surface.

Viewing distance: A parameter for creating a perspective view, which represents the distance between the viewer and the surface.

Viewshed: Areas of the land surface that are visible from an observation point, or observation points.

VIP: An elevation point selection algorithm, which evaluates the importance of an elevation point by measuring how well its value can be estimated from the neighboring point values.

Watershed: An area that drains water and other substances to a common outlet.

Watershed analysis: Analysis that involves derivation of flow direction, watershed boundaries, and stream networks.

Z-scale: A parameter for creating a perspective view, which is the ratio between the vertical scale and the horizontal scale. It is also called the *vertical exaggeration factor*.

APPLICATIONS: TERRAIN MAPPING AND ANALYSIS

This applications section includes six tasks. Task 1 uses DEM data for terrain mapping. To create a 3-D perspective view in Task 1, you will use Arc-Scene, accessible through the 3D Analyst extension. Tasks 2 and 3 perform viewshed analysis, first with two given lookout locations and then with three lookout locations of your choice. Task 4 lets you run watershed analysis and introduces you to a new feature in ArcGIS 8.2 that supports multiple line statements in the Raster Calculator. Task 5 lets you build and modify a TIN in ArcMap. In

Task 6, you will build a TIN in ArcInfo Workstation and derive a slope and an aspect map from the TIN.

Task 1: Use DEM for Terrain Mapping

What you need: *plne*, an elevation grid; *streams. shp*, a stream shapefile.

The elevation grid *plne* is imported from a USGS 7.5-minute DEM. The shapefile *streams.shp* shows major streams in the study area. Task 1 covers terrain mapping and analysis using Spatial Analyst and 3D Analyst in ArcMap.

1.1 Create a contour theme

1. Start ArcCatalog, and make connection to the Chapter 12 database. Launch ArcMap. Add *plne* to Layers, and rename Layers Task 1. Select Extensions from the Tools menu and check the boxes for Spatial Analyst and 3D Analyst. Then click the View menu, point to Toolbars, and make sure that the boxes for Spatial Analyst and 3D Analyst are both checked.

2. Click the Spatial Analyst dropdown arrow, point to Surface Analysis, and select Contour. In the Contour dialog, select *plne* for the Input surface, enter 100 (meters) as the Contour interval and 800 (meters) as the Base contour, and save the Output features as *ctour.shp*. Click OK to dismiss the dialog.

3. *Ctour.shp* appears on the map. Do the following to label the contour lines. Select Properties from the context menu of *ctour*. Under the Labels tab, check the box to Label Features in this layer and select CONTOUR from the Label Field dropdown list. Click OK to dismiss the dialog. The contour lines are now labeled. (To remove the contour labels, right-click *ctour* and uncheck Label Features.)

1.2 Create a vertical profile

1. Add *streams.shp* to Task 1. Select Open Attribute Table from the context menu of *streams*. Click the Options dropdown arrow and choose Select by Attributes. Enter the following SQL statement in the expression box: "USGH_ID" = 167. Click Apply. Close the *streams* attribute table. Use the Zoom In tool to zoom in the selected stream.

2. Make sure that the 3D Analyst toolbar is available. Click the Interpolate Line tool on the 3D Analyst toolbar. Use the mouse pointer to digitize points along the selected stream. Double-click the last point to finish

digitizing. A rectangle with handles appears around the digitized stream.

3. Click the Create Profile Graph tool on the 3D Analyst toolbar. A vertical profile appears with a default title and subtitle. Right-click the title bar of the graph and select Properties. The Graph Properties dialog allows you to enter a new title and subtitle and to choose other advanced design options.

4. The digitized stream becomes a graphic element on the map. You can delete it by using the Select Elements tool to first select it. To unselect the stream, choose Clear Selected Features from the Selection menu.

1.3 Create a hillshade theme

1. Click the Spatial Analyst dropdown arrow, point to Surface Analysis, and select Hillshade. In the Hillshade dialog, select *plne* for the Input surface. Take the default values of 315 for Azimuth, 45 for Altitude, 1 for Z factor, and 30 for Output cell size. Opt for a Temporary Output raster. Click OK to dismiss the dialog. *Hillshade of plne* is added to the Table of Contents.

2. Try different values of azimuth and altitude to see how these two parameters affect hill shading. For example, a hillshade map will look darker with a lower altitude.

1.4 Create a perspective view

1. Make sure that the 3D Analyst toolbar is available. Click the ArcScene tool on the 3D Analyst toolbar. The ArcScene application opens. Add *plne* and *stream.shp* to view. By default, *plne* is displayed in a planimetric view, without the 3-D effect. Select Properties from the context menu of *plne*. Click the Base Heights tab. Click the radio button next to "Obtain heights for layer from surface," and select *plne* for the surface. Click OK to dismiss the dialog.

2. *Plne* is now displayed in a 3-D perspective view. The next step is to superimpose

streams on the surface. Select Properties from the context menu of *streams*. Click the Base Heights tab. Click the radio button next to "Obtain heights for layer from surface," and select *plne* for the surface. Click OK to dismiss the dialog.

3. Using the properties of *plne* and *streams*, you can change the look of the 3-D view. For example, you can change the color scheme for displaying *plne*. Select Properties from the context menu of *plne*. Click the Symbology tab. Right-click the Color Ramp box and uncheck Graphic View. Click the Color Ramp dropdown arrow and select Elevation #1. Click OK. Elevation #1 uses the conventional color scheme to display the 3-D view of *plne*. Click the symbol for *streams* in the Table of Contents. Select the River symbol from the Symbol Selector, and click OK.

4. You can tone down the color symbols for *plne* so that *streams* can stand out more. Select Properties from the context menu of *plne*. Click the Display tab, enter 40 (%) in the box next to Transparent, and click OK.

5. The 3D Analyst toolbar has tools for you to navigate, zoom in or out, center on target, zoom to target, and to perform other manipulations. For example, the navigate tool allows you to rotate the 3-D surface. Exit ArcScene when you are done.

1.5 Create a slope theme

1. Click the Spatial Analyst dropdown arrow, point to Surface Analysis, and select Slope. In the Slope dialog, select *plne* for the Input surface and opt for a temporary Output raster.

2. *Slope of plne* shows a degree slope map in a default classification. Select Properties from the context menu of *Slope of plne*. Click the Symbology tab, and click Classify. Click the Classes dropdown arrow and change the number of classes from 9 to 5. Click the first cell under Break Values, and enter 10. Enter 20, 30, 40, and 53 in the next four cells.

Click OK to dismiss the dialogs. The slope map now has a new classification.

3. Reclassify in the Spatial Analyst dropdown list can also be used for re-classification. The difference is that Reclassify creates a new integer grid based on the new classification.

1.6 Create an aspect theme

1. Click the Spatial Analyst dropdown arrow, point to Surface Analysis, and select Aspect. In the Aspect dialog, select *plne* for the Input surface and opt for a temporary Output raster. Click OK to dismiss the dialog

2. *Aspect of plne* shows an aspect map with the eight principal directions and flat area. But the aspect grid is actually a floating-point grid and does not have an attribute table. You can use Reclassify to convert the aspect grid to an integer grid or to change the aspect classification, for example, from the eight to four principal directions.

3. Select Reclassify from the Spatial Analyst dropdown list. Select *Aspect of plne* for the Input raster. Click Classify. Select a different method than Manual from the Method dropdown list to activate Classes. Click the Classes dropdown arrow and choose 10. Click the first cell under Break Values and enter −1. Enter 22.5, 67.5, 112.5, 157.5, 202.5, 247.5, 292.5, 337.5, and 360 in the following nine cells. Click OK to dismiss the dialog.

4. Old values in the Reclassify dialog are now updated with the break values you have entered. Now you have to change new values. Click the first cell under New values and enter −1. Click and enter 1, 2, 3, 4, 5, 6, 7, 8, and 1 in the following 9 cells. The last cell has a value of 1 because the cell (337.5°–360°) and the second cell (−1°–22.5°) make up the north aspect. Click OK to dismiss the Reclassify dialog. *Reclass of Aspect of plne* is an integer aspect grid with the eight principal directions and flat (−1).

Task 2. Perform Viewshed Analysis

What you need: *plne*, an elevation grid; and *lookouts.shp*, a lookout location shapefile.

The elevation grid *plne* is the same as in Task 1. The lookout location shapefile contains two points, labeled 1 and 2. A viewshed analysis can determine areas in *plne* that are visible from the two lookout locations and areas that are not visible. To better visualize the relationship between visibility and terrain, you will also use a hillshade map of *plne*.

1. Select Data Frame from the Insert menu in ArcMap. Rename the new data frame Tasks 2&3, and add *plne* and *lookouts.shp* to Tasks 2&3. First, create a hillshade map of *plne*. Click the Spatial Analyst dropdown arrow, point to Surface Analysis, and select Hillshade. Select *plne* for the input surface and take the default values for the other parameters. Click OK to dismiss the dialog. *Hillshade of plne* is added to the map.

2. Now you will run a viewshed analysis. Click the Spatial Analyst dropdown arrow, point to Surface Analysis, and select Viewshed. Make sure that the Input surface is *plne* and the Observer points are from *lookouts*. Opt for a temporary Output raster. Click OK to dismiss the Viewshed dialog.

3. *Viewshed of lookouts* separates the visible areas from not visible areas. Portions of the visible area are visible to only one observer point, whereas others are visible to both observer points. Right-click *Viewshed of lookouts* and select Properties. Under the Symbology tab, click Unique Values in the Show box. Under the Display tab, enter 50 next to Transparent. Click OK to dismiss the dialog. *Viewshed of lookouts* now shows three classes: 0 for not visible, 1 for visible from one observer point, and 2 for visible from both observer points.

4. Right-click *Viewshed of lookouts* and select Open Attribute Table. The table shows cell counts for the three classes of 0, 1, and 2.

Task 3. Create a New Lookout Shapefile for Viewshed Analysis

What you need: *plne* and *lookouts.shp*, same as in Task 2.

Task 3 asks you to digitize three new lookout locations before running a viewshed analysis.

1. Select Copy from the context menu of *lookouts*. Select Paste Layer from the context menu of Tasks 2&3. The copied shapefile is also named *lookouts*. Right-click the first *lookouts* in the Table of Contents, and select Properties. Under the General tab, change Layer Name from *lookouts* to *newpoints*. *Newpoints* contains two observer points from *lookouts*, which will be removed in this task and replaced with three new observer points.

2. Click the Editor Toolbar button to open the Editor toolbar. Click the Editor menu and select Start Editing. Make sure that the Task is to Create New Feature and the Target is *newpoints*. Click the Edit tool on the Editor toolbar. Click an observer point in *newpoints*. When the point is shown in cyan, delete it. Click the other observer point and delete it.

3. Next add new observer points. To find suitable observer point locations, you can use *Hillshade of plne* as a guide and the Zoom In tool for close-up looks. You can also use *plne* and the Identify tool to find elevation data. When you are ready to add an observer point, click the Create New Feature tool first and then click the intended location of the observer point. Add two more observer points. Click the Editor menu and select Stop Editing. Save the edits. You are ready to use *newpoints* for viewshed analysis.

4. Click the Spatial Analyst dropdown arrow, point to Surface Analysis, and select Viewshed. Make sure that the Input surface is *plne* and the Observer points are from *newpoints*. Opt for a temporary Output raster. Click OK to dismiss the dialog.

5. *Viewshed of newpoints* shows visible and not visible areas. The attribute table of *Viewshed*

of newpoints provides cell counts of visible from one point, visible from two points, and visible from three points.

6. To save *newpoints* as a shapefile, you can right-click *newpoints*, point to Data, and select Export Data. In the Export Data dialog, specify the path and name of the output shapefile.

Task 4. Delineate Watersheds

What you need: *emidafill*, a filled DEM; *pourpoints.shp*, a point shapefile; and *emidastrm.shp*, a stream shapefile.

Task 4 shows you the process of delineating watersheds using a DEM as the data source. Watershed delineation can be areawide or for specific points. *Emidafill* is a filled elevation grid because depressions have been removed from the original DEM. *Pourpoints.shp* contains three points, each of which will be used to derive a specific watershed. *Emidastrm.shp* provides a reference. ArcGIS does not offer a hydrologic modeling extension. Therefore, you will use Raster Calculator to perform the task.

1. Select Data Frame from the Insert menu in ArcMap. Rename the new data frame Task 4, and add *emidafill*, *pourpoints.shp*, and *emidastrm.shp* to Task 4.

2. Make sure that the Spatial Analyst extension is checked and visible. Click the Spatial Analyst dropdown arrow and select Raster Calculator. The first step in watershed delineation is to create a flow direction grid. In the Raster Calculator's expression box, type flowdirection([emidafill]). Flowdirection in the expression is a function, and *emidafill* is the argument or input. Click Evaluate. The output grid is named *Calculation*, a generic output name. Rename *Calculation Flowdirection* and make it a permanent grid. *Flowdirection* shows the flow direction of each cell in *emidafill*.

3. Next create a flow accumulation grid. Type flowaccumulation([flowdirection]) in the

Raster Calculator's expression box. Click Evaluate. Rename the output grid *flowaccumu* and make it permanent.

4. Next create a source grid, which will be used as the input later for watershed delineation. Creating a source grid involves two steps. First, you select from (or threshold) *flowaccumu* those cells that have more than 500 cells flowing into them. Type con([flowaccumu] > 500,1) in the Raster Calculator's expression box. (Make sure that you have a space before and after >.) Click Evaluate. Rename the output grid *net* and make it permanent. Second, you assign unique values to sections of *net* between junctions (intersections). Type streamlink([net],[flowdirection]) in the Raster Calculator. Click Evaluate. Rename the output grid *source* and make it permanent.

5. Now you have the necessary input grids for watershed delineation. Type watershed([flowdirection],[source]) in the Raster Calculator's expression box. Click Evaluate. Rename the output grid *watershed* and make it permanent. Change the symbology of *watershed* to that of unique values so that you can see individual watersheds.

6. ArcGIS 8.2 has a new functionality that supports multiple lines in the Raster Calculator's expression box. Therefore, you can string together the statements in steps 2 to 5 and their output grids as follows:

[flowdirection] = flowdirection([emidafill])

[flowaccumu] = flowaccumulation ([flowdirection])

[net] = con([flowaccumu] > 500,1)

[source] = streamlink([net],[flowdirection])

[watershed] = watershed([flowdirection], [source])

Click Evaluate. Spatial Analyst executes the above statements and adds the output grids to

ArcMap. This new functionality saves time for an experienced user.

7. The second part of the task is to derive a specific watershed for each point in *pourpoints.shp*. Because Raster Calculator works only with grids, you must first convert *pourpoints.shp* to a grid. Select Options in the Spatial Analyst dropdown menu. In the Options dialog, click the Extent tab and select Same as Layer "flowdirection" for the Analysis Extent. Click the Cell Size tab and select Same as Layer "flowdirection" for the analysis cell size. Click OK to dismiss the Options dialog. Click the Spatial Analyst dropdown arrow, point to Convert, and select Features to Raster. In the Features to Raster dialog, select *pourpoints* for the input features and save the output raster as *pointgd*.

8. Click the Spatial Analyst dropdown arrow and select Raster Calculator. Type watershed([flowdirection],[pointgd]) in the Raster Calculator's expression box. Click Evaluate. Rename the output grid *ptwsheds* and make it permanent.

Task 5. Build and Display a TIN in ArcMap

What you need: *emidalat*, an elevation grid; and *emidastrm.shp*, a stream shapefile.

Task 5 shows you how to construct a TIN from an elevation grid and to modify the TIN with *emidastrm.shp* as breaklines. You will also display different features of the TIN.

1. Select Data Frame from the Insert menu in ArcMap. Rename the new data frame Task 5, and add *emidalat* and *emidastrm.shp* to Task 5.

2. Make sure that the 3D Analyst toolbar is available. Click the 3D Analyst menu, point to Convert, and select Raster to TIN. In the Convert Raster to TIN dialog, make sure that *emidalat* is the Input raster. Enter 10 as the z-tolerance. Specify the Output TIN as *emidatin*. Click OK to dismiss the dialog.

3. *Emidatin* is added to the map. The next step is to modify *emidatin* with *emidastrm*, which contains streams. Click the 3D Analyst menu, point to Create/Modify TIN, and select Add Features to TIN. In the Add Features to TIN dialog, make sure that *emidatin* is the Input TIN. Check *emidastrm* in the Layers box. Then select None for the Height source and Triangulate as hard line. Click OK to dismiss the dialog.

4. You can view *emidatin* in a variety of ways. Select Properties from the context menu of *emidatin*. Click the Symbology tab. Click the Add button below the Show frame. An Add Renderer scroll list appears with choices related to the display of edges, faces, or nodes that make up *emidatin*. Click Faces with the same symbol in the list, click Add, and click Dismiss. Uncheck all the boxes in the Show frame except Faces. Make sure that the box to show hillshade illumination effect in 2-D display is checked. Click OK on the Layer Properties dialog. With its faces in the same symbol, *emidatin* can be used as a background in the same way as a hillshade map for displaying map features such as streams, vegetation, and so on.

Task 6. Process TIN in ArcInfo Workstation

What you need: *emidalat*, an elevation grid; and *breakstrm*, a stream coverage.

In Task 6, you will build a TIN from *emidalat* and *breakstrm* in ArcInfo Workstation and then derive slope and aspect maps from the TIN as polygon coverages.

1. The VIP algorithm selects "very important" points from an elevation grid to be included in a TIN. ArcInfo Workstation allows users to specify a percentage of elevation points to be selected. The default is 10%.

Arc: vip emidalat emidavip /*emidavip is the output point coverage

2. The CREATETIN command in ArcInfo Workstation can create a TIN using data from different sources. For this task, you will use two data sources: (1) elevation points selected by the VIP algorithm and (2) the hard breakline of *emidastrm*. Elevation along streams in *emidastrm* will be derived from the elevation grid *emidalat*. CREATETIN is a dialog command. You will use subcommands in the dialog to enter the input data.

Arc: createtin emidatin2 /* *emidatin2* is the output tin
Createtin: cover emidavip point /* input elevation points from *emidavip*
Createtin: lattice emidalat breakstrm line hardline /* input *emidastrm* as hardline
Createtin: end /* exit createtin

3. The next step converts *emidatin2* to a polygon coverage. Each triangle in *emidatin2* becomes a polygon with its slope and aspect

measures. Slopes can be measured in percent or degrees.

Arc: tinarc emidatin2 emidapoly poly percent /* *emidapoly* is the polygon coverage

4. *Emidapoly.pat* contains the items of percent slope and aspect. Typically, percent slope values are grouped into classes so that slopecode 1 = 0–20%, slopecode 2 = 20–40%, and so on. After the slope classification is completed in Tables, *emidapoly* can be dissolved into a slope (polygon) map by using slopecode as the dissolve item. Likewise, *emidapoly* can be dissolved into an aspect (polygon) map by using aspectcode as the dissolve item. The following shows the command to dissolve *emidapoly* into a slope map called *emidaslope*:

Arc: dissolve emidapoly emidaslope slopecode poly

REFERENCES

Bolstad, P. V., and T. Stowe. 1994. An Evaluation of DEM Accuracy: Elevation, Slope, and Aspect. *Photogrammetric Engineering and Remote Sensing* 60: 1327–32.

Campbell, J. 1984. *Introductory Cartography*. Englewood Cliffs, NJ: Prentice Hall.

Carter, J. R. 1989. Relative Errors Identified in USGS Gridded DEMs. *Proceedings, AUTO–CARTO* 9, pp. 255–65.

Carter, J. R. 1992. The Effect of Data Precision on the Calculation of Slope and Aspect Using Gridded DEMs. *Cartographica* 29: 22–34.

Chang, K., and Z. Li. 2000. Modeling Snow Accumulation

with a Geographic Information System. *International Journal of Geographical Information Science* 14: 693–707.

Chang, K., and B. Tsai. 1991. The Effect of DEM Resolution on Slope and Aspect Mapping. *Cartography and Geographic Information Systems* 18: 69–77.

Chen, Z. T., and J. A. Guevara. 1987. Systematic Selection of Very Important Points (VIP) from Digital Terrain Model for Constructing Triangular Irregular Networks. *Proceedings, AUTO-CARTO* 8, pp. 50–56.

Clarke, K. C. 1995. *Analytical and Computer Cartography*, 2d ed. Englewood Cliffs, NJ: Prentice Hall.

Eyton, J. R. 1991. Rate-of-Change Maps. *Cartography and Geographic Information Systems* 18: 87–103.

Fisher, P. F. 1991. First Experiments in Viewshed Uncertainty: The Accuracy of the Viewshed Area. *Photogrammetric Engineering and Remote Sensing* 57: 1321–27.

Fisher, P. F. 1993. Algorithm and Implementation Uncertainty in Viewshed Analysis. *International Journal of Geographical Information Systems* 7: 331–47.

Fisher, P. R. 1996. Extending the Applicability of Viewsheds in Landscape Planning. *Photogrammetric Engineering and Remote Sensing* 62: 1297–1302.

Fleming, M. D., and R. M. Hoffer. 1979. *Machine Processing of Landsat MSS Data and DMA Topographic Data for Forest Cover Type Mapping*. LARS Technical Report 062879. Laboratory for Applications of Remote Sensing, Purdue University, West Lafayette, IN.

Flood, M. 2001. Laser Altimetry: From Science to Commercial LIDAR Mapping. *Photogrammetric Engineering and Remote Sensing* 67: 1209–17.

Florinsky, I. V. 1998. Accuracy of Local Topographic Variables Derived from Digital Elevation Models. *International Journal of Geographical Information Systems* 12: 47–61.

Franklin, S. E. 1987. Geomorphometric Processing of Digital Elevation Models. *Computers & Geosciences* 13: 603–9.

Gallant J. C., and J. P. Wilson. 2000. Primary Topographic Attributes. In J. P. Wilson and J. C. Gallant, eds., *Terrain Analysis: Principles and Applications,* pp. 51–85. New York: Wiley.

Gao, J. 1998. Impact of Sampling Intervals on the Reliability of Topographic Variables Mapped from Grid DEMs at a Micro-Scale. *International Journal of Geographical Information Systems* 12: 875–90.

Gertner, G., G. Wang, S. Fang, and A. B. Anderson. 2002. Effect and Uncertainty of Digital Elevation Model Spatial Resolutions on Predicting the Topographical Factor for Soil Loss Estimation. *Journal of Soil and Water Conservation* 57: 164–74.

Hill, J. M, L. A. Graham, and R. J. Henry. 2000. Wide-Area Topographic Mapping and Applications Using Airborne Light Detection and Ranging (LIDAR) Technology. *Photogrammetric Engineering and Remote Sensing* 66: 908–14.

Hodgson, M. E. 1998. Comparison of Angles from Surface Slope/Aspect Algorithms. *Cartography and Geographic Information Systems* 25: 173–85.

Horn, B. K. P. 1981. Hill Shading and the Reflectance Map. *Proceedings of the IEEE* 69 (1): 14–47.

Isaacson, D. L., and W. J. Ripple. 1990. Comparison of 7.5-Minute and 1-Degree Digital Elevation Models. *Photogrammetric Engineering and Remote Sensing* 56: 1523–27.

Jenson, S. K. 1991. Applications of Hydrologic Information Automatically Extracted from Digital Elevation Models. In K. J. Bevan and I. D. Moore, eds., *Terrain Analysis and Distributed Modeling in Hydrology,* pp. 35–48. Chichester, England: Wiley.

Jenson, S. K., and J. O. Domingue. 1988. Extracting Topographic Structure from Digital Elevation Data for Geographic Information System Analysis. *Photogrammetric Engineering and Remote Sensing* 54: 1593–1600.

Jones, K. H. 1998. A Comparison of Algorithms Used to Compute Hill Slope as a Property of the DEM. *Computers & Geosciences* 24: 315–23.

Jones, T. A., D. E. Hamilton, and C. R. Johnson. 1986. *Contouring Geologic Surfaces with the Computer.* New York: Van Nostrand Reinhold.

Kumler, M. P. 1994. An Intensive Comparison of Triangulated Irregular Networks (TINs) and Digital Elevation Models (DEMs). *Cartographica* 31 (2): 1–99.

Lane, S. N., K. S. Richards, and J. H. Chandler, eds., 1998. *Landform Monitoring, Modelling and Analysis.* Chichester, England: Wiley.

Lee, J. 1991. Comparison of Existing Methods for Building Triangular Irregular Network Models of Terrain from Grid Digital Elevation Models. *International Journal of Geographical Information Systems* 5: 267–85.

Maloy, M. A., and D. J. Dean. 2001. An Accuracy Assessment of Various GIS-Based Viewshed Delineation Techniques. *Photogrammetric Engineering and Remote Sensing* 67: 1293–98.

Miller, D. 2001. A Method for Estimating Changes in the Visibility of Land Cover. *Landscape and Urban Planning* 54: 93–106.

Moore, I. D. 1996. Hydrological Modeling and GIS. In M. F. Goodchild, L. T. Steyaert, B. O. Parks, C. Johnston, D. Maidment, M. Crane, and S. Glendinning, eds., *GIS and Environmental Modeling: Progress and Research Issues,* pp. 143–48. Fort Collin, CO: GIS World Books.

Moore, I. D., R. B. Grayson, and A. R. Ladson. 1991. Digital Terrain Modelling: A Review of

Hydrological, Geomorphological, and Biological Applications. *Hydrological Process* 5: 3–30.

O'Callaghan, J. F., and D. M. Mark. 1984. The Extraction of Drainage Networks from Digital Elevation Data. *Computer Vision, Graphics and Image Processing* 28: 323–44.

Ritter, P. 1987. A Vector-Based Slope and Aspect Generation Algorithm. *Photogrammetric Engineering and Remote Sensing* 53: 1109–11.

Saunders, W. 1999. Preparation of DEMs for Use in Environmental Modeling Analysis. *Proceedings of the 1999 ESRI User Conference* (**http://gis.esri.com/ library/userconf/proc99/procee d/papers/pap802/p802.htm**).

Sharpnack, D. A., and G. Akin. 1969. An Algorithm for Computing Slope and Aspect from Elevations.

Photogrammetric Engineering 35: 247–48.

Skidmore, A. K. 1989. A Comparison of Techniques for Calculating Gradient and Aspect from a Gridded Digital Elevation Model. *International Journal of Geographical Information Systems* 3: 323–34.

Srinivasan, R., and B. A. Engel. 1991. Effect of Slope Prediction Methods on Slope and Erosion Estimates. *Applied Engineering in Agriculture* 7: 779–83.

Thelin, G. P., and R. J. Pike. 1991. *Landforms of the Conterminous United States: A Digital Shaded-Relief Portrayal,* map I-2206, scale 1:3,500,000. Washington, DC: U.S. Geological Survey.

Tsai, V. J. D. 1993. Delaunay Triangulations in TIN Creation: An Overview and Linear Time Algorithm. *International Journal of Geographical Information Systems* 7: 501–24.

Wilson, J. P., and J. C. Gallant, eds. 2000. *Terrain Analysis: Principles and Applications.* New York: Wiley.

Wing, M. G., and R. Johnson. 2001. Quantifying Forest Visibility with Spatial Data. *Environmental Management* 27: 411–20.

Zar, J. H. 1984. *Biostatistical Analysis*, 2d ed. Englewood Cliffs, NJ: Prentice Hall.

Zevenbergen, L. W., and C. R. Thorne. 1987. Quantitative Analysis of Land Surface Topography. *Earth Surface Processes and Landforms* 12: 47–56.

Zhang, W., and D. R. Montgomery. 1994. Digital Elevation Model Grid Size, Landscape Representation, and Hydrologic Simulations. *Water Resources Research* 30: 1019–28.

13

SPATIAL INTERPOLATION

13.1 INTRODUCTION

Besides the land surface covered in Chapter 12, GIS users also work with another type of surface, which may not be physically present but can be visualized in the same way as the land surface. Cartographers have called this type of surface the statistical surface (Robinson et al. 1995). Examples of the statistical surface include precipitation, snow accumulation, water table, and population density.

How can one construct a statistical surface? The answer is similar to that for the land surface except that input data are typically limited to a sample of point data. To make a precipitation map, for example, one will not find a regular array of weather stations like a digital elevation model (DEM). Therefore, a process of filling in data between the sampled points is required.

Spatial interpolation is the process of using points with known values to estimate values at other points. These points with known values are called known points, control points, sampled points, or observations. For example, one can estimate the precipitation value at a location with no recorded data through interpolation of known precipitation readings at nearby weather stations. In GIS applications, spatial interpolation is typically applied to a grid with estimates made for all cells. Spatial interpolation is therefore a means of converting point data to surface data so that the surface data can be used with other surfaces for analysis and modeling. Many GIS packages provide spatial interpolation functionalities (Box 13.1).

This chapter has four main sections. Section 13.2 reviews the elements of spatial interpolation including control points and type of spatial interpolation. Sections 13.3 and 13.4 cover global methods and local methods, respectively. Section 13.5 discusses comparison of interpolation methods and comparison measures. Perhaps more than any other topic in GIS, spatial interpolation depends on the computing algorithm. Worked examples are included in this chapter to show how spatial interpolation is carried out mathematically.

This chapter does not include isopleth mapping or areal interpolation. Isopleth mapping uses assigned points such as polygon centroids in interpolation (Robinson et al. 1995). The result is not nearly as accurate as spatial interpolation that uses actual points such as weather stations. Areal interpolation (Chapter 10) transfers known data from one set of polygons to another and does not involve point or surface data.

13.2 ELEMENTS OF SPATIAL INTERPOLATION

13.2.1 Control Points

Control Points are points with known values. They provide the data for the development of an **interpolator,** usually a mathematical equation, for spatial interpolation. The number and distribution

---

 Box 13.1 **A Survey of Spatial Interpolation among GIS Packages**

Spatial interpolation is usually included in a raster-based GIS package. The following lists some of the interpolation methods covered in this chapter and the GIS packages that carry them.

Trend surface: ArcGIS, IDRISI, GRASS, ILWIS

Regression: ArcGIS, IDRISI, GRASS (linear regression)

Inverse distance weighted interpolation: ArcGIS, IDRISI, GRASS, ILWIS, MFworks, SPANS, Vertical Mapper

Spline: ArcGIS, GRASS (regularized splines with tension)

Kriging: ArcGIS, IDRISI, ILWIS, GRASS, MFworks, SPANS, Vertical Mapper

of control points can greatly influence the accuracy of spatial interpolation. A basic assumption in spatial interpolation is that the value to be estimated at a point is more influenced by nearby known points than those that are farther away. To be effective for estimation, control points should be well distributed within the study area. But this ideal situation is rare in real-world applications because a study area often contains data-poor areas.

Figure 13.1 is a map of 105 weather stations and their 30-year average annual precipitation data in Idaho. The map clearly shows two major data-poor areas in the state: the central part (Clearwater Mountains, Salmon River Mountains, and Lemhi Range), and the southwestern corner (Owyhee Mountains). These data-poor areas can cause problems in spatial interpolation.

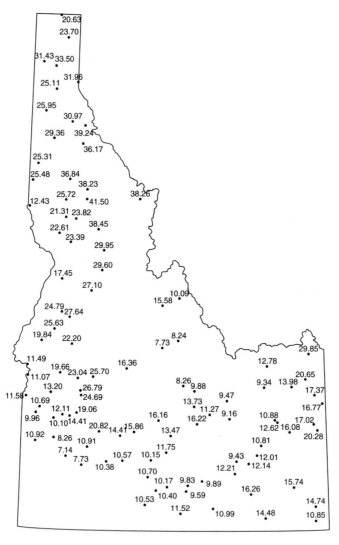

Figure 13.1

A map of 105 weather stations in Idaho and their 30-year average annual precipitation values.

13.2.2 Type of Spatial Interpolation

Spatial interpolation methods can be categorized in several ways. First, they can be grouped into global and local methods. A **global interpolation** method uses every known point available to estimate an unknown value. A **local interpolation** method, on the other hand, uses a sample of known points to estimate an unknown value. Because the difference between the two groups lies in the number of control points used in estimation, one may view the scale from global to local as a continuum. Typically a local method is chosen for two reasons: (1) faraway known points have little influence on the value to be estimated at a point, thus justifying interpolation based on a neighborhood; and (2) a local method requires much less computation than a global method.

Second, spatial interpolation methods can be grouped into exact and inexact interpolation. **Exact interpolation** predicts a value at the point location that is the same as its known value. In other words, exact interpolation generates a surface that passes through the control points. In contrast, **inexact interpolation**, or approximate interpolation, predicts a value at the point location that differs from its known value.

Third, spatial interpolation methods may be deterministic or stochastic. A **deterministic interpolation** method provides no assessment of errors with predicted values. A **stochastic interpolation** method, on the other hand, offers assessment of prediction errors with estimated variances. The assumption of a random process is normally required for a stochastic method.

Table 13.1 shows a classification of spatial interpolation methods covered in this chapter. Notice that the two global methods can also be used in local operations.

13.3 GLOBAL METHODS

13.3.1 Trend Surface Analysis

An inexact interpolation method, **trend surface analysis** approximates points with known values with a polynomial equation (Davis 1986; Bailey and Gatrell 1995). The equation or the interpolator can then be used to estimate values at other points. A linear or first-order trend surface uses the equation

(13.1)

$$z_{x,y} = b_0 + b_1 x + b_2 y$$

where the attribute value z is a function of x and y coordinates. The b coefficients are estimated from the known points (Box 13.2). Because the trend surface model is computed by the least-squares method, the "goodness of fit" of the surface can be measured. Also, the deviation or the residual between the observed and the estimated values can be computed for each known point.

TABLE 13.1   A Classification of Spatial Interpolation Methods

Global		Local	
Deterministic	*Stochastic*	*Deterministic*	*Stochastic*
Trend surface (inexact)*	Regression (inexact)	Thiessen (exact)	Kriging (exact)
		Density estimation (inexact)	
		Inverse distance weighted (exact)	
		Splines (exact)	

*Given some required assumptions, trend surface analysis can be treated as a special case of regression analysis and thus a stochastic method (Griffith and Amrhein 1991).

Box 13.2   A Worked Example of Trend Surface Analysis

Figure 13.2 shows five weather stations with known values around point 0 with an unknown value. The table below shows the x-, y-coordinates of the points, measured in row and column of a grid with a cell size of 2000 meters, and their known values.

Point	x	y	Value
1	69	76	20.820
2	59	64	10.910
3	75	52	10.380
4	86	73	14.600
5	88	53	10.560
0	69	67	?

This example shows how we can use Eq. (13.1), or a linear trend surface, to interpolate the unknown value at point 0. The least-squares method is commonly used to solve for the coefficients of b_0, b_1, and b_2 in Eq. (13.1). Therefore, the first step is to set up three normal equations, similar to those for a regression analysis:

$$\Sigma z = b_0 n + b_1 \Sigma x + b_2 \Sigma y$$
$$\Sigma xz = b_0 \Sigma x + b_1 \Sigma x^2 + b_2 \Sigma xy$$
$$\Sigma yz = b_0 \Sigma y + b_1 \Sigma xy + b_2 \Sigma y^2$$

The equations can be rewritten in matrix form as

$$\begin{bmatrix} n & \Sigma x & \Sigma y \\ \Sigma x & \Sigma x^2 & \Sigma xy \\ \Sigma y & \Sigma xy & \Sigma y^2 \end{bmatrix} \cdot \begin{bmatrix} b_0 \\ b_1 \\ b_2 \end{bmatrix} = \begin{bmatrix} \Sigma z \\ \Sigma xz \\ \Sigma yz \end{bmatrix}$$

Using the values of the five known points, we can calculate the statistics and substitute the statistics into the equation

$$\begin{bmatrix} 5 & 377 & 318 \\ 377 & 29,007 & 23,862 \\ 318 & 23,862 & 20,714 \end{bmatrix} \cdot \begin{bmatrix} b_0 \\ b_1 \\ b_2 \end{bmatrix} = \begin{bmatrix} 67.270 \\ 5043.650 \\ 4445.800 \end{bmatrix}$$

We can then solve the b coefficients by multiplying the inverse of the first matrix on the left (shown with four decimal digits because of the very small numbers) by the matrix on the right:

$$\begin{bmatrix} 23.2102 & -0.1631 & -0.1684 \\ -0.1631 & 0.0018 & 0.0004 \\ -0.1684 & 0.0004 & 0.0021 \end{bmatrix} \cdot \begin{bmatrix} 67.270 \\ 5043.650 \\ 4445.800 \end{bmatrix} = \begin{bmatrix} -10.094 \\ 0.020 \\ 0.347 \end{bmatrix}$$

Using the coefficients, the unknown value at point 0 can be estimated by

$$z_0 = -10.094 + (0.020)(69) + (0.347)(67) = 14.535$$

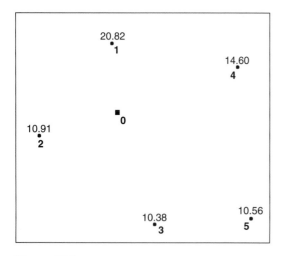

Figure 13.2
The unknown value at point 0 is interpolated by five surrounding stations with known values.

The distribution of most natural phenomena is usually more complex than an inclined plane surface from a first-order model. Higher-order trend surface models are required to approximate more complex surfaces. A cubic or third-order model, for example, includes hills and valleys. A cubic trend surface is based on the equation

(13.2)

$$z_{x,y} = b_0 + b_1 x + b_2 y + b_3 x^2 + b_4 xy + b_5 y^2 + b_6 x^3 + b_7 x^2 y + b_8 xy^2 + b_9 y^3$$

A GIS package may offer up to 12th-order trend surface models. Figure 13.3 shows an isoline map derived from a third-order trend surface of annual precipitation in Idaho created from 105 data points with a cell size of 2000 meters. An isoline map is

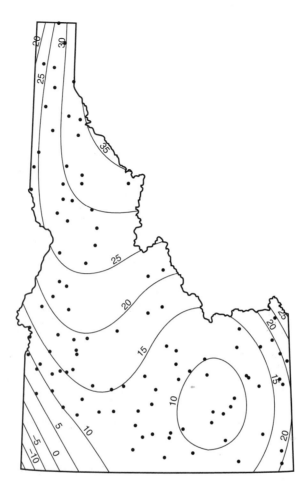

Figure 13.3
An isoline map of a third-order trend surface created from 105 control points with annual precipitation values.

like a contour map, useful for visualization as well as measurement. An obvious problem with the map is the negative values in the data-poor southwest corner of the state.

There are variations to the above trend surface analysis. Logistic trend surface analysis uses known points with binary data (i.e., 0 and 1) and produces a probability surface. **Local polynomial interpolation** uses a sample of known points to estimate the unknown value of a cell. Conversion of a triangulated irregular network (TIN) to a DEM (Chapter 12), for instance, uses local polynomial interpolation.

13.3.2 Regression Models

A **regression model** relates a dependent variable to a number of independent variables in a linear equation (an interpolator), which can then be used for prediction or estimation. Many regression models use nonspatial attributes such as income and education and are not considered methods for spatial interpolation. But exceptions can be made for regression models that use spatial variables such as distance to a river (Burrough and McDonnell 1998; Rogerson 2001).

A watershed-level regression model for snow accumulation developed by Chang and Li (2000) is an example. The model uses snow water equivalent (SWE) as the dependent variable and location and topographic variables as the independent variables. One of their watershed models takes the form of

(13.3)

$$SWE = b_0 + b_1 EASTING + b_2 SOUTHING + b_3 ELEV + b_4 PLAN1000$$

where EASTING and SOUTHING correspond to the column number and the row number in an elevation grid, ELEV is the elevation value, and PLAN1000 is a surface curvature measure. With the b coefficients in Eq. (13.3) estimated in a regression analysis using known values from snow courses, the model can then estimate SWE for all cells in the watershed and produce a continuous SWE surface.

PRISM (parameter-elevation regressions on independent slopes model), a popular tool for generating precipitation estimates in mountainous regions is another example (Daly et al. 1994). The input to PRISM consists of point measurements of climate data and a DEM, and the output is a predicted map in raster format. The estimate for each cell is based on a separate regression equation using data from nearby climate stations, which are in turn weighted by the variables of distance, eleva-

tion, vertical layer, topographic facet, and coastal proximity. The Natural Resources Conservation Service (NRCS) has used PRISM to map precipitation for all 50 states of the United States **(http://www.ftw.nrcs.usda.gov/prism/prismdata.html/).**

13.4 LOCAL METHODS

Because local interpolation uses a sample of known points, it is important to know how to select a sample. The first issue in sampling is the number of points to be used in estimation. GIS packages typically let users specify the number of points or use a default number (e.g., 7 to 12 points). One might assume that more points would result in more accurate estimates. But the validity of this assumption depends on the distribution of known points relative to the cell to be estimated, the extent of spatial autocorrelation, and the quality of data (Yang and Hodler 2000). More points usually imply more generalized estimations.

After the number of points is determined, the next task is to search for those known points (Figure 13.4). A simple option is to use the closest known points to the point to be estimated. Another option is to select known points within a radius, the size of which may depend on the distribution

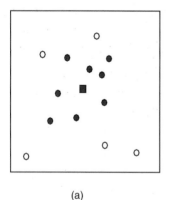
(a)

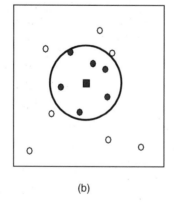
(b)

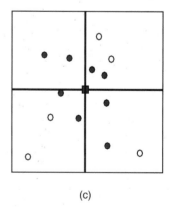
(c)

Figure 13.4
Three search methods for sample points: (a) find the closest points to the point to be estimated, (b) find points within a radius, and (c) find points within each of the four quadrants.

of known points. Other search options may incorporate the component of direction such as a quadrant or octant requirement (Davis 1986). A quadrant requirement means selecting known points from each of the four quadrants around a cell to be estimated. An octant requirement means using eight sectors. In some cases, known points are chosen from an ellipse representing the movement of air or water pollutants.

13.4.1 Thiessen Polygons

Thiessen polygons assume that any point within a polygon is closer to the polygon's known point than any other known points. Thiessen polygons were originally proposed to estimate areal averages of precipitation by making sure that any point within a polygon is closer to the polygon's weather station than any other station (Tabios and Salas 1985).

Thiessen polygons do not use an interpolator but require initial triangulation for connecting known points. Because different ways of connecting points can form different sets of triangles, the Delaunay triangulation—the same method for constructing a TIN—is often used in preparing Thiessen polygons (Davis 1986). The Delaunay triangulation ensures that each known point is connected to its nearest neighbors, and that triangles

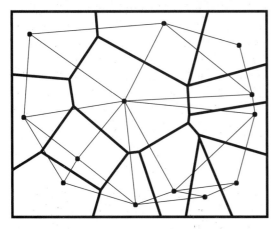

Figure 13.5
The diagram shows known points, Delaunay triangulation in thinner lines, and Thiessen polygons in thicker lines.

are as equilateral as possible. After triangulation, Thiessen polygons can be easily constructed by connecting lines drawn perpendicular to the sides of each triangle at their midpoints (Figure 13.5). Thiessen polygons are also called *Voronoi polygons.*

13.4.2 Density Estimation

Density estimation measures densities in a grid by using a sample of known points. There are simple and kernel density estimation methods.

To use the simple density estimation method, one can place a grid on a point distribution, tabulate points that fall within each cell, sum the point values, and estimate the cell's density by dividing the total point value by the cell size. Figure 13.6 shows the input and output of an example of simple density estimation. The input is a distribution of sighted deer locations plotted with a 50-meter interval to accommodate the resolution of telemetry. Each deer location has a count value measuring how many times a deer was sighted at the location. The output is a density grid, which has a cell size of 10,000 square meters or 1 hectare and a density measure of number of sightings per hectare. A circle, rectangle, wedge, or ring based at the center of a cell may replace the cell in the calculation.

Kernel estimation associates each known point with a kernel function for the purpose of estimation (Silverman 1986; Scott 1992; Bailey and Gatrell 1995). Expressed as a bivariate probability density function, a kernel function looks like a "bump," centering at a known point and tapering off to 0 over a defined bandwidth or window area (Silverman 1986) (Figure 13.7). The kernel function and the bandwidth determine the shape of the bump, which in turn determines the amount of smoothing in estimation. The kernel density estimator at point x is then the sum of bumps placed at the known points x_i within the bandwidth:

(13.4)

$$\hat{f}(x) = \frac{1}{nh^d} \sum_{i=1}^{n} K\left[\frac{1}{h}(x - x_i)\right]$$

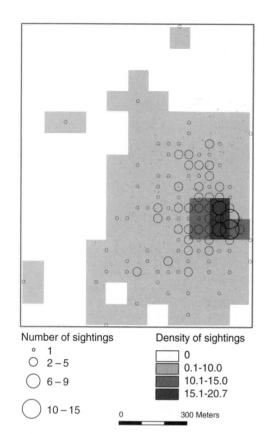

Number of sightings
- ° 1
- ○ 2 – 5
- ◯ 6 – 9
- ◯ 10 – 15

Density of sightings
- □ 0
- ▨ 0.1-10.0
- ▩ 10.1-15.0
- ■ 15.1-20.7

0 ——— 300 Meters

Figure 13.6
The simple density estimation method is used to compute the number of deer sightings per hectare from the point data.

where $K( )$ is the kernel function, h is the bandwidth, n is the number of known points within the bandwidth, and d is the data dimensionality. For two-dimensional data ($d = 2$), the kernel function is usually given by

(13.5)

$$K(x) = 3\pi^{-1}(1 - X^{T}X)^{2}, \text{ if } X^{T}X < 1$$

$$K(x) = 0, \text{ otherwise}$$

By substituting Eq. (13.5) for $K( )$, Eq. (13.4) can be rewritten as

(13.6)

$$\hat{f}(x) = \frac{3}{nh^{2}\pi} \sum_{i=1}^{n} \{1 - \frac{1}{h^{2}}[(x - x_{i})^{2} + (y - y_{i})^{2}]\}^{2}$$

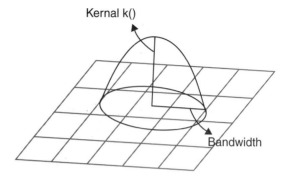

Kernal k()

Bandwidth

Figure 13.7
A kernel function, which represents a probability density function, looks like a "bump" above a grid.

where π is a constant, and $(x - x_{i})$ and $(y - y_{i})$ are the deviations in x-, y-coordinates between the point x and the observations x_{i} within the bandwidth.

Using the same input as for the simple estimation method, Figure 13.8 shows the output density grid from kernel estimation. Density values in the grid are expected values rather than probabilities (Box 13.3). Kernel estimation usually produces a smoother density surface than the simple estimation method.

13.4.3 Inverse Distance Weighted Interpolation

Inverse distance weighted (IDW) interpolation is an exact method that enforces that the estimated value of a point is influenced more by nearby known points than those farther away. The degree of influence is expressed by the inverse of the distance between points raised to a power. A power of 1.0 means a constant rate of change in value between points (linear interpolation). A power of 2.0 or higher suggests that the rate of change in values is higher near a known point and levels off away from it. One characteristic of IDW interpolation is that all predicted values are within the range of maximum and minimum values of the known points. The method has been popular in computer-assisted mapping since the 1970s (Monmonier 1982).

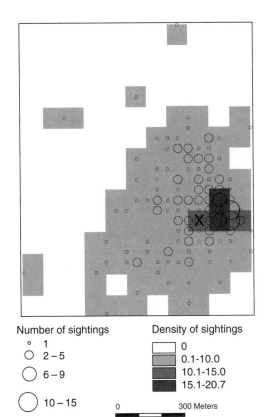

Number of sightings

° 1
○ 2 – 5
◯ 6 – 9
◯ 10 – 15

Density of sightings

□	0
▨	0.1-10.0
▨	10.1-15.0
■	15.1-20.7

0   300 Meters

Figure 13.8
The kernel estimation method is used to compute the number of sightings per hectare from the point data. The letter X marks the cell, which is used as an example in Box 13.3.

The general equation for the IDW method is

(13.7)

$$z_0 = \frac{\sum\limits_{i=1}^{s} z_i \dfrac{1}{d_i^k}}{\sum\limits_{i=1}^{s} \dfrac{1}{d_i^k}}$$

where z_0 is the estimated value at point 0, z_i is the z value at known point i, d_i is the distance between point i and point 0, s is the number of known points used in estimation, and k is the specified power.

Figure 13.9 shows an annual precipitation surface in Idaho created by the IDW method with a power of 2 from 105 known points (Box 13.4). Figure 13.10 shows an isoline map of the surface, which is also called an isohyet map. Small, enclosed isolines are typical of IDW interpolation. Irregular isolines in the southwest corner of the state are due to the absence of known points.

13.4.4 Thin-Plate Splines

Splines for spatial interpolation are conceptually similar to splines for line generalization (Chapter 5), except that in spatial interpolation they apply to surfaces rather than lines. **Thin-plate splines** create a surface that passes through the control points and has the least possible change in slope at all points (Franke 1982). In other words, thin-plate

---

 Box **13.3** **A Worked Example of Kernel Estimation**

Τhis example shows how the value of the cell marked X in Figure 13.8 is derived. The window area is defined as a circle with a radius of 100 meters (h). Therefore, only points within the 100-meter radius of the center of the cell can influence the estimation of the cell density. Using the 10 points within the cell's neighborhood, we can compute the cell density by

$$\frac{3}{\pi} \sum_{i=1}^{10} n_i \left\{ 1 - \frac{1}{h^2} [(x - x_i)^2 + (y - y_i)^2] \right\}^2$$

where n_i is the number of sightings at point i, x_i and y_i are the x-, y-coordinates of point i, and x and y are the x-, y-coordinates of the center of the cell to be estimated. Because the density is measured per 10,000 square meters or hectare, h^2 in Eq. (13.6) is canceled out. Also, because the output shows an expected value rather than a probability, n in Eq. (13.6) is not needed. The computation shows the cell density to be 11.421.

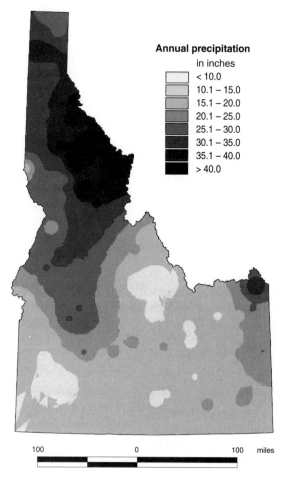

Figure 13.9
An annual precipitation surface map created by the inverse distance weighted method.

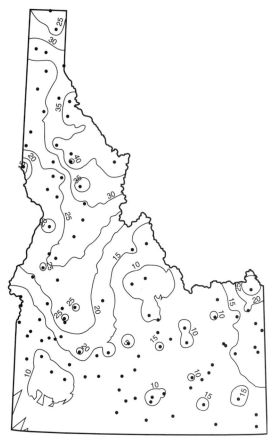

Figure 13.10
An isohyet map created by the inverse distance weighted method.

splines fit the control points with a minimum-curvature surface. The approximation of thin-plate splines is of the form

(13.8)

$$Q(x,y) = \sum A_i d_i^2 \log d_i + a + bx + cy$$

where x and y are the x-, y-coordinates of the point to be interpolated, $d_i^2 = (x - x_i)^2 + (y - y_i)^2$, and x_i and y_i are the x-, y-coordinates of control point i. Thin-plate splines consist of two components: $(a + bx + cy)$ represents the local trend function,

which has the same form as a linear or first-order trend surface, and $d_i^2 \log d_i$ represents a basis function, which is designed to obtain minimum curvature surfaces (Watson 1992). The coefficients A_i, and a, b, and c are determined by a linear system of equations (Franke 1982)

(13.9)

$$\sum_{i=1}^{n} A_i d_i^2 \log d_i + a + bx + cy = f_i$$

(13.10)

$$\sum_{i=1}^{n} A_i = 0$$

Box 13.4 A Worked Example of Inverse Distance Weighted Estimation

This example uses the same data set as in Box 13.2, but interpolates the unknown value at point 0 by the IDW method. The table below shows the distances in thousands of meters between point 0 and the five known points:

Between Points	Distance
0,1	18.000
0,2	20.880
0,3	32.310
0,4	36.056
0,5	47.202

We can substitute the known values and the distances into Eq. (13.7) and estimate z_0:

$$\sum z_i \, 1/d_i^2 = (20.820)(1/18.000)^2 +$$
$$(10.910)(1/20.880)^2 + (10.380)(1/32.310)^2 +$$
$$(14.600)(1/36.056)^2 + (10.560)(1/47.202)^2 = 0.1152$$

$$\sum 1/d_i^2 = (1/18.000)^2 + (1/20.880)^2 +$$
$$(1/32.310)^2 + (1/36.056)^2 + (1/47.202)^2 = 0.0076$$

$$z_0 = 0.1152/0.0076 = 15.158$$

(13.11)

$$\sum_{i=1}^{n} A_i x_i = 0$$

(13.12)

$$\sum_{i=1}^{n} A_i y_i = 0$$

where n is the number of control points, and f_i is the known value at control point i. The estimation of the coefficients requires $n + 3$ simultaneous equations.

Other algorithms may be used for creating minimum-curvature surfaces. For example, instead of the basis function $d^2 \log d$ in Eq. (13.9), the bi-harmonic Green function, $d^2(\log d - 1)$, may be used (Watson 1992; Middleton 2000).

Unlike the IDW method, predicted values from thin-plate splines are not limited within the range of maximum and minimum values of the known points. In fact, a major problem with thin-plate splines is the steep gradients in data-poor areas, often referred to as overshoots. Different methods for correcting overshoots have been proposed. Thin-plate splines with tension, for example, allow the user to control the tension to be

pulled on the edges of the surface (Franke 1985; Mitas and Mitasova 1988). Other methods include regularized splines (Mitas and Mitasova 1988) and regularized splines with tension (Mitasova and Mitas 1993). All these methods belong to a diverse group called **radial basis functions** (Box 13.5).

The approximation of **regularized splines** has the same local trend function as thin-plate splines, but the basis function has a different form

(13.13)

$$\frac{1}{2\pi}\left\{\frac{d^2}{4}\left[\ln\left(\frac{d}{2\tau}\right) + c - 1\right] + \tau^2\left[K_0\left(\frac{d}{\tau}\right) + c + \ln\left(\frac{d}{2\pi}\right)\right]\right\}$$

where τ is the weight to be used with the splines method, d is the distance between the point to be interpolated and control point i, c is a constant of 0.577215, and $K_0(d/\tau)$ is the modified zeroth-order Bessel function, which can be approximated by a polynomial equation (Abramowitz and Stegun 1964). The τ value is usually set between 0 and 0.5 because higher τ values tend to result in overshoots in data-poor areas. ArcInfo Workstation uses the default τ value of 0.1.

Box 13.5 Radial Basis Functions

Radial basis functions (RBF) refer to a large group of interpolation methods. All of them are exact interpolators. The selection of a basis function or equation determines how the surface will fit between the control points. Both ArcInfo Workstation and the Spatial Analyst extension to ArcGIS offer thin-plate splines with tension and regularized splines. The Geostatistical Analyst extension to ArcGIS, on the other hand, has a menu choice of five RBF methods: thin-

plate spline, spline with tension, completely regularized spline, multiquadric function, and inverse multiquadric function. Each RBF method also has a parameter that controls the smoothness of the generated surface. Although each combination of an RBF method and a parameter value can create a new surface, the difference between the surfaces is usually small.

The **thin-plate splines with tension** method has the following form:

(13.14)

$$a + \sum_{i=1}^{n} A_i R(d_i)$$

where a represents the trend function, and the basis function $R(d)$ is

(13.15)

$$-\frac{1}{2\pi\varphi^2}\left[\ln\left(\frac{d\varphi}{2}\right) + c + K_0(d\varphi)\right]$$

where φ is the weight to be used with the tension method. If the weight φ is set close to 0, then the approximation with tension is similar to the basic thin-plate splines method. A larger φ value reduces the stiffness of the plate and thus the range of interpolated values so that the interpolated surface resembles the shape of a membrane passing through the control points (Franke 1985). ArcInfo Workstation uses the default φ value of 0.1.

Thin-plate splines and their variations are recommended for smooth, continuous surfaces such as elevation and water table. Splines have also been used for interpolating mean rainfall surface (Hutchinson 1995) and land demand surface (Wickham et al. 2000). Figures 13.11 and 13.12, respectively, show annual precipitation surfaces created by the regularized splines method and the

splines with tension method (Box 13.6). The isolines in both figures are very smooth, compared to isolines in Figure 13.10, which were created by the IDW method. Interpolation by splines in data-poor areas can be excessive, as exemplified by the 45-inch isohyet in north Idaho in Figure 13.11.

13.4.5 Kriging

Kriging (after the South African mining engineer, D. G. Krige) is a geostatistical method for spatial interpolation. The technique of kriging assumes that the spatial variation of an attribute such as changes in grade within an ore body is neither totally random (stochastic) nor deterministic (Davis 1986; Isaaks and Srivastava 1989; Webster and Oliver 1990; Cressie 1991; Bailey and Gatrell 1995; Webster and Oliver 2001). Instead, the spatial variation may consist of three components: a spatially correlated component, representing the variation of the regionalized variable; a "drift" or structure, representing a trend; and a random error term. The presence or absence of a drift and the interpretation of the regionalized variable have led to development of different kriging methods for spatial interpolation.

13.4.5.1 Ordinary Kriging

Assuming the absence of a drift, **ordinary kriging** focuses on the spatially correlated component. The

Figure 13.11
An isohyet map created by the regularized splines method.

Figure 13.12
An isohyet map created by the splines with tension method.

measure of the degree of spatial dependence among the sampled points is the average **semi-variance:**

(13.16)

$$\gamma(h) = \frac{1}{2n} \sum_{i=1}^{n} [z(\mathrm{x}_i) - z(x_i + h)]^2$$

where h is the distance or lag between sampled points; n is the number of pairs of sampled points separated by h; and z is the attribute value. The computation of $\gamma(h)$ follows two steps. First, pairs of sampled points are grouped by distance. For example, if the distance is 2000 meters, then pairs of points separated by less than 2000 meters are grouped into 0–2000, pairs of points separated between 2000 and 4000 meters are grouped into 2000–4000, and so on. Second, the average distance h and the average semivariance $\gamma(h)$ are calculated for each group. If spatial dependence exists among the points, then pairs of points that are closer in distance will have more similar values than pairs that are farther apart. Thus, $\gamma(h)$ is expected to increase as h increases in the presence of spatial dependence.

A **semivariogram** is a plot that has $\gamma(h)$ along the y-axis and h along the x-axis (Figure 13.13). A semivariogram can be dissected into three possible elements: nugget, range, and sill. The nugget is the semivariance at the distance of 0, representing the spatially uncorrelated noise. The range is the

Box 13.6 A Worked Example of Thin-Plate Splines with Tension

This example uses the same data set as in Box 13.2 but interpolates the unknown value at point 0 by the splines with tension method. The method first involves calculation of $R(d)$ in Eq. (13.15) using the distances between the point to be estimated and the known points, distances between the known points, and the φ value of 0.1. The following table shows the $R(d)$ values along with the distance values.

Points	0,1	0,2	0,3	0,4	0,5
Distance	18.000	20.880	32.310	36.056	47.202
$R(d)$	-7.510	-9.879	-16.831	-18.574	-22.834
Points	1,2	1,3	1,4	1,5	2,3
Distance	31.240	49.476	34.526	59.666	40.000
$R(d)$	-16.289	-23.612	-17.879	-26.591	-20.225
Points	2,4	2,5	3,4	3,5	4,5
Distance	56.920	62.032	47.412	26.076	40.200
$R(d)$	-25.843	-27.214	-22.868	-13.415	-20.305

The next step is to solve for A_i in Eq. (13.14). We can substitute the calculated $R(d)$ values into Eq. (13.14)

and rewrite the equation and the constraint about A_i in matrix form

$$\begin{bmatrix} 1 & 0 & -16.289 & -23.612 & -17.879 & -26.591 \\ 1 & -16.289 & 0 & -20.225 & -25.843 & -27.214 \\ 1 & -23.612 & -20.225 & 0 & -22.868 & -13.415 \\ 1 & -17.879 & -25.843 & -22.868 & 0 & -20.305 \\ 1 & -26.591 & -27.214 & -13.415 & -20.305 & 0 \\ 0 & 1 & 1 & 1 & 1 & 1 \end{bmatrix} \cdot \begin{bmatrix} a \\ A_1 \\ A_2 \\ A_3 \\ A_4 \\ A_5 \end{bmatrix} = \begin{bmatrix} 20.820 \\ 10.910 \\ 10.380 \\ 14.600 \\ 10.560 \\ 0 \end{bmatrix}$$

The matrix solutions are

$$a = 13.203 \quad A_1 = 0.396 \quad A_2 = -0.226$$
$$A_3 = -0.058 \quad A_4 = -0.047 \quad A_5 = -0.065$$

Now we can calculate the value at point 0 by

$$z_0 = 13.203 + (0.396)(-7.510) + (-0.226)(-9.879)$$
$$+ \qquad (-0.058)(-16.831) \qquad +$$
$$(-0.047)(-18.574) + (-0.065)(-22.834) = 15.795$$

The splines with tension method is one of the thin-plate splines methods discussed in the text. Using the same data set, the estimation of z_0 by other methods is as follows: 16.350 by thin-plate splines and 15.015 by regularized splines (using the τ value of 0.1).

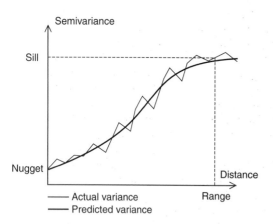

Figure 13.13

A semivariogram shows semivariances along the *y*-axis and distances along the *x*-axis.

spatially correlated portion of the semivariogram that shows increases in the semivariance with the distance. Beyond the range, the semivariance levels off to a relatively constant value. The semivariance, at which the leveling takes place, is called the sill.

A semivariogram may be used alone as a measure of spatial correlation in the same way as spatial autocorrelation (Chapter 11). But to be used as an interpolator in kriging, the semivariogram must be fitted with a mathematical function or model. The fitted semivariogram can then be used for estimating the semivariance at any given distance. Fitting a model to a semivariogram is a difficult and often controversial task in geostatistics (Webster and Oliver 2001). One reason for the difficulty is the number of models to choose from. For example, the Geostatistical Analyst extension to

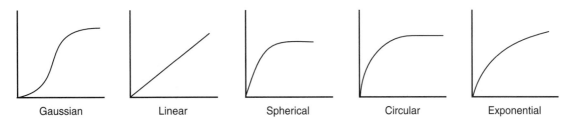

Figure 13.14
Five mathematical models for fitting semivariograms: Gaussian, linear, spherical, circular, and exponential.

ArcGIS offers 11 models including spherical, circular, exponential, and Gaussian (Figure 13.14). The other reason is the lack of a standardized procedure for comparing the models. Webster and Oliver (2001) have recommended a procedure that combines visual inspection and cross-validation. Cross-validation, as discussed later in Section 13.4.6, is a method for comparing interpolation methods.

Ordinary kriging uses the fitted semivariogram directly in spatial interpolation. The general equation for estimating the z value at a point is

(13.17)

$$z_0 = \sum_{i=1}^{s} z_x W_x$$

where z_0 is the estimated value, z_x is the known value at point x, W_x is the weight associated with point x, and s is the number of sampled points used in estimation. The weights can be derived from solving a set of simultaneous equations. For example, the following equations are needed for a point (0) to be estimated from three known points (1, 2, 3):

(13.18)

$$W_1\gamma(h_{11}) + W_2\gamma(h_{12}) + W_3\gamma(h_{13}) + \lambda = \gamma(h_{10})$$
$$W_1\gamma(h_{21}) + W_2\gamma(h_{22}) + W_3\gamma(h_{23}) + \lambda = \gamma(h_{20})$$
$$W_1\gamma(h_{31}) + W_2\gamma(h_{32}) + W_3\gamma(h_{33}) + \lambda = \gamma(h_{30})$$
$$W_1 + W_2 + W_3 + 0 = 1.0$$

where $\gamma(h_{ij})$ is the semivariance between known points i and j, $\gamma(h_{i0})$ is the semivariance between the ith known point and the point to be estimated, and λ is a Lagrange multiplier, which is added to ensure the minimum possible estimation error. The above equations can be rewritten in matrix form

$$\begin{bmatrix} \gamma(h_{11}) & \gamma(h_{12}) & \gamma(h_{13}) & 1 \\ \gamma(h_{21}) & \gamma(h_{22}) & \gamma(h_{23}) & 1 \\ \gamma(h_{31}) & \gamma(h_{32}) & \gamma(h_{33}) & 1 \\ 1 & 1 & 1 & 0 \end{bmatrix} \cdot \begin{bmatrix} W_1 \\ W_2 \\ W_3 \\ \lambda \end{bmatrix} = \begin{bmatrix} \gamma(h_{10}) \\ \gamma(h_{20}) \\ \gamma(h_{30}) \\ 1 \end{bmatrix}$$

Once the weights are solved, Eq. (13.17) can be used to estimate z_0

$$z_0 = z_1 W_1 + z_2 W_2 + z_3 W_3$$

The above example shows that weights used in kriging involve not only semivariances between the point to be estimated and the known points but also those between the known points. This is different from the IDW method, which uses only weights applicable to the point to be estimated and the known points. Another important difference between kriging and other local methods is that kriging produces a variance measure for each estimated point to indicate the reliability of the estimation. For the above example, the variance estimation can be calculated by

(13.19)

$$s^2 = W_1\gamma(h_{10}) + W_2\gamma(h_{20}) + W_3\gamma(h_{30}) + \lambda$$

Figure 13.15 shows an annual precipitation surface created by ordinary kriging with the linear model (Box 13.7). Figure 13.16 shows the distribution of the standard deviation of the kriged surface. (The

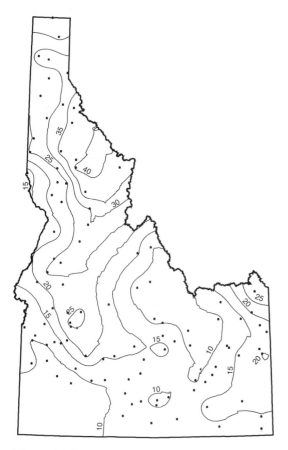

Figure 13.15
An isohyet map created by ordinary kriging with the linear model.

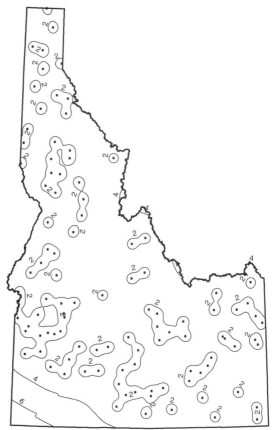

Figure 13.16
The map shows the standard deviation of the annual precipitation surface created by ordinary kriging with the linear model.

standard error of the estimate is another term for the standard deviation.) As expected, the standard deviation is highest in data-poor areas, such as the southwest corner of Idaho.

13.4.5.2 Universal Kriging

Universal kriging assumes that the spatial variation in z values has a drift or a trend in addition to the spatial correlation between the sampled points. Typically, universal kriging incorporates a first-order (plane surface) or a second-order (quadratic surface) polynomial in the kriging process. A first-order polynomial is

(13.20)

$$M = b_1 x_i + b_2 y_i$$

where M is the drift, x_i and y_i are the x-, y-coordinates of sampled point i, and b_1 and b_2 are the drift coefficients. A second-order polynomial is

(13.21)

$$M = b_1 x_i + b_2 y_i + b_3 x_i^2 + b_4 x_i y_i + b_5 y_i^2$$

Higher-order polynomials are usually not recommended for two reasons. First, kriging is

Box 13.7 A Worked Example of Ordinary Kriging Estimation

This worked example uses ordinary kriging for spatial interpolation. The first step is to construct a semivariogram by using point data of average annual precipitation values at 105 weather stations in Idaho (Figure 13.17). The distance interval in the semivariogram is measured in meters. The next step is to fit the semivariogram with a mathematical function. This example chooses the linear model, which is defined by

$$\gamma(h) = C_0 + C\,(h/a),\; 0 < h <= a$$
$$\gamma(h) = C_0 + C,\; h > a$$
$$\gamma(0) = 0$$

where $\gamma(h)$ is the semivariance at distance h, C_0 is the semivariance at distance 0, a is the range, and C is the sill, or the semivariance at a. The output from ArcInfo Workstation shows: $C_0 = 0$, $C = 112.475$, and $a = 458,000$.

Now we can use the model for spatial interpolation. The scenario is the same as in Box 13.2: using five points with known values to estimate an unknown value. This step begins with the computation of distances between points and the semivariances at those distances based on the linear model. The results are shown in the following table, with distances in thousands of meters:

Points ij	0,1	0,2	0,3	0,4	0,5
h_{ij}	18.000	20.880	32.310	36.056	47.202
$\gamma(h_{ij})$	4.420	5.128	7.935	8.855	11.592
Points ij	1,2	1,3	1,4	1,5	2,3
h_{ij}	31.240	49.476	34.526	59.666	40.000
$\gamma(h_{ij})$	7.672	12.150	8.479	14.653	9.823
Points ij	2,4	2,5	3,4	3,5	4,5
h_{ij}	56.920	62.032	47.412	26.076	40.200
$\gamma(h_{ij})$	13.978	15.234	11.643	6.404	9.872

Using the semivariances, we can write the simultaneous equations for solving the weights in matrix form

$$
\begin{bmatrix}
0 & 7.672 & 12.150 & 8.479 & 14.653 & 1 \\
7.672 & 0 & 9.823 & 13.978 & 15.234 & 1 \\
12.150 & 9.823 & 0 & 11.643 & 6.404 & 1 \\
8.479 & 13.978 & 11.643 & 0 & 9.872 & 1 \\
14.653 & 15.234 & 6.404 & 9.872 & 0 & 1 \\
1 & 1 & 1 & 1 & 1 & 0
\end{bmatrix}
\cdot
\begin{bmatrix}
W_1 \\ W_2 \\ W_3 \\ W_4 \\ W_5 \\ \lambda
\end{bmatrix}
=
\begin{bmatrix}
4.420 \\ 5.128 \\ 7.935 \\ 8.855 \\ 11.592 \\ 1
\end{bmatrix}
$$

The matrix solutions are

$$W_1 = 0.397 \quad W_2 = 0.318 \quad W_3 = 0.182$$
$$W_4 = 0.094 \quad W_5 = 0.009 \quad \lambda = -1.161$$

Using Eq. (13.17), we can estimate the unknown value at point 0 by

$$z_0 = (0.397)(20.820) + (0.318)(10.910) +$$
$$(0.182)(10.380) + (0.094)(14.600) +$$
$$(0.009)(10.560) = 15.091$$

We can also estimate the variance at point 0 by

$$s^2 = (4.420)(0.397) + (5.128)(0.318) +$$
$$(7.935)(0.182) + (8.855)(0.094) +$$
$$(11.592)(0.009) - 1.161 = 4.605$$

In other words, the standard deviation (s), or the standard error of estimate, at point 0 is 2.146.

performed on the residuals after the trend is removed. A higher-order polynomial will leave little variation in the residuals for assessing uncertainty. Second, a higher-order polynomial means a larger number of the b_i coefficients, which must be esti-

mated along with the weights, and a larger set of simultaneous equations to be solved.

Figure 13.18 shows an annual precipitation surface created by universal kriging with the linear (first-order) drift, and Figure 13.19 shows the

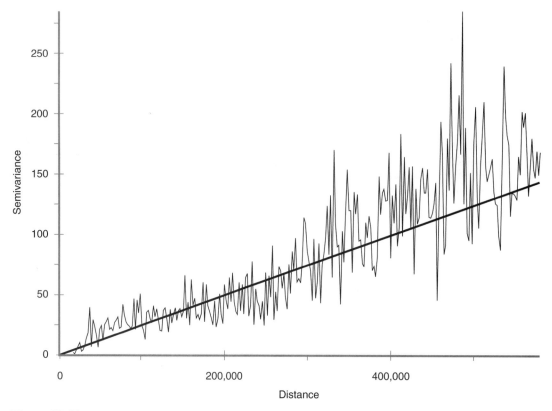

Figure 13.17
A semivariogram constructed from annual precipitation values at 105 weather stations in Idaho. The linear model provides the trend line.

distribution of the standard deviation of the kriged surface (Box 13.8). Both maps are similar to those created by ordinary kriging based on the linear model. The main difference is in the southwest corner of Idaho with the presence of negative values, a result of extrapolation by the linear drift of the universal kriging model.

13.4.5.3 Other Kriging Methods

Besides ordinary kriging and universal kriging, other kriging methods such as block kriging and cokriging have been proposed in the literature (Bailey and Gatrell 1995; Burrough and McDonnell 1998; Webster and Oliver 2001). Block kriging estimates the average value of a variable over

some small area or block rather than at a point. Cokriging uses one or more secondary variables, which are correlated with the primary variable of interest, in interpolation. It assumes that the correlation between the variables can be used to improve the prediction of the value of the primary variable. For example, better results in precipitation interpolation have been reported by including elevation as an additional variable in cokriging (Martinez-Cob 1996).

13.4.6 Comparison of Spatial Interpolation Methods

Using the same data but different methods, one can expect to find different interpolation results. Like-

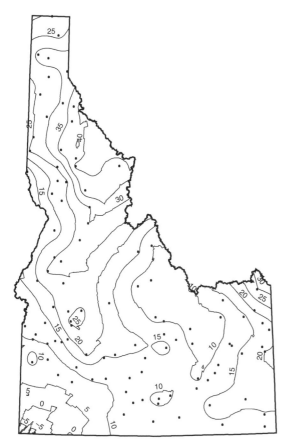

Figure 13.18
An isohyet map created by universal kriging with the linear drift.

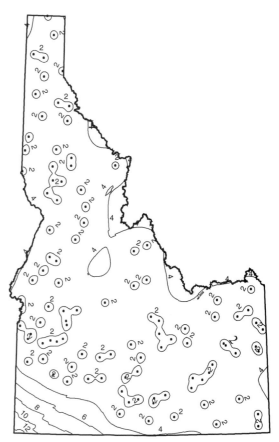

Figure 13.19
The map shows the standard deviation of the annual precipitation surface created by universal kriging with the linear drift.

wise, different predicted values can occur by using the same method but different parameter values. GIS packages such as ArcGIS offer a variety of spatial interpolation methods (Box 13.9). How can we compare these methods? This section describes the methods and useful statistics for comparison.

Figure 13.20 shows the difference of annual precipitation estimates between regularized splines and IDW. Figure 13.21 shows the difference between ordinary kriging with the linear model and regularized splines. The two maps have similar general patterns, although the difference between ordinary kriging and regularized splines is smaller

than that between regularized splines and IDW. Notice that on both maps, the data-poor areas have the largest difference (i.e., more than 3 inches either positively or negatively). This suggests the importance of having more control points in the data-poor area. No matter which method is used, spatial interpolation can never substitute for observed data. But if the option of adding more points is not feasible, how can one tell which interpolation method, or which parameter value, is better?

Comparison of local methods is usually based on statistical measures, although some studies

Box 13.8 A Worked Example of Universal Kriging Estimation

This example uses universal kriging to estimate the unknown value at point 0 (Box 13.2) and assumes that (1) the drift is linear and (2) the semivariogram is fitted with a linear model. Because of the additional drift component, this example uses eight simultaneous equations:

$$W_1\gamma(h_{11}) + W_2\gamma(h_{12}) + W_3\gamma(h_{13}) + W_4\gamma(h_{14}) + W_5\gamma(h_{15}) + \lambda + b_1x_1 + b_2y_1 = \gamma(h_{10})$$

$$W_1\gamma(h_{21}) + W_2\gamma(h_{22}) + W_3\gamma(h_{23}) + W_4\gamma(h_{24}) + W_5\gamma(h_{25}) + \lambda + b_1x_2 + b_2y_2 = \gamma(h_{20})$$

$$W_1\gamma(h_{31}) + W_2\gamma(h_{32}) + W_3\gamma(h_{33}) + W_4\gamma(h_{34}) + W_5\gamma(h_{35}) + \lambda + b_1x_3 + b_2y_3 = \gamma(h_{30})$$

$$W_1\gamma(h_{41}) + W_2\gamma(h_{42}) + W_3\gamma(h_{43}) + W_4\gamma(h_{44}) + W_5\gamma(h_{45}) + \lambda + b_1x_4 + b_2y_4 = \gamma(h_{40})$$

$$W_1\gamma(h_{51}) + W_2\gamma(h_{52}) + W_3\gamma(h_{53}) + W_4\gamma(h_{54}) + W_5\gamma(h_{55}) + \lambda + b_1x_5 + b_2y_5 = \gamma(h_{50})$$

$$W_1 + W_2 + W_3 + W_4 + W_5 + 0 + 0 + 0 = 1$$

$$W_1x_1 + W_2x_2 + W_3x_3 + W_4x_4 + W_5x_5 + 0 + 0 + 0 = x_0$$

$$W_1y_1 + W_2y_2 + W_3y_3 + W_4y_4 + W_5y_5 + 0 + 0 + 0 = y_0$$

where x_0 and y_0 are the x-, y-coordinates of the point to be estimated, and x_i and y_i are the x-, y-coordinates of known point i; otherwise, the notations are the same as in Box 13.7. The x-, y-coordinates are actually rows and columns in the output grid using a cell size of 2000 meters.

Similar to ordinary kriging, semivariance values for the equations can be derived from the semivariogram and the linear model. The next step is to rewrite the equations in matrix form:

$$\begin{bmatrix} 0 & 7.672 & 12.150 & 8.479 & 14.653 & 1 & 69 & 76 \\ 7.672 & 0 & 9.823 & 13.978 & 15.234 & 1 & 59 & 64 \\ 12.150 & 9.823 & 0 & 11.643 & 6.404 & 1 & 75 & 52 \\ 8.479 & 13.978 & 11.643 & 0 & 9.872 & 1 & 86 & 73 \\ 14.653 & 15.234 & 6.404 & 9.872 & 0 & 1 & 88 & 53 \\ 1 & 1 & 1 & 1 & 1 & 0 & 0 & 0 \\ 69 & 59 & 75 & 86 & 88 & 0 & 0 & 0 \\ 76 & 64 & 52 & 73 & 53 & 0 & 0 & 0 \end{bmatrix} \cdot \begin{bmatrix} W_1 \\ W_2 \\ W_3 \\ W_4 \\ W_5 \\ \lambda \\ b_1 \\ b_2 \end{bmatrix} = \begin{bmatrix} 4.420 \\ 5.128 \\ 7.935 \\ 8.855 \\ 11.592 \\ 1 \\ 69 \\ 67 \end{bmatrix}$$

The solutions are

$$W_1 = 0.387 \quad W_2 = 0.311 \quad W_3 = 0.188 \quad W_4 = 0.093$$
$$W_5 = 0.021 \quad \lambda = -1.154 \quad b_1 = 0.009 \quad b_2 = -0.010$$

The estimated value at point 0 is

$$z_0 = (0.387)(20.820) + (0.311)(10.910) + (0.188)(10.380) + (0.093)(14.600) + (0.021)(10.560) = 14.981$$

And, the variance at point 0 is

$$s^2 = (4.420)(0.387) + (5.128)(0.311) + (7.935)(0.188) + (8.855)(0.093) + (11.592)(0.021) - 1.154 = 4.710$$

The standard deviation (s) at point 0 is 2.170. These results from universal kriging are very similar to those from ordinary kriging.

---

have also suggested the importance of the visual quality of generated surfaces such as preservation of distinct spatial pattern and visual pleasantness and faithfulness (Declercq 1996; Yang and Hodler 2000). Cross-validation and validation are two common statistical techniques for comparison.

Cross-validation compares the interpolation methods by repeating the following procedure for each interpolation method to be compared:

1. Remove a known point from the data set.
2. Use the remaining points to estimate the value at the point previously removed.
3. Calculate the predicted error of the estimation by comparing the estimated with the known value.

After completing the procedure for each known point, one can calculate diagnostic statistics to

Box 13.9 Spatial Interpolation Using ArcGIS

ArcInfo Workstation has the command TREND for running trend surface analysis from first to 12th order and REGRESSION for running regression analysis. The commands for local interpolation in ArcInfo Workstation are: THIESSEN for Thiessen polygons, POINTDENSITY for density estimation (simple and kernel), IDW for the inverse distance weighted interpolation method, SPLINE for thin-plate splines with tension and regularized splines, and KRIGING for ordinary kriging and universal kriging.

The Spatial Analyst extension to ArcGIS has menu access to density estimation, IDW, splines (splines with tension and regularized splines), and Kriging

(ordinary and universal). Most users, however, will opt to run spatial interpolation using the Geostatistical Analyst extension. Geostatistical Analyst's main menu has the selections of explore data, geostatistical wizard, and create subsets. The explore data selection offers histogram, semivariogram, QQ plot, and others for exploratory data analysis (Chapter 9). The geostatistical wizard offers a large variety of global and local interpolation methods including IDW, trend surface, local polynomial, radial basis function, kriging, and cokriging. The wizard also provides cross-validation measures. The create subsets selection is designed for model validation.

---

assess the accuracy of the interpolation method. Two common diagnostic statistics are the root mean square (RMS) and the standardized RMS:

(13.22)

$$RMS = \sqrt{\frac{1}{n} \sum_{i=1}^{n} (z_{i,\,act} - z_{i,\,est})^2}$$

(13.23)

$$Standardized\ RMS = \sqrt{\frac{1}{n} \sum_{i=1}^{n} \frac{(z_{i,\,act} - z_{i,\,est})^2}{s^2}}$$

$$= \frac{RMS}{s}$$

where n is the number of points, $z_{i,act}$ is the known value of point i, $z_{i,est}$ is the estimated value of point i, s^2 is the variance, and s is the standard deviation.

The RMS statistic is available for all exact local methods. But the standardized RMS is only available for kriging because the variance is required for the computation. The interpretation of the statistics is as follows:

- A better interpolation method should yield a smaller RMS.
- A better kriging method should yield a smaller RMS and a standardized RMS closer to 1.

If the standardized RMS is 1, it means that the RMS equals s. Therefore, the estimated standard deviation is a reliable measure of the uncertainty of predicted values.

Cross-validation has shown, for example, the use of elevation as an additional variable to the x, y location in elevation-detrended kriging or cokriging yields better results than kriging (Phillips et al. 1992; Garen et al. 1994; Carroll and Cressie 1996). The Applications section of this chapter uses RMS statistics to find the optimal parameter value for an interpolation method.

The **validation** technique compares the interpolation methods by first dividing known points into two samples: one sample for developing the models for each interpolation method to be compared and the other sample for testing the accuracy of the models. The diagnostic statistics of RMS and standardized RMS derived from the test sample can then be used to compare the methods. Validation may not be a feasible option if the number of known points is too small to be split into two samples.

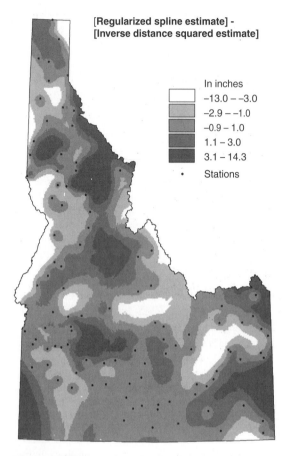

Figure 13.20
The map shows the difference between surfaces generated from the regularized splines method and the inverse distance weighted method. A local operation, in which one surface grid was subtracted from the other, created the map.

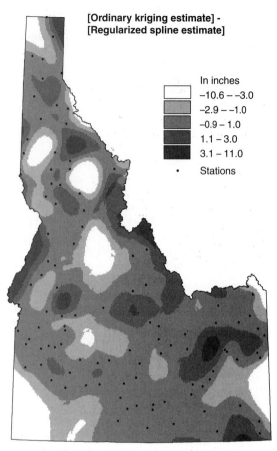

Figure 13.21
The map shows the difference between surfaces generated from the ordinary kriging with linear model method and the regularized splines method.

KEY CONCEPTS AND TERMS

Control points: Points with known values in spatial interpolation. Also called *known points*, *sampled points*, or *observations*.

Cross-validation: A technique for comparing interpolation methods, which removes one control point at a time and uses the remaining points to estimate the value at the location of the point removed.

Density estimation: A local interpolation method, which measures densities in a grid based on a distribution of points and point values.

Deterministic interpolation: A spatial interpolation method that provides no assessment of errors with predicted values.

Exact interpolation: An interpolation method that predicts a value that is the same as the known value of a control point.

Global interpolation: An interpolation method that uses every control point available in estimating an unknown value.

Inexact interpolation: An interpolation method that predicts a value that differs from the known value of a control point.

Interpolator: A mathematical equation for spatial interpolation.

Inverse distance weighted interpolation: A local interpolation method, which assumes that the unknown value of a point is influenced more by nearby points than those farther away.

Kernel estimation: A local interpolation method, which associates each known point with a kernel function in the form of a bivariate probability density function.

Kriging: A local interpolation method, which assumes that the spatial variation of an attribute includes a spatially correlated component, representing the variation of the regionalized variable.

Local interpolation: An interpolation method that uses a sample of known points in estimating an unknown value.

Local polynomial interpolation: A local interpolation method that uses a sample of points with known values and a polynomial equation to estimate the unknown value of a point.

Ordinary kriging: A kriging method, which assumes the absence of a drift or trend and focuses on the spatially correlated component.

Radial basis functions: A diverse group of methods for spatial interpolation including thin-plate splines, thin-plate splines with tension, and regularized splines.

Regression model: A global interpolation method that uses a number of independent variables to estimate a dependent variable.

Regularized splines: A variation of thin-plate splines for spatial interpolation.

Semivariance: A measure of the degree of spatial dependence among points used in kriging.

Semivariogram: A diagram relating the semivariance to the distance between points used in kriging.

Spatial interpolation: The process of using points with known values to estimate unknown values at other points.

Stochastic interpolation: A spatial interpolation method that offers assessment of prediction errors with estimated variances.

Thiessen polygons: A local interpolation method, which ensures that every unsampled point within a polygon is closer to the polygon's known point than any other known points.

Thin-plate splines: A local interpolation method, which creates a surface passing through points with the least possible change in slope at all points.

Thin-plate splines with tension: A variation of thin-plate splines for spatial interpolation.

Trend surface analysis: A global interpolation method that uses points with known values and a polynomial equation to approximate a surface.

Universal kriging: A kriging method, which assumes that the spatial variation of an attribute has a drift or a structural component in addition to the spatial correlation between sampled points.

Validation: A technique for comparing interpolation methods, which splits control points into two samples, one for developing the model and the other for testing the accuracy of the model.

APPLICATIONS: SPATIAL INTERPOLATION

This applications section covers seven tasks. Task 1 covers the global method of trend surface analysis. Tasks 2 through 7 deal with the local interpolation methods of kernel density estimation, IDW, thin-plate spline, and kriging. Kriging includes ordinary kriging, universal kriging, and cokriging. You can access the local interpolation methods of IDW, spline, and kriging through Spatial Analyst, 3D Analyst, and Geostatistical Analyst. You will use Geostatistical Analyst in this section whenever possible because Geostatistical Analyst provides more information and better user interface than the other extensions. For example, Geostatistical Analyst computes the RMS, a cross-validation statistic that shows how closely a model predicts the known values at control points. Therefore, the RMS can be used as one of the statistics in comparing models.

Task 1: Use Trend Surface Analysis for Global Interpolation

What you need: *stations.shp*, a shapefile containing 105 weather stations in Idaho; *idoutl.shp*, an Idaho outline shapefile; and *idoutlgd*, an Idaho outline grid.

In Task 1 you will explore the average annual precipitation from 1961 to 1990 stored in the attribute called ann_prec in *stations.shp*, before running a trend surface analysis.

1. Start ArcCatalog, and make connection to the Chapter 13 database. Launch ArcMap. Add *stations.shp*, *idoutl.shp*, and *idoutlgd* to Layers and rename Layers Task 1. Select Extensions from the Tools menu and make sure that both Geostatistical Analyst and Spatial Analyst are checked. Click the View menu, point to Toolbars, and make sure that both Geostatistical Analyst and Spatial Analyst are checked.

2. Click the Geostatistical Analyst dropdown arrow, point to Explore Data, and select Trend Analysis. (Ignore the dialog that says that all values in the selected attribute are

identical.) At the bottom of the Trend Analysis dialog, click the dropdown arrow to select *stations* for Layer and ANN_PREC for Attribute.

3. Maximize the Trend Analysis dialog. The 3-D diagram shows two Trend Projections: The YZ Plane dips from north to south, and the XZ plane dips initially from west to east and then rises. The north–south trend is much stronger than the east–west trend, suggesting that the general precipitation pattern in Idaho decreases from north to south. Close the dialog.

4. Click the Geostatistical Analyst dropdown arrow, and select Geostatistical Wizard. You will first choose the geostatistical method and input data. In the Methods frame, click Global Polynomial Interpolation. Click the Input Data dropdown arrow and select *stations*. Click the Attribute dropdown arrow and select ANN_PREC. The next panel lets you choose the power of the trend surface model. The Power list provides the choice from 1 to 10.

5. Select 1 for the Power. The next panel shows scatter plots (Predicted versus Measured values, and Error versus Measured values) and statistics related to the first-order trend surface model. The RMS statistic measures the overall fit of the trend surface model. In this case, it has a value of 5.94. Click Back and change the Power to 2, and click Next. The RMS statistic has a value of 6.03. Repeat the same procedure with other power numbers. The trend surface model with the lowest RMS statistic is considered the best overall model for this task. For ANN_PREC, the best overall model has the power of 5 and the RMS statistic of 3.94. Change the Power to 5, and click Finish. Click OK in the Output Layer Information dialog.

6. Added to the Table of Contents is *Global Polynomial Interpolation Prediction Map*,

which is a Geostatistical Analyst output layer and has the same extent as *stations*. To clip *Global Polynomial Interpolation Prediction Map* to fit Idaho, you must first convert the map to a raster grid and then clip the grid using the Spatial Analyst extension.

7. Right-click *Global Polynomial Interpolation Prediction Map*, point to Data, and select Export to Raster. In the Export to Raster dialog, enter 2000 (meters) for the Cell Size and specify *trend5_grid* for the Output Raster. Click OK to dismiss the dialog. Add *trend5_grid* to the map. (Notice that *trend5_grid* has cell values ranging from -8 to 1242. Most of those extreme cell values are located outside the state border.)

8. Now you are ready to clip *trend5_grid*. Click the Spatial Analyst dropdown arrow and select Options. Under the General tab, select the Chapter 13 database for the working directory and select *idoutlgd* for the analysis mask. Click OK to dismiss the Options dialog. Click the Spatial Analyst dropdown arrow and select Raster Calculator. Double-click *trend5_grid* in the Layers frame so that *trend5_grid* appears in the expression box. Click Evaluate. *Calculation* is the clipped *trend5_grid*.

9. You can generate isolines from *Calculation* for data visualization. Click the Spatial Analyst dropdown arrow, point to Surface Analysis, and select Contour. In the Contour dialog, make sure that *Calculation* is the input surface, enter 5 for the contour interval and 0 for the base contour, and specify *trend5ctour.shp* for the output features. (Here contour lines are not lines connecting points of equal elevations but equal precipitation values.) Click OK. Right-click *trend5ctour* in the Table of Contents and select Properties. Click the Labels tab. Check the box next to Label Features in this layer, select CONTOUR from the Label Field dropdown list, and click OK. The map now shows isoline labels.

10. *Calculation* is a temporary grid. You can save it as a permanent grid by going through the following steps: right-click *calculation* and select Make Permanent, and enter the name for the permanent grid.

Task 2: Use Kernel Density Estimation for Local Interpolation

What you need: *deer.shp*, a point shapefile showing deer locations.

Task 2 uses the kernel estimation method to compute the average number of deer sightings per hectare from *deer.shp*. Deer location data have a 50-meter minimum discernible distance; therefore, some locations have multiple sightings.

1. Select Data Frame from the Insert menu in ArcMap. Rename the new data frame Task 2, and add *deer.shp* to Task 2. Make sure that the Spatial Analyst toolbar is available.

2. Select Properties from the context menu of *deer*. Click the Symbology tab. Select Quantities and Graduated symbols in the Show box, and select SIGHTINGS from the Value dropdown list. Click OK. The map shows deer sightings at each location in graduated symbols.

3. Because you have used *idoutlgd* as the analysis mask in Task 1, you must remove it before proceeding with the next step. Click the Spatial Analyst dropdown arrow and select Options. Under the General tab, select None from the Analysis mask dropdown list. Click OK to dismiss the Options dialog.

4. Click the Spatial Analyst dropdown arrow, and click Density. You need to work with several parameters in the Density dialog. Make sure that *deer* is the input data. Select SIGHTINGS from the Population field dropdown list. Make sure that Density type is Kernel. Enter 100 for the search radius. Select Hectares from the Area units dropdown list. Enter 100 for the output cell size. Click OK to dismiss the dialog. *Density of deer* shows deer sighting densities computed by the kernel estimation method.

5. To view *density of deer* on top of *deer*, you can make *density of deer* transparent. Right-click *density of deer* in the Table of Contents and select Properties. Under the Display tab, enter 30 next to Transparent and click OK.

Task 3: Use IDW for Local Interpolation

What you need: *stations.shp* and *idoutlgd*, same as in Task 1.

This task lets you create a precipitation grid using the IDW method.

1. Select Data Frame from the Insert menu in ArcMap. Rename the new data frame Tasks 3&4, and add *stations.shp* and *idoutlgd* to Tasks 3&4. Make sure that the Geostatistical Analyst and Spatial Analyst toolbars are both available.

2. Click the Geostatistical Analyst dropdown arrow and select Geostatistical Wizard. Make sure that the Input Data is *stations*. Select ANN_PREC from the Attribute dropdown list. Click Inverse Distance Weighting in the Methods frame. Click Next.

3. The Step 1 panel includes a graphic frame and a method frame for specifying IDW parameters. The default IDW method uses the power of 2, 15 neighbors (control points), and a circular area from which control points are selected. The graphic frame shows *stations* and the points and their weights (shown in percentages and color symbols) used in deriving the estimated value for a test location. You can click the "Predict/Estimate Value and Identify Neighbors in the Center of Ellipse" tool and then click any point within the graphic frame to see how the point's estimated value is derived.

4. The Optimize Power Value button is included in the Step 1 panel. Because a change of the power value will change the estimated value at a point location, you can click the button and ask Geostatistical Wizard to find the optimal power value while holding other parameter values constant. Geostatistical

Wizard employs the cross-validation technique to find the optimal power value. Click the Optimal button, and the Power field shows a value of 4.2639. Click Next.

5. The panel for Step 2 allows you to draw scatter plots relating predicted to measured values, or relating error to measured values. The panel also shows the statistics of prediction errors. For example, the RMS statistic is 3.322 using the optimal power value of 4.2639, compared to 3.678 using a power value of 2. Click Finish. Click OK in the Output Layer Information dialog.

6. You can follow the same steps as in Task 1 to convert *Inverse Distance Weighting Prediction Map* to a grid, to clip the grid by using *idoutlgd* as the analysis mask, and to create isolines from the clipped grid.

Task 4: Compare Two Splines Methods

What you need: *stations.shp* and *idoutlgd*.

Task 4 compares the results from two thin-plate splines methods. The task has three parts: one, create an interpolated grid using the regularized splines method; two, create an interpolated grid using the thin-plate splines with tension method; and three, use a local operation to compare the two grids. The result can show the difference between the two interpolation methods.

1. Click the Geostatistical Analyst dropdown arrow and select Geostatistical Wizard. Make sure that the Input Data is *stations*. Select ANN_PREC from the Attribute dropdown list. Click Radial Basis Functions in the Methods frame. Click Next.

2. In the Step 1 panel, click the Kernel Functions dropdown arrow and select Completely Regularized Spline. The panel contains other information: the optimal parameter value of 0.00010205, the selection method for neighbors (control points), and a graphic frame showing how the estimated value for a test location is derived. Click Next.

3. The Step 2 panel shows scatter plots relating predicted to measured values or relating error to measured values and statistics of prediction errors. For example, the RMS statistic is 3.045 using the optimal parameter value of 0.00010205, compared to 3.187 using a parameter value of 0.0002. Click Finish. Click OK in the Output Layer Information dialog.

4. *Radial Basis Functions Prediction Map* shows the result of the regularized splines method. Right-click *Radial Basis Functions Prediction Map*, point to Data, and select Export to Raster. In the Export to Raster dialog, change the Cell Size to 2000 (meters) and save the Output Raster as *regularized*. Click OK to dismiss the dialog.

5. Now you will create an output raster using the spline with tension method. Click the Geostatistical Analyst dropdown arrow and select Geostatistical Wizard. Make sure that that the input data is *stations*, the attribute is ANN_PREC, and Radial Basis Functions is the method. Click Next.

6. In the Step 1 panel, select Spline with Tension from the Kernel Functions dropdown list. Notice that the optimal parameter value is 0.000027474. Click Next.

7. The Step 2 panel shows the RMS statistic of 2.875. Click Finish. Click OK in the Output Layer Information dialog.

8. Right-click *Radial Basis Functions_2 Prediction Map*, point to Data, and select Export to Raster. In the Export to Raster dialog, change the Cell Size to 2000 (meters) and save the Output Raster as *tension*. Click OK to dismiss the dialog.

9. Click the Spatial Analyst dropdown arrow and select Options. Make sure that, under the General tab, the Chapter 13 database is the working directory and *idoutlgd* is the analysis mask. Click OK to dismiss the dialog.

10. Select Raster Calculator from the Spatial Analyst dropdown list. Prepare the following

expression in the expression box: [regularized] − [tension]. Click Evaluate.

11. *Calculation* shows the difference in cell values between the two grids, *regularized* and *tension*. Right-click *Calculation* and select Properties. Click the Symbology tab. Click Classify. In the Classification dialog, select 3 in the Classes dropdown list. Click the first cell under Break Values and enter −2. Enter 2 and 2.6 in the next two cells. Click OK in the Classification dialog and the Layer Properties dialog. *Calculation* now shows the difference between *regularized* and *tension* in three classes (< −2, −2-2, and > 2).

Task 5: Use Ordinary Kriging for Local Interpolation

What you need: *stations.shp* and *idoutlgd*.

Task 5 lets you run ordinary kriging in ArcMap. Kriging is a complex topic and requires expert knowledge in selecting the proper model for the data to be interpolated. For this task as well as the following two tasks, you will use the RMS statistic to select the mathematical model.

1. Select Data Frame from the Insert menu in ArcMap. Rename the new data frame Tasks 5-7, and add *stations.shp* and *idoutlgd* to Tasks 5-7. Click the Geostatistical Analyst dropdown arrow and select Geostatistical Wizard. In the first panel, make sure that the Input Data is *stations*. Select ANN_PREC from the Attribute dropdown list. Click Kriging in the Methods frame. Click Next.

2. The Step 1 panel lets you choose the kriging method and the map output. Click Ordinary Kriging Prediction Map in the Geostatistical Methods frame. Click Next.

3. The Step 2 panel shows the semivariogram/covariance view, and the choice of the mathematical models for approximating the semivariogram/covariance. Choose the Exponential model (this model results in the lowest RMS of 2.918). The mathematical equation for the model is

shown in the lower left of the panel. Click Next.

4. The Step 3 panel shows the selection method for neighbors (control points), and a graphic frame showing how the estimated value for a test location is derived. Click Next.

5. The Step 4 panel shows scatter plots (Predicted versus Measured values, Error versus Measured values, Standardized Error versus Measured values, and Quantile-Quantile plot for Standardized Error values), and statistics related to the model including the RMS. Click Finish. Click OK in the Output Layer Information dialog.

6. *Ordinary Kriging Prediction Map* is added to the map. To derive a prediction standard error map, you will click the Ordinary Kriging Prediction Standard Error Map in Step 1 and repeat Steps 2 to 4.

7. You can follow the same steps as in Task 1 to convert *Ordinary Kriging Prediction Map* and *Ordinary Kriging Prediction Standard Error Map* to grids, to clip the grids by using *idoutlgd* as the mask grid, and to create isolines from the clipped grids.

Task 6: Use Universal Kriging for Local Interpolation

What you need: *stations.shp* and *idoutlgd*.

In Task 6 you will run universal kriging in ArcMap. The trend to be removed from the kriging process is the first-order trend surface.

1. Click the Geostatistical Analyst dropdown arrow and select Geostatistical Wizard. Make sure that the Input Data is *stations* and the attribute is ANN_PREC. Click Kriging in the Methods frame. Click Next.

2. In the Step 1 panel, click Universal Kriging Prediction Map in the Geostatistical Methods frame. Select First from the Order of Trend dropdown list. Click Next.

3. The Step 2 panel shows the first-order trend that will be removed from the kriging process. Click Next.

4. The Step 3 panel shows the semivariogram/covariance view, and the choice of the mathematical models for approximating the semivariogram/covariance. Choose the Tetraspherical model (this model results in the lowest RMS statistic of 3.019). The mathematical equation for the model is shown in the lower left of the panel. Click Next.

5. The Step 4 panel shows the selection method for neighbors (control points), and a graphic frame showing how the estimated value for a test location is derived. Click Next.

6. The Step 5 panel shows scatter plots (Predicted versus Measured values, Error versus Measured values, Standardized Error versus Measured values, and Quantile-Quantile plot for Standardized Error values), and statistics related to the model including the RMS. Click Finish. Click OK in the Output Layer Information dialog.

7. *Universal Kriging Prediction Map* is an interpolated map from universal kriging. To derive a prediction standard error map, you will click the Ordinary Kriging Prediction Standard Error Map in Step 1 and repeat Steps 2 to 5.

8. You can follow the same steps as in Task 1 to convert *Universal Kriging Prediction Map* and *Universal Kriging Prediction Standard Error Map* to grids, to clip the grids by using *idoutlgd* as the mask grid, and to create isolines from the clipped grids.

Task 7: Use Cokriging for Local Interpolation

What you need: *stations.shp* and *idoutlgd*.

In Task 7 you will run cokriging in ArcMap by using the variables of annual precipitation and elevation.

1. Click the Geostatistical Analyst dropdown arrow and select Geostatistical Wizard. Click cokriging in the Methods frame. Under the Dataset 1 tab, select *stations* for the Input Data and ANN_PREC for the Attribute. Under the Dataset 2 tab, select *stations* for the Input Data and ELEVATION for the Attribute. Click Next.

2. In the Step 1 panel, click Ordinary Cokriging Prediction Map. Click Next.

3. The Step 2 panel shows the semivariogram/covariance view, and the choice of the mathematical models for approximating the semivariogram/covariance. Choose the Pentaspherical model (this model results in the lowest RMS statistic of 3.363). The mathematical equation for the model is shown in the lower left of the panel. Click Next.

4. The Step 3 panel shows the selection method for neighbors (control points), and a graphic frame showing how the estimated value for a test location is derived. Click Next.

5. The Step 4 panel shows scatter plots (Predicted versus Measured values, Error versus Measured values, Standardized Error versus Measured values, and Quantile-Quantile plot for Standardized Error values), and statistics related to the model including the RMS. Click Finish. Click OK in the Output Layer Information dialog.

6. *Ordinary Cokriging Prediction Map* is an interpolated map from cokriging using ANN_PREC and ELEVATION. To derive a prediction standard error map, you will click the Ordinary Kriging Prediction Standard Error Map in Step 2 and repeat Steps 3 to 5.

7. You can follow the same steps as in Task 1 to convert *Ordinary Cokriging Prediction Map* and *Ordinary Cokriging Prediction Standard Error Map* to grids, to clip the grids by using *idoutlgd* as the mask grid, and to create isolines from the clipped grids.

REFERENCES

Abramowitz, M., and I. A. Stegun. 1964. *Handbook of Mathematical Functions*. New York: Dover.

Bailey, T. C., and A. C. Gatrell. 1995. *Interactive Spatial Data Analysis*. Harlow, England: Longman Scientific & Technical.

Burrough, P. A., and R. A. McDonnell. 1998. *Principles of Geographical Information Systems*. Oxford, England: Oxford University Press.

Carroll, S. S., and N. Cressie. 1996. A Comparison of Geostatistical Methodologies Used to Estimate Snow Water Equivalent. *Water Resources Bulletin* 32: 267–78.

Chang, K., and Z. Li. 2000. Modeling Snow Accumulation with a Geographic Information System. *International Journal of Geographical Information Science* 14: 693–707.

Cressie, N. 1991. *Statistics for Spatial Data*. Chichester, England: Wiley.

Daly, C., R. P. Neilson, and D. L. Phillips. 1994. A Statistical-Topographic Model for Mapping Climatological Precipitation over Mountainous Terrain. *Journal of Applied Meteorology* 33: 140-58.

Davis, J. C. 1986. *Statistics and Data Analysis in Geology*, 2d ed. New York: Wiley.

Declercq, F. A. N. 1996. Interpolation Methods for Scattered Sample Data: Accuracy, Spatial Patterns, Processing Time. *Cartography and Geographic Information Science* 23: 128–44.

Franke, R. 1982. Smooth Interpolation of Scattered Data by Local Thin Plate Splines.

Computers and Mathematics with Applications 8: 273–81.

Franke, R. 1985. Thin Plate Spline with Tension. *Computer-Aided Geometrical Design* 2: 87–95.

Garen, D. C., G. L. Johnson, and C. J. Hanson. 1994. Mean Areal Precipitation for Daily Hydrologic Modeling in Mountainous Regions. *Water Resources Bulletin* 30: 481–91.

Griffith, D. A., and C. G. Amrhein. 1991. *Statistical Analysis for Geographers.* Englewood Cliffs, NJ: Prentice Hall.

Hutchinson, M. F. 1995. Interpolating Mean Rainfall Using Thin Plate Smoothing Splines. *International Journal of Geographical Information Systems* 9: 385–403.

Issaks, E. H., and R. M. Srivastava. 1989. *An Introduction to Applied Geostatistics.* Oxford, England: Oxford University Press.

Martinez-Cob, A. 1996. Multivariate Geostatistical Analysis of Evapotranspiration and Precipitation in Mountainous Terrain. *Journal of Hydrology* 174: 19–35.

Middleton, G. V. 2000. *Data Analysis in the Earth Sciences Using Matlab.* Upper Saddle River, NJ: Prentice Hall.

Mitas, L., and H. Mitasova. 1988. General Variational Approach to the Interpolation Problem. *Computers and Mathematics with Applications* 16: 983–92.

Mitasova, H., and L. Mitas. 1993. Interpolation by Regularized Spline with Tension: I. Theory and Implementation. *Mathematical Geology* 25: 641–55.

Monmonier, M. S. 1982. *Computer-Assisted Cartography: Principles and Prospects.* Englewood Cliffs, NJ: Prentice Hall.

Phillips, D. L., J. Dolph, and D. Marks. 1992. A Comparison of Geostatistical Procedures for Spatial Analysis of Precipitation in Mountainous Terrain. *Agricultural and Forest Meteorology* 58: 119–41.

Robinson, A. H., J. L. Morrison, P. C. Muehrcke, A. J. Kimerling, and S. C. Guptill. 1995. *Elements of Cartography,* 6th ed. New York: Wiley.

Rogerson, P. A. 2001. *Statistical Methods for Geography.* London: Sage Publications.

Scott, D. W. 1992. *Multivariate Density Estimation: Theory, Practice, and Visualization.* New York: Wiley.

Silverman, B. W. 1986. *Density Estimation.* London: Chapman and Hall.

Tabios, G. Q., III, and J. D. Salas. 1985. A Comparative Analysis of Techniques for Spatial Interpolation of Precipitation. *Water Resources Bulletin* 21: 365–80.

Watson, D. F. 1992. *Contouring: A Guide to the Analysis and Display of Spatial Data.* Oxford: Pergamon Press.

Webster, R., and M. A. Oliver. 1990. *Statistical Methods in Soil and Land Resource Survey.* Oxford, England: Oxford University Press.

Webster, R., and M. A. Oliver. 2001. *Geostatistics for Environmental Scientists.* Chichester, England: Wiley.

Wickham, J. D., R. V. O'Neill, and K. B. Jones. 2000. A Geography of Ecosystem Vulnerability. *Landscape Ecology* 15: 495–504.

Yang, X., and T. Hodler. 2000. Visual and Statistical Comparisons of Surface Modeling Techniques for Point-Based Environmental Data. *Cartography and Geographic Information Science* 27: 165–75.

GIS MODELS AND MODELING

14.1 INTRODUCTION

The previous chapters have presented tools for exploring, manipulating, and analyzing vector data and raster data. One of many uses of these tools is to build models. What is a model? A **model** is a simplified representation of a phenomenon or a system. Several types of models have already been covered in this book. A map is a model. So are the vector and raster data models for representing spatial features and the relational database model for representing a database system. A model helps us better understand a phenomenon or a system by retaining the significant features and relationships of reality.

This chapter discusses use of GIS in building models. Two points must be clarified at the start. First, the chapter deals with models using geographically referenced data or geospatial data. Some researchers have used the term "spatially explicit models" to describe these models. Second, the emphasis is the use of GIS in modeling rather than the models. Although this chapter covers a number of models, the intent is simply to use them as examples. A basic requirement in modeling is the modeler's interest and knowledge of the system to be modeled (Hardisty et al. 1993). This is why many models are discipline specific. For example, models of the environment usually consist of atmospheric, hydrologic, land surface/subsurface, and ecological models. It would be impossible for a GIS book to discuss each of these environmental models, not to mention models from other disciplines.

This chapter is divided into the following five sections. Section 14.2 discusses the basic elements of GIS modeling. Sections 14.3 and 14.4 cover binary models and index models, respectively. Section 14.5 deals with regression models, both linear and logistic. Section 14.6 introduces process models including soil erosion and other environmental models. Although these four types of models differ in degree of complexity, they share two common elements: a set of selected spatial variables, and the functional or mathematical relationship between the variables.

14.2 BASIC ELEMENTS OF GIS MODELING

14.2.1 Classification of GIS Models

It is difficult to classify many models used by GIS users. DeMers (2002), for example, classified models by purpose, methodology, and logic. But the boundary between the classification criteria is not always clear. Rather than proposing an exhaustive classification, this section covers some broad categories of models. The purpose is to provide an introduction to the models to be discussed later.

A model may be **descriptive** or **prescriptive.** A descriptive model describes the existing conditions of spatial data, and a prescriptive model offers a prediction of what the conditions could be or should be. If we use maps as analogies, a vegetation map would represent a descriptive model and a potential natural vegetation map, a prescriptive model. The vegetation map shows existing vegetation, whereas the potential natural vegetation map predicts the vegetation that could occupy a site without disturbance or climate change.

A model may be **deterministic** or **stochastic.** Both deterministic and stochastic models are mathematical models represented by equations with parameters and variables. A stochastic model considers the presence of some randomness in one or more of its parameters or variables, but a deterministic model does not. As a result of random processes, the predictions of a stochastic model can have measures of error or uncertainty, typically expressed in probabilistic terms. This is why a stochastic model may also be called a probabilistic or statistical model. Among the local interpolation methods covered in Chapter 13, for instance, only kriging represents a stochastic model. Besides producing a prediction map, a kriging interpolator can generate the variance for each predicted value.

A model may be **static** or **dynamic.** A dynamic model emphasizes the changes of spatial data and the interactions between variables, whereas a static model deals with the state of spatial data at a given time. Time is important to show the process of changes in a dynamic model (Peuquet 1994). Simulation is a technique that can generate different states of spatial data over time. Many environmental models such as groundwater pollution and soil water distribution are best studied as dynamic models (Rogowski and Goyne 2002).

A model may be **deductive** or **inductive.** A deductive model represents the conclusion derived from a set of premises. These premises are often based on scientific theories or physical laws. An inductive model represents the conclusion derived from empirical data and observations. To assess the potential for a landslide, for example, one can use a deductive model based on laws in physics or use an inductive model based on recorded data from past landslides.

14.2.2 The Modeling Process

The development of a model follows a series of steps. The first step is to define the goals of the model. This is analogous to defining a research problem. What is the phenomenon to be modeled? Why is the model necessary? What spatial and time scales are appropriate for the model? One can use a conceptual or schematic model to show the essential structure of the model.

The second step is to break down the model into elements and to define the properties of each element and the interactions between the elements. A flowchart is a useful tool for linking the elements. Also at this step, one will gather mathematical equations of the model and commands in a GIS (or the computer codes) to carry out the computation.

The third step is the implementation and calibration of the model. Data are needed for running and calibrating the model. The objective of model calibration is to improve the performance of the model by adjusting the numeric values of the parameters in an iterative process. Each time, the output from the model is compared to the observed data before adjusting the parameter values. Uncertainties in model prediction are a major problem in calibrating a deterministic model. Sensitivity analysis is a technique that can quantify these uncertainties by measuring the effects of input changes on the output.

A calibrated model is a tool ready for prediction. But the model must be validated before it can be generally accepted. Model validation assesses the model's ability to predict under conditions that are different from those used in the calibration phase. Model validation therefore requires a different set of data from those used for developing the model. One can split observed data into two subsets: one subset for developing the model and the other subset for model validation (Chang and Li 2000). But in many cases the required additional data set presents a problem and forces the modeler to forgo the validation step. A model that has not been validated is likely to be ignored by other researchers (Brooks 1997).

14.2.3 The Role of GIS in Modeling

GIS can assist the modeling process in several ways. First, a GIS is a tool that can process, display, and integrate different data sources including maps, digital elevation models (DEMs), GPS (global positioning system) data, images, and tables. These data are needed for the implementation, calibration, and validation of a model. A GIS can function as a database management tool and, at the same time, is useful for modeling-related tasks such as exploratory data analysis and data visualization.

Second, models built with a GIS can be vector-based or raster-based. The choice depends on the nature of the model, data sources, and the computing algorithm. A raster-based model is preferred if the spatial phenomenon to be modeled varies continuously over the space such as soil erosion and snow accumulation. A raster-based model is also preferred if satellite images and DEMs constitute a major portion of the input data, or if the modeling involves intense and complex computations. But raster-based models are not recommended for studies of travel demand, for example, because travel demand modeling requires the use of a topology-based road network (Chang et al. 2002). Vector-based models are generally recommended for spatial phenomena that involve well-defined locations and shapes.

Third, the distinction between raster-based and vector-based models does not preclude GIS users from integrating both types of data in the modeling process. Algorithms for conversion between vector and raster data are easily available in GIS packages. The decision as to which data format to use in analysis should be based on the efficiency and the expected result, rather than the format of the original data. For instance, if a vector-based model requires a precipitation map as the input, it would be easier to interpolate a precipitation grid from known points and then to derive the precipitation map from the grid.

Fourth, the process of modeling may take place in a GIS or require the linking of a GIS to other computer programs. Many GIS packages including ArcGIS, GRASS, IDRISI, SPANS, ILWIS, Vertical Mapper, and MFworks have extensive analytical functions for modeling. But a GIS package cannot accommodate statistical analysis as well as a statistical analysis package can, or perform dynamic simulation efficiently. In those cases, one would link a GIS to a statistical analysis package or a simulation program.

14.2.4 Integration of GIS and Other Modeling Programs

There are three scenarios for linking a GIS to other computer programs (Corwin et al. 1997). GIS users may encounter all three scenarios in the modeling process, depending on the tasks to be accomplished.

A **loose coupling** involves transfer of data files between the GIS and other programs. For example, one would export data to be run in a statistical analysis package from the GIS and import results from statistical analysis back to the GIS for data visualization or display. Under this scenario, GIS users must create and manipulate data files to be exported or imported unless the interface has already been established between the GIS and the targeted computer program. A **tight coupling** gives the GIS and other programs a common user interface. For instance, the GIS can have a menu selection to run a simulation program on soil erosion. An **embedded system** bundles the GIS and other programs with shared memory and a common interface. The Geostatistical Analyst extension to ArcGIS is an example of having geostatistical functions embedded into a GIS environment. The other option is to embed selected GIS functions into a statistical analysis environment (Zhang and Griffith 2000).

Among the four types of models included in this chapter, regression and process models usually require the coupling of a GIS with other programs. Binary and index models, on the other hand, can be built entirely in a GIS.

14.3 BINARY MODELS

A **binary model** uses logical expressions to select map features from a composite map or multiple grids. The output of a binary model is in binary format: 1 (true) for map features that meet the selection criteria and 0 (false) for map features that do not. One might consider a binary model as an extension of data query or map query.

A vector-based binary model requires that map overlay operations be performed to combine geometries and attributes to be used in data query into a composite map (Figure 14.1). In contrast, a raster-based binary model can be derived directly from querying multiple grids, with each grid representing a criterion (Figure 14.2).

14.3.1 Applications of the Binary Model

Change detection is a simple application of the binary model. By overlaying two maps representing land covers at two different points in time, one can query the attribute data of the composite map to find, for example, where forested land has been converted to housing development. This type of operation can also be performed with raster data by querying two grids for a specific change. Or, one can use the local operation of combine to get unique combinations of cell values from two grids onto the output first, and then query the output grid for a specific change.

Siting analysis is probably the most common application of the binary model. A siting analysis determines if a unit area (i.e., a polygon or a cell) meets a set of selection criteria for locating a landfill, a ski resort, or a university campus. There are at least two approaches to running a siting analysis. One is to evaluate a set of nominated or preselected sites, and the other is to evaluate all potential sites.

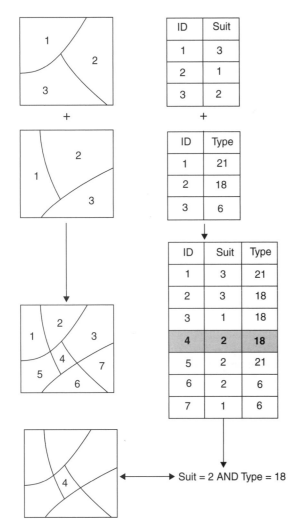

Figure 14.1
This diagram illustrates a vector-based binary model. First, overlay the two maps so that their spatial features and attributes (Suit and Type) are combined. Then, use the query statement, Suit = 2 AND Type = 18, to select polygon 4 and save it to the output.

Although the two approaches may use different sets of selection criteria (e.g., more stringent criteria for evaluating preselected sites), they follow the same procedure for evaluation. This section therefore does not distinguish between the two approaches.

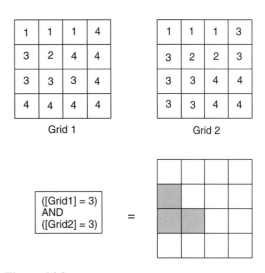

Figure 14.2
This diagram illustrates a raster-based binary model. Use the query statement, [Grid1] = 3 AND [Grid2] = 3, to select three cells (shaded) and save them to the output grid.

Suppose a county government wants to select potential industrial sites that meet the following criteria:

- At least 5 acres in size
- Commercial zones
- Vacant or for sale
- Not subject to flooding
- Not more than 1 mile from a heavy-duty road
- Less than 10 percent slope

Operationally, the task involves the following four steps:

1. Gather all digital maps relevant to the selection criteria. Some of these maps may require preprocessing. For instance, one may derive a commercial zone map from a land-use map. (The alternative is to use the land-use map directly and to query the commercial zone later.)
2. Create a 1-mile buffer zone map by buffering heavy-duty roads.
3. Perform a series of INTERSECT operations to combine the road buffer zone map and other maps.

4. Query the composite map to find which parcels are potential industrial sites.

The above procedure uses INTERSECT for map overly because INTERSECT, a Boolean operation that uses the AND connector (Chapter 10), can limit the output to only those parcels that meet the criteria. Each time the INTERSECT operation reduces the size of areas that may contain the potential sites. The procedure also uses well-defined or "crisp" threshold values to remove land from consideration. The road buffer is exactly 1 mile and the minimum parcel size is exactly 5 acres. These threshold values automatically exclude parcels that are more than 1 mile from heavy-duty roads or are smaller than 5 acres.

Government programs such as the Conservation Reserve Program (Box 14.1) are often characterized by their detailed and explicit guidelines. Sharp threshold values simplify the process of siting analysis. But one might question if they are too restrictive or arbitrary in real-world applications. In an example cited by Steiner (1983), local residents questioned whether any land could meet the criteria of a county comprehensive plan on rural housing. To mitigate the suspicion, a study was made to show that there were indeed sites available. An alternative to crisp threshold values is to use fuzzy sets (Hall et al. 1992; Hall and Arnberg 2002; Stoms et al. 2002). Fuzzy sets do not use sharp boundaries. Rather than being placed in a class (e.g., outside the road buffer), a unit area is associated with a group of membership grades, which suggest the extents to which the unit area belongs to different classes. Therefore, fuzziness is a way to handle uncertainty and complexity in multicriteria evaluation.

14.4 INDEX MODELS

An **index model** calculates the index value for each unit area and produces a ranked map based on the index values. An index model is similar to a binary model in that both involve multicriteria evaluation and both depend on map overlay operations in data processing. But an index model produces for each unit area an index value rather than a simple yes or no.

Box 14.1   The Conservation Reserve Program

The Conservation Reserve Program (CRP) is a voluntary program administered by the Farm Service Agency (FSA) of the U.S. Department of Agriculture (**http://www.fsa.usda.gov/dafp/cepd/crp.htm**). Its main goal is to reduce soil erosion on marginal croplands (Osborne 1993). Land eligible to be placed in the CRP includes cropland that is planted to an agricultural commodity during two of the five most recent crop years. Additionally, the cropland must meet the following criteria:

- Have an erosion index of 8 or higher or be considered highly erodible land
- Be considered a cropped wetland
- Be devoted to any of a number of highly beneficial environmental practices, such as filter strips, riparian buffers, grass waterways, shelter belts, wellhead protection areas, and other similar practices
- Be subject to scour erosion
- Be located in a national or state CRP conservation priority area, or be cropland associated with or surrounding noncropped wetlands

The difficult part of implementing the CRP in a GIS is putting together the necessary map layers, unless they are already available in a statewide database (Wu et al. 2002).

14.4.1 The Weighted Linear Combination Method

The primary consideration in developing an index model, either vector- or raster-based, is the method for computing the index value. The **weighted linear combination** method is probably the most common method for computing the index value (Saaty 1980; Banai-Kashani 1989; Malczewski 2000). Following the analytic hierarchy process (AHP) proposed by Saaty (1980), weighted linear combination involves evaluation at three levels (Figure 14.3).

First, the relative importance of each criterion, or factor, is evaluated against other criteria. Many studies have used expert-derived paired comparison for evaluating criteria (Saaty 1980; Banai-Kashani 1989; Pereira and Duckstein 1993; Jiang and Eastman 2000; Basnet et al. 2001). This method involves performing ratio estimates for each pair of criteria. For instance, if criterion A is considered to be three times more important than criterion B, then 3 is recorded for A/B and 1/3 is recorded for B/A. Using a criterion matrix of ratio estimates and their reciprocals as the input, the paired comparison method derives a weight for each criterion. The criterion weights are expressed in percentages, with the total equaling 100 percent or 1.0. Paired comparison is available in various software packages (e.g., Expert Choice).

Second, data for each criterion are standardized. A common method for data standardization is linear transformation. For example, the following formula can convert interval data into a standardized scale of 0.0 to 1.0:

(14.1)

$$S_i = \frac{X_i - X_{\min}}{X_{\max} - X_{\min}}$$

where S_i is the standardized value for the original value X_i, $X_{\min}$ is the lowest original value, and $X_{\max}$ is the highest original value. One cannot use Eq. (14.1) if the original data are nominal or ordinal data. In those cases, a ranking procedure based on expertise and knowledge can convert the data into a standardized range such as 0 to 1, 1 to 5, or 0 to 100.

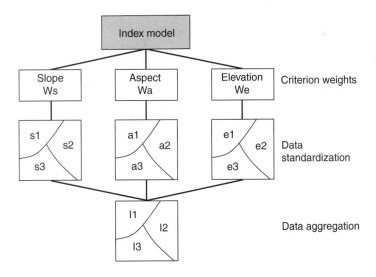

Figure 14.3
To build an index model with the selection criteria of slope, aspect, and elevation, the weighted linear combination method involves evaluation at three levels. The first level of evaluation determines the criterion weights (e.g., W_s for slope). The second level of evaluation determines standardized values for each criterion (e.g., $s1$, $s2$, and $s3$ for slope). The third level of evaluation determines the index (aggregate) value for each unit area.

Third, the index value is calculated for each unit area by summing the weighted criterion values and dividing the sum by the total of the weights:

(14.2)

$$I = \frac{\sum_{i=1}^{n} w_i x_i}{\sum_{i=1}^{n} w_i}$$

where I is the index value, n is the number of criteria, w_i is the weight for criterion i, and x_i is the standardized value for criterion i.

Figure 14.4 shows the procedure for developing a vector-based index model, and Figure 14.5 a raster-based index model. As long as criterion weighting and data standardization are well defined, it is not difficult to use the weighted linear combination method to build an index model in a GIS. But one must document standardized values and criterion weights in detail. User interface to simplify the process of building an index model is therefore welcome (Box 14.2).

14.4.2 Other Methods

There are many alternatives to the above weighted linear combination method. These alternatives mainly deal with the issues of independence of factors, criterion weights, data aggregation, and data standardization.

Weighted linear combination cannot deal with interdependence between factors (Hopkins 1977). A land suitability model may include soils and slope in a linear equation and treat them as independent factors. But in reality soils and slope are dependent of one another. One solution to the interdependence problem is to use a nonlinear function and express the relationship between factors mathematically. But a nonlinear function is usually limited to two factors rather than multiple factors as required in an index model. Another solution is the rule of combination method proposed by Hopkins (1977). Using rules of combination, one would assign suitability values to sets of combinations of environmental factors and express them through verbal logic instead of numeric terms. The rule of combination method has been widely

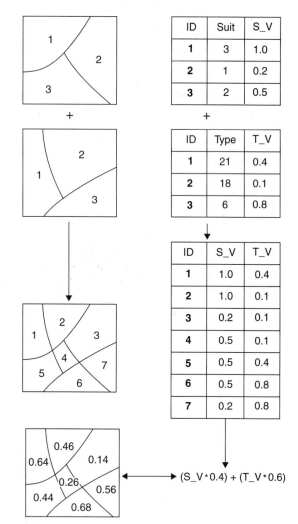

Figure 14.4
This diagram illustrates a vector-based index model. First, standardize the Suit and Type values of the two input maps into a scale of 0.0 to 1.0. Second, overlay the two maps. Third, assign a weight of 0.4 to the map with Suit and a weight of 0.6 to the map with Type. Finally, calculate the index value for each polygon in the output by summing the weighted criterion values. For example, Polygon 4 has an index value of 0.26 (0.5*0.4 + 0.1*0.6).

adopted in land suitability studies (e.g., Steiner 1983), but the method can become unwieldy given a large variety of criteria and data types (Pereira and Duckstein 1993).

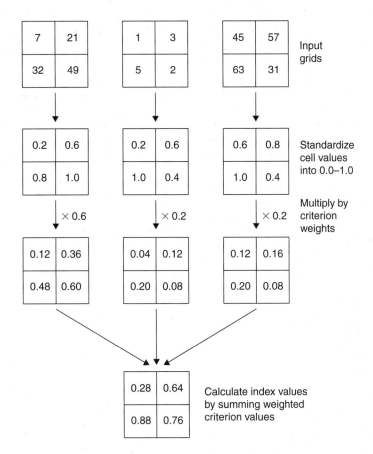

Figure 14.5
This diagram illustrates a raster-based index model. First, standardize the cell values of each input grid into a scale of 0.0 to 1.0. Second, multiply each input grid by its criterion weight. Finally, calculate the index values in the output grid by summing the weighted cell values. For example, the index value of 0.28 is calculated by: 0.12 + 0.04 + 0.12, or 0.2*0.6 + 0.2*0.2 + 0.6*0.2.

Paired comparison for determining criterion weights is sometimes called direct assessment. An alternative to direct assessment is trade-off weighting (Hobbs and Meier 1994; Xiang 2001). Trade-off weighting determines the criterion weights by asking participants to state how much of one criterion they are willing to give up to obtain a given improvement in another criterion. In other words, trade-off weighting is based on the degree of compromise one is willing to make between two criteria when an ideal combination of the two criteria is not attainable. Although realistic in some real-world applications, trade-off weighting has shown

to be more difficult to understand and use than direct assessment (Hobbs and Meier 1994).

Data aggregation refers to the derivation of the index value. Weighted linear combination calculates the index value by summing the weighted criterion values. One alternative is to skip the computation entirely and assign the lowest value, the highest value, or the most frequent value among the criteria to the index value (Chrisman 2001). Another alternative is the ordered weighted averaging (OWA) operator, which uses ordered weights instead of criterion weights in computing the index value (Yager 1988; Jiang and Eastman

Box 14.2 | **ModelBuilder**

Available in ArcView 3.2, ModelBuilder is a tool for building a raster-based index model. One can use the tool to first build a model diagram (flowchart) by stringing together a series of the input, the spatial function to be operated on the input, and the output. ModelBuilder then executes the operations specified in the model diagram in sequence to create a ranked map using the weighted or unweighted linear combination method. Once a model is built, one can rerun the model by changing any of the inputs or weights in the model diagram. ModelBuilder is therefore useful for sensitivity analysis. ModelBuilder is not available in ArcGIS 8.x, but ESRI Inc. plans to bring it back in ArcGIS 9.

Among GIS packages, IDRISI is well known for its multicriteria evaluation tools, including tools for weighted linear combination, ordered weighted averaging, paired comparison, and fuzzy data standardization. Some of these tools are discussed in Section 14.4.2. IDRISI32 has a new Decision Wizard that can guide the user through the process of multicriteria decision-making. The design of the Decision Wizard is similar to that of ModelBuilder.

2000). Suppose that a set of weights is {0.6, 0.4}. If the ordered position of the criteria at location 1 is {A, B}, then criterion A has the weight of 0.6 and criterion B has the weight of 0.4. If the ordered position of the criteria at location 2 is {B, A}, then B has the weight of 0.6 and A has the weight of 0.4. Ordered weighted averaging is therefore more flexible than weighted linear combination. Flexibility in data aggregation can also be achieved by use of fuzzy sets, which allow a unit area to belong to different classes with membership grades (Hall et al. 1992; Stoms et al. 2002).

Data standardization converts the values of each criterion into a standardized scale. A common method is linear transformation as shown in Eq. (14.1), but there are other quantitative methods. In their habitat evaluation study, Pereira and Duckstein (1993) used expert-derived value functions for data standardization, with a specific, often non-linear, function for each criterion. Jiang and Eastman (2000) used a fuzzy set membership function to transform data into fuzzy measures for their analysis of industrial allocation.

Criterion weights, data aggregation, and data standardization are the same issues that a spatial decision support system (SDSS) must deal with. Designed to work with spatial data, an SDSS can assist the decision maker in making a choice from a set of alternatives according to given evaluation criteria (Densham 1991; Jankowski 1995; Malczewski 1999). Although its emphasis is on decision making, an SDSS shares the same methodology as an index model. An index model developer can therefore benefit from becoming familiar with the SDSS literature.

14.4.3 Applications of the Index Model

Index models are commonly used for suitability analysis and vulnerability analysis. Here we will look at six examples from different disciplines.

Example 1 In 1981 the Natural Resources Conservation Service (NRCS), known then as the Soil Conservation Service, proposed the Land Evaluation and Site Assessment (LESA) system, a tool intended to be used by state and local planners in determining the conditions that justify conversion of agricultural land to other uses (Wright et al. 1983). LESA consists of two sets of factors: LE measures the inherent soil-based qualities of land for agricultural use, and SA measures demands for nonagricultural uses. Many state and local governments have developed, or are in the process of developing, their own LESA models. As an example, the California LESA model completed in 1997 (**http://www.consrv.ca.gov/DLRP/qh_lesa.htm**) uses the following factors and factor weights (in parentheses):

1. Land evaluation factors
   - Land capability classification (25 percent)
   - Storie index rating (25 percent)

2. Site assessment factors
   - Project size (15 percent)
   - Water resource availability (15 percent)
   - Surrounding agricultural lands (15 percent)
   - Surrounding protected resource lands (5 percent)

For a given location, one would first rate (standardize) each of the factors on a 100-point scale and then sum the weighted factor scores to derive a single index value.

Example 2 The U.S. Environmental Protection Agency developed the DRASTIC model for evaluating groundwater pollution potential (Aller et al. 1987). The acronym DRASTIC stands for the seven factors used in weighted linear combination: Depth to water, net Recharge, Aquifer media, Soil media, Topography, Impact of the vadose zone, and hydraulic Conductivity. The use of DRASTIC involves rating each parameter, multiplying the rating by a weight, and summing the total score by

(14.3)

$$\text{Total score} = \sum_{i=1}^{7} W_i P_i$$

where P_i is factor (input parameter) i and W_i is the weight applied to P_i.

A critical review of the DRASTIC model in terms of the selection of factors and the interpretation of numeric scores and weights is available in Merchant (1994).

Example 3 Lathrop and Bognar (1998) prioritized areas for conservation protection in Sterling Forest on the New York–New Jersey border by using the following five factors:

- Development limitations due to soil conditions/steep slopes/flooding
- Nonpoint source pollution potential due to proximity to water/wetlands
- Habitat fragmentation potential due to distance from existing roads and development
- Sensitive wildlife habitat areas
- Aesthetic impact (visibility from the Appalachian and Sterling Ridge Trails)

After making a grid for each factor, they ranked the environmental cost/development constraints for each grid from 1 to 5, with 1 being very slight and 5 being very severe. They then assigned the maximum value of any of the five factors to each cell in the output grid. Cells with low values were suitable for conservation protection.

Example 4 Habitat Suitability Index (HSI) models typically evaluate habitat quality by using weighted linear combination and factors considered to be important to the wildlife species (Brooks 1997). The following is an HSI model for pine marten developed by Kliskey et al. (1999):

(14.4)

$$\text{HSI} = \text{sqrt} \left\{ [(3SR_{BSZ} + SR_{SC} + SR_{DS}) / 6] \atop [(SR_{CC} + SR_{SS}) / 2] \right\}$$

where SR_{BSZ}, SR_{SC}, SR_{DS}, SR_{CC}, and SR_{SS} are the ratings for biogeoclimatic zone, site class, dominant species, canopy closure, and seral stage, respectively. The model was scaled so that the HSI values ranged from 0 for unsuitable habitat to 1 for optimal habitat.

Example 5 Chuvieco and Congalton (1989) constructed a forest fire hazard index model for a study area in the Mediterranean coast of Spain by using the following five factors:

- Vegetation species, classified according to fuel class, stand conditions, and site (v)
- Elevation (e)
- Slope (s)
- Aspect (a)

- Proximity to roads and trails, campsites, or housing (*r*)

They standardized the values of each factor by using 0, 1, and 2 for high, medium, and low fire hazard, respectively. They then used the following weighted linear equation to compute the hazard index *H:*

(14.5)

$$H = 1 + 100v + 30s + 10a + 5r + 2e$$

To make the forest fire hazard index model more operational, a follow-up study included new variables such as weather data, more objective criteria for weighting variables, and a new scheme for integrating the variables (Chuvieco and Salas 1996).

Example 6 Finco and Hepner (1999) built a model of human vulnerability to chemical accident by including the following demographic factors

- Total population
- Number of people younger than 18 years of age and older than 65 years
- Economic status as measured by household income
- Proximity to sensitive institutions such as schools, hospitals, and health clinics

They rated (standardized) each factor from 0 to 10, with 10 being most vulnerable. The total population and sensitive population factors were positively related to vulnerability, whereas the economic status factor was inversely related to vulnerability. They assumed 100 meters as the zone of influence for sensitive institutions. They then combined the four factors into a single measure of vulnerability.

14.5 REGRESSION MODELS

A **regression model** relates a dependent variable to a number of independent (explanatory) variables in an equation, which can then be used for prediction or estimation (Rogerson 2001). Like an index model, a regression model can use map

overlay operations in a GIS to combine all the variables needed for the analysis. There are two types of regression model: linear regression and logistic regression. Some GIS packages are capable of performing linear or logistic regression analysis. Both ArcInfo Workstation and IDRISI have commands to build raster-based linear or logistic models. GRASS has a command to build linear regression models. Unlike statistical analysis packages, these GIS commands do not offer choices of methods for running linear or logistic regression analysis.

14.5.1 Linear Regression Models

A multiple linear regression model is defined by

(14.6)

$$y = a + b_1x_1 + b_2x_2 + \ldots + b_nx_n$$

where *y* is the dependent variable, x_i is the independent variable *i*, and $b_1, \ldots, b_n$ are the regression coefficients. All variables in the equation are numeric variables. They can also be the transformation of some variables. Common transformations include square, square root, and logarithmic.

The primary purpose of linear regression is to predict values of *y* from values of x_i. But linear regression requires several assumptions about the error, or residual, between the predicted value and the actual value (Miles and Shevlin 2001):

- The errors have a normal distribution for each set of values of the independent variables.
- The errors have the expected (mean) value of zero.
- The variance of the errors is constant for all values of the independent variables.
- The errors are independent of one another.

An additional assumption in the case of multiple linear regression is that the correlation among the independent variables should not be high.

Numerous linear regression models are available in the literature. As an example, Chang and Li (2000) used linear regression to model snow accumulation: Snow water equivalent (SWE) was the dependent variable and location and topographic variables derived from a DEM were the independent variables. One of their watershed-level models was expressed as

(14.7)

$$SWE = a + b_1 \text{ EASTING} + b_2 \text{ SOUTHING} + b_3 \text{ ELEV}$$

where a, b_1, b_2, and b_3 are the regression coefficients, EASTING is the column number of a grid cell, SOUTHING is the row number of a cell, and ELEV is the elevation value of a cell.

14.5.2 Logistic Regression Models

Logistic regression is used when the dependent variable is categorical (e.g., presence or absence) and the independent variables are categorical, numeric, or both (Menard 2002). Although having the same form as linear regression, logistic regression uses the logit of y as the dependent variable:

(14.8)

$$logit(y) = a + b_1x_1 + b_2x_2 + b_3x_3 + \dots$$

The logit of y is the natural logarithm of the odds (also called odds ratio):

(14.9)

$$logit(y) = \ln(p/(1 - p))$$

where ln is the natural logarithm, $p/(1 - p)$ is the odds, and p is the probability of the occurrence of y. To convert logit (y) back to the odds or the probability, Eq. (14.9) can be rewritten as

(14.10)

$$p/(1 - p) = e^{(a + b_1x_1 + b_2x_2 + b_3x_3 + \dots)}$$

(14.11)

$$p = e^{(a + b_1x_1 + b_2x_2 + b_3x_3 + \dots)} / [1 + e^{(a + b_1x_1 + b_2x_2 + b_3x_3 + \dots)}]$$

or,

(14.12)

$$p = 1/[1 + e^{-(a + b_1x_1 + b_2x_2 + b_3x_3 + \dots)}]$$

where e is the exponent.

The main advantage of using logistic regression is that it does not require the assumptions needed for linear regression. Many logistic regression models are also available in the literature. A habitat suitability model for red squirrel developed by Pereira and Itami (1991) is based on the following model:

(14.13)

$$logit(y) = 0.002 \text{ elevation} - 0.228 \text{ slope} + 0.685 \text{ canopy1} + 0.443 \text{ canopy2} + 0.481 \text{ canopy3} + 0.009 \text{ aspectE-W}$$

where canopy1, canopy2, and canopy3 represent three categories of canopy.

Another example is a logistic regression model built by Mladenoff et al. (1995) to estimate the amount and spatial distribution of favorable gray wolf habitat. They first created wolf pack areas (presence) and nonpack areas (absence) in polygon coverages. Wolf pack areas were home ranges derived from telemetry data of wolf location points. Nonpack areas were randomly located in the study area at least 10 kilometers from known pack territories. Both pack and nonpack areas were then overlaid with the landscape coverages of human population density, prey density, road density, land cover, and land ownership. Using the composite coverage, stepwise logistic regression analysis converged on the following model:

(14.14)

$$logit(y) = 6.5988 - 14.6189 R$$

where R is road density. This logistic regression model, which was based on data from Wisconsin, was then applied to a three-state region (Wisconsin, Minnesota, and Michigan) to map the amount and distribution of favorable wolf habitat. The same model was used in a subsequent study to predict habitat suitable for wolves in the Northeast (Mladenoff and Sickley 1998).

14.6 PROCESS MODELS

A **process model** integrates existing knowledge about the environmental processes in the real world into a set of relationships and equations for quantifying the processes (Beck et al. 1993). Modules or submodels are often needed to cover different components of a process model. Some of these modules may use mathematical equations derived from empirical data, whereas others may use equations derived from laws in physics. A process model offers both a predictive capability and an explanation that is inherent in the proposed processes (Hardisty et al. 1993). Therefore process models are by definition predictive and dynamic models.

Examples of process models covered in this section all fall under the category of environmental models. Environmental models are typically process models because they must deal with the interaction of many variables including physical variables such as climate, topography, vegetation, and soils as well as cultural variables such as land management. As to be expected, environmental models are complex and data-intensive and usually face issues of uncertainty to a greater extent than traditional natural science or social science models (Couclelis 2002).

14.6.1 Soil Erosion Models

Soil erosion is an environmental process that involves climate, soil properties, topography, soil surface conditions, and human activities. A well-known model of soil erosion is the Revised Universal Soil Loss Equation (RUSLE), the updated version of the Universal Soil Loss Equation (USLE) (Wischmeier and Smith 1965, 1978; Renard et al. 1997). RUSLE predicts the average soil loss carried by runoff from specific field slopes in specified cropping and management systems and from rangeland.

RUSLE is a multiplicative model with six factors:

(14.15)

$$A = R K L S C P$$

where A is the average soil loss, R is the rainfall-runoff erosivity factor, K is the soil erodibility factor, L is the slope length factor, S is the slope steepness factor, C is the crop management factor, and P is the support practice factor. Box 14.3 includes a more detailed description of the six factors in RUSLE.

To use RUSLE, one must first understand the basic structure of the model:

- The dimension of A is expressed in ton per acre per year (ton.acre^{-1}.yr^{-1}) in the United States, which is determined by the dimensions of R and K.
- L and S are combined into a single topographic factor LS.
- LS, C, and P are dimensionless, and their values represent deviations from a standard derived by RUSLE developers from experimental data.
- RUSLE considers the interactions between the factors. For example, both C and P are affected by the rainfall condition.
- RUSLE can compute a soil loss value for each time period over which the crop and soil parameters as well as the climate variables are assumed to remain constant.
- USLE was originally developed from analyses of field data and precipitation records for cropland east of the Rocky Mountains. RUSLE has incorporated modifications of the original equations for areas in the West. RUSLE has also included estimation of the K factor using regression analysis with soil and soil profile data.

Among the six factors in RUSLE, the slope length factor L poses more questions than other factors (Renard et al. 1997). Slope length is defined as the horizontal distance from the point of origin of overland flow to the point where either the slope gradient decreases enough that deposition begins or the flow is concentrated in a defined channel (Wischmeier and Smith 1978). RUSLE developers recommend that slope length be measured from samples taken in the field. But it is a problem to measure slope length when the topography is

The rainfall–runoff erosivity factor (*R*) quantifies the effect of raindrop impact and also measures the amount and rate of runoff likely to be associated with the rain. In the United States, the *R* factor can be interpolated from the isoerodent map compiled by RUSLE developers from analysis of rainfall data. Separate equations are available for calculating *R* for cropland in the Northwestern wheat and range region and other special areas.

The soil erodibility factor (*K*) measures change in the soil per unit of applied external force or energy. Specifically, it is the rate of soil loss per rainfall erosion index unit as measured on an experimental plot, which has a length of 72.6 feet and has a slope of 9 percent. RUSLE developers recommend that the *K* factor be obtained from direct measurements on natural runoff plots. But studies have shown that *K* factors can be estimated from soil and soil profile parameters. In the United States, the *K* factor is available in the Soil Survey Geographic (SSURGO) database.

The topographic factor (*LS*) represents the ratio of soil loss on a given slope length (*L*) and steepness (*S*) to soil loss from a slope that has a length of 72.6 feet and a steepness of 9 percent. RUSLE developers recommend that *L* be measured from samples taken in the field. The procedure for converting *L* and *S* into the *LS* factor varies, depending on whether the slope is uniform, irregular, or segmented.

The crop management factor (*C*) represents the effect of cropping and management practices on soil erosion. It uses an area under clean-tilled continuous-fallow conditions as a standard to measure soil loss under actual conditions. The calculation of *C* considers the change of crop and soil parameters over time as well as the changing climate variables.

The support practice factor (*P*) is the ratio of soil loss with a specific support practice to the corresponding loss with upslope and downslope tillage. The support practices for cultivated land include contouring, strip cropping, terracing, and subsurface drainage.

complex and irregular (Desmet and Govers 1996b), or to select "representative" transects in a farm field (Busacca et al. 1993).

Moore and Burch (1986) have proposed a method for estimating *LS*. Based on the unit stream power theory, the method uses the equation

(14.16)

$$LS = (A_s/22.13)^m(\sin \beta/0.0896)^n$$

where A_s is the upslope contributing area, β is the slope angle, *m* is the slope length exponent, and *n* is the slope steepness exponent. The exponent *m* is estimated to be 0.6, and *n* 1.3. When implemented in a raster-based GIS, the *LS* factor for each grid cell can be calculated from the slope and the catchment area of the cell (Moore and Wilson 1992; Moore et al.1993; Moore and Wilson 1994; Gertner et al. 2002).

RUSLE developers, however, recommend that the *L* and *S* components be separated in the computational procedure for the *LS* factor (Foster 1994; Renard et al. 1997). The equation for *L* is

(14.17)

$$L = (\lambda / 72.6)^m$$

where λ is the measured slope length, and *m* is the slope length exponent. The exponent *m* is calculated by

$$m = \beta / (1 + \beta)$$
$$\beta = (\sin \theta / 0.0896) / [3.0(\sin \theta)^{0.8} + 0.56]$$

where β is the ratio of rill erosion (caused by flow) to interrill erosion (principally caused by raindrop impact), and θ is the slope angle. The equation for *S* is

(14.18)

$S = 10.8 \sin \theta + 0.03$, for slopes of less than 9%

$S = 16.8 \sin \theta - 0.50$, for slopes of 9% or steeper

Both L and S also need to be adjusted for special conditions, such as the adjustment of the slope length exponent m for the erosion of thawing, cultivated soils by surface flow, and the use of a different equation than Eq. (14.17) for slopes shorter than 15 feet.

One can follow the same procedure and equations as proposed by RUSLE developers but still automate the estimation of L and S using a GIS. One method proposed by Hickey et al. (1994) uses DEMs as the input and estimates for each grid cell the slope steepness by computing the maximum downhill slope and the slope length by iteratively calculating the cumulative downhill slope length. The maximum downhill slope is the maximum slope from the center cell to its eight neighbors, similar to the way the flow direction is calculated (Chapter 12). Another method proposed by Desmet and Govers (1996a) also uses DEMs as the input and estimates for each cell the slope steepness by using a quadratic surface-fitting method and the slope length by calculating the unit-contributing area.

USLE and RUSLE have evolved over the past 50 years. This soil erosion model has gone through numerous cycles of model development, calibration, and validation. The process continues. A new, updated model called WEPP (Water Erosion Prediction Project) is expected to replace RUSLE (Laflen 1991). In addition to modeling soil erosion on hillslopes, WEPP can model the deposition in the channel system. Integration of WEPP and GIS is desirable because a GIS can be used to extract hillslopes and channels as well as identify the watershed (Cochrane and Flanagan 1999). A description and download site of WEPP is available at **http://topsoil.nserl.purdue.edu/nserlweb/weppmain/wepp.html.**

14.6.2 Other Process Models

The AGNPS (Agricultural Nonpoint Source) model analyzes nonpoint source pollution and

estimates runoff water quality from agricultural watersheds (Young et al. 1987). AGNPS is event-based and operates on a cell basis. Using various types of input data, the model simulates runoff, sediment, and nutrient transport. For example, AGNPS uses a modified form of USLE to estimate upland erosion for single storms (Young et al. 1989):

(14.19)

$$SL = (EI)\ K\ LS\ C\ P\ (SSF)$$

where SL is the soil loss, EI is the product of the storm total kinetic energy and maximum 30-minute intensity, K is the soil erodibility, LS is the topographic factor, C is the cultivation factor, P is the supporting practice factor, and SSF is a factor to adjust for slope shape within the cell. Detached sediment calculated from the equation is routed through the cells according to yet another equation based on the characteristics of the watershed.

The SWAT (Soil and Water Assessment Tool) model predicts the impact of land management practices on water quality and quantity, sediment, and agricultural chemical yields in large complex watersheds (Srinivasan and Arnold 1994). SWAT is a process-based continuous simulation model. Inputs to SWAT include land management practices such as crop rotation, irrigation, fertilizer use, and pesticide application rates, as well as the physical characteristics of the basin and subbasins such as precipitation, temperature, soils, vegetation, and topography. The creation of the input data files requires substantial knowledge at the subbasin level. Model outputs include simulated values of surface water flow, groundwater flow, crop growth, sediment, and chemical yields.

Other process models on nonpoint source pollutants in the vadose zone (Corwin et al. 1997), landslides (Montgomery et al. 1998), and groundwater contamination (Loague and Corwin 1998) are also available in the literature.

14.6.3 GIS and Process Models

Process models are typically raster-based. The role of a GIS in building a process model depends on the complexity of the model. A simple process model may be prepared and run entirely within a GIS. But more often a GIS is delegated to the role of performing modeling-related tasks such as data visualization, database management, and exploratory data analysis. The GIS is then linked to other computer programs for complex and dynamic analysis.

Commercial GIS packages do not offer commands for building process models. GRASS has commands for preparing input variables to AG-NPS, running the model, and viewing the model.

GRASS also has a command that can produce the *LS* and *S* factors in RUSLE. The NRCS has a demonstration project using the Soil Survey Geographic (SSURGO) database to develop SWAT models of five small watersheds in Iowa (**http://waterhome.tamu.edu/NRCSdata/SWAT_SSURGO**). SWAT is included in the Better Assessment Science Integrating point and Nonpoint Sources (BASINS) system developed by the U.S. Environmental Protection Agency, which can be downloaded at **http://www.epa.gov/waterscience/BASINS/**.

KEY CONCEPTS AND TERMS

Binary model: A GIS model that uses logical expressions to select map features from a composite map or multiple grids.

Deductive model: A model that represents the conclusion derived from a set of premises.

Descriptive model: A model that describes the existing conditions of spatial data.

Deterministic model: A mathematical model that does not consider randomness.

Dynamic model: A model that emphasizes the changes of spatial data and the interactions between variables.

Embedded system: GIS is bundled with other computer programs in a system with shared memory and a common interface.

Index model: A GIS model that uses the index value calculated from a composite map or multiple grids to produce a ranked map.

Inductive model: A model that represents the conclusion derived from empirical data and observations.

Loose coupling: The process for linking GIS and other computer programs through transfer of data files.

Model: A simplified representation of a phenomenon or a system.

Prescriptive model: A model that offers a prediction of what the conditions of spatial data could be or should be.

Process model: A GIS model that integrates existing knowledge into a set of relationships and equations for quantifying the physical processes.

Regression model: A GIS model that uses a dependent variable and a number of independent variables in a regression equation for prediction or estimation.

Static model: A model that deals with the state of spatial data at a given time.

Stochastic model: A mathematical model that considers the presence of some randomness in one or more of its parameters or variables.

Tight coupling: The process for linking GIS and other computer programs through a common user interface.

Weighted linear combination: A method that computes the index value for each unit area by summing the products of the standardized value and the weight for each criterion.

APPLICATIONS: GIS MODELS AND MODELING

This applications section covers four tasks. Tasks 1 and 2 let you build binary models using vector data and raster data, respectively. Tasks 3 and 4 let you build index models using vector data and raster data, respectively.

Task 1: Build a Vector-Based Binary Model

What you need: *elevzone.shp*, an elevation zone shapefile; *stream.shp*, a stream shapefile.

Task 1 asks you to locate the potential habitats of a plant species. Both *elevzone.shp* and *stream.shp* are measured in meters and spatially registered. The field zone in e*levzone.shp* shows three elevation zones. The potential habitats must meet the following criteria: (1) in elevation zone 2 and (2) within 200 meters of streams.

1. Start ArcCatalog, and make connection to the Chapter 14 database. Launch ArcMap. Add *stream.shp* and *elevzone.shp* to Layers, and rename the data frame Tasks 1&2. You will first buffer *streams* with a buffer distance of 200 meters. Select Buffer Wizard from the Tools menu. In the first panel, make sure that you want to buffer features of *stream*. In the second panel, specify 200 (meters) as the buffer distance. In the third panel, take the default of dissolving barriers between buffers and specify *strmbuf.shp* for the output shapefile. Click Finish.

2. The next step is to overlay *elevzone* and *strmbuf*. Select GeoProcessing Wizard from the Tools menu. In the first panel, click the radio button to intersect two layers. In the second panel, select *strmbuf* from list 1 for the input layer to intersect, select *elevzone* from list 2 for the polygon overlay layer, and specify *pothab.shp* for the output shapefile. Click Finish.

3. Now you want to query *pothab* and select areas in elevation zone 2. Select Open Attribute Table from the context menu of *pothab*. Click the Options dropdown arrow

and choose Select by Attributes. Enter the following SQL statement in the expression box: "ZONE" = 2, and click Apply. The potential habitats are highlighted in both the attribute table and the map.

4. Oftentimes you want to save a binary model into a new shapefile so that you can use it for reference or further analysis. To convert the selected polygons in *pothab* into a new shapefile, right-click *pothab*, point to Data, and select Export Data. In the Export Data dialog, choose selected features for export, specify *final.shp* for the output shapefile, and click OK.

Task 2: Build a Raster-Based Binary Model

What you need: *elevzone_gd*, an elevation zone grid; *stream_gd*, a stream grid.

Task 2 tackles the same problem as Task 1 except that Task 2 uses raster data. Both *elevzone_gd* and *stream_gd* have the cell resolution of 30 meters. The cell value in *elevzone_gd* corresponds to the elevation zone. The cell value in *stream_gd* corresponds to the stream ID.

1. Add *stream_gd* and *elevzone_gd* to Tasks 1&2. Make sure that the Spatial Analyst extension is selected and its toolbar is checked. The first step is to create continuous distance measures from *stream_gd*. Click the Spatial Analyst dropdown arrow, point to Distance, and select Straight Line. Make sure that *stream_gd* is the raster to calculate the distance to, enter 30 (meters) for the output cell size, and opt for a temporary output raster. Click OK. *Distance to stream_gd* is the temporary output raster.

2. Now you can query *elevzone_gd* and *Distance to stream_gd* to locate the potential habitats. Select Raster Calculator from the Spatial Analyst dropdown list. Enter the following expression in the expression box:

[Distance to stream_gd] <= 200 AND [elevzone_gd] = 2. Click Evaluate.

3. *Calculation* shows the potential habitats with the value of 1. Compare *Calculation* with *final* from Task 1. They should cover the same areas.

Task 3: Build a Vector-Based Index Model

What you need: *soil.shp*, a soil shapefile; *landuse.shp*, a land-use shapefile; *depwater.shp*, a depth to water shapefile.

Task 3 simulates a project on mapping groundwater vulnerability. The project assumes that groundwater vulnerability is related to three variables: soil characteristics, depth to water, and land use. Each variable has been rated on a scoring system from 0 to 50. For example, scores of 50, 35, 20, and 10 have been assigned to the depth to water classes of 1–25, 26–50, 51–100, and 101–250 feet, respectively. Soilrate shows the scores in *soil.shp*, dwrate in *depwater.shp*, and lurate in *landuse.shp*. The score of 99 is assigned to areas such as urban and built-up areas in *landuse.shp*, which should not be included in the model. The project also assumes that the soil factor is more important than the other two factors and is assigned a weight of 3, compared to 1 for the other two factors. The index model can therefore be expressed as Index value = 3 x soilrate + lurate + dwrate.

1. Select Data Frame from the Insert menu in ArcMap. Rename the new data frame Tasks 3&4, and add *soil.shp*, *landuse.shp*, and *depwater.shp* to Tasks 3&4. The main part of Task 3 is to overlay all three shapefiles. You can only overlay two at a time. Therefore, you need to perform overlay (INTERSECT) twice for three shapefiles.

2. Select GeoProcessing Wizard from the Tools menu. In the first panel, click the radio button to intersect two layers. In the second panel, select *landuse* from list 1 as the input layer to intersect, select *soil* from list 2 as the polygon overlay layer, and specify

landsoil.shp for the output shapefile. Click Finish.

3. Next, INTERSECT *landsoil* with *depwater*. Repeat the above step but use *landsoil* and *depwater* as the two layers to be overlaid. Specify the output as *vulner.shp*. Click Finish.

4. Select Open Attribute Table from the context menu of *vulner*. The table has all three rates needed for computing the index value. But you must go through a couple of steps before computation. You need to add a new field to the attribute table for the index value. Then you need to exclude areas with the lurate value of 99 from computation.

5. Before adding a new field to *vulner* in ArcCatalog, you must remove *vulner* from ArcMap. Right-click *vulner* in the Table of Contents of ArcMap and select Remove. Then select Properties from the context menu of *vulner* in ArcCatalog. (Select Refresh from the View menu if *vulner* is not listed in the Catalog tree.) Click the Fields tab. Click the first empty cell under Field Name and enter TOTAL. Click the cell next to TOTAL and select Float. Click OK to dismiss the dialog.

6. Add *vulner.shp* back to Tasks 3&4 in ArcMap. Select Open Attribute Table from the context menu of *vulner*. TOTAL appears in the table with 0s. Click the Options dropdown arrow and choose Select by Attributes. Enter the following SQL statement in the Select by Attributes dialog: "LURATE" <> 99. Click Apply. Click Selected in the Attributes of vulner table so that only selected records (nonurban areas) are shown. Right-click TOTAL and select Calculate Values. Click Yes in the Field Calculator message box. Enter the following expression in the Field Calculator dialog: 3 * [SOILRATE] + [LURATE] + [DWRATE]. Click OK to dismiss the dialog. The field TOTAL is populated with the calculated index values. To see the range of the TOTAL

values, right-click TOTAL and select Statistics. The Selection Statistics of vulner dialog shows a minimum of 145.2 and a maximum of 250.

7. You can assign a TOTAL value of −99 to urban areas. Click All in the Attributes of vulner table. Click the Options dropdown arrow and select Switch Selection. Click Selected so that only the selected records (urban areas) are shown. Right-click TOTAL and select Calculate Values. Enter −99 in the expression box, and click OK. Click All in the Selected Attributes of vulner table. Then click the Options dropdown arrow and select Clear Selection. Close the table.

8. You probably want to display the index values of *vulner* in the map. Select Properties from the context menu of *vulner*. Under the Symbology tab, choose Quantities and Graduated colors in the Show box. Click the Value dropdown arrow and select TOTAL. To distinguish urban areas, which are not included in the analysis, you can click Classify and enter 0, 175, 200, 225, and 250 as Break Values in the Classification dialog. You can also double-click the default symbol for urban areas (range −99 – 0) and change it to a Hollow symbol for differentiation.

9. Once the index value map is made, you can modify the classification so that the grouping of index values may represent a rank order such as very severe (5), severe (4), moderate (3), slight (2), and very slight (1). You can then convert the index value map into a ranked map by doing the following: save the rank of each class under a new field called rank, and then use the dissolve operation (available in GeoProcessing Wizard) to remove boundaries of polygons that fall within the same rank. The ranked map should look much simpler than the index value map.

Task 4: Build a Raster-Based Index Model

What you need: *soil_gd*, a soils grid; *landuse_gd*, a land-use grid; *depwater_gd*, a depth to water grid.

Task 4 performs the same analysis as Task 3 but uses raster data. All three grids have the cell resolution of 90 meters. The cell value in *soil_gd* corresponds to soilrate, the cell value in *landuse_gd* corresponds to lurate, and the cell value in *depwater_gd* corresponds to dwrate.

1. Add *soil_gd*, *landuse_gd*, and *depwater_gd* to Tasks 3&4. First, reclassify 99 in *landuse_gd* as NoData because the cell value of 99 represents urban areas, which should not be included in the model. Click the Spatial Analyst dropdown arrow and choose Reclassify. In the Reclassify dialog, select *landuse_gd* for the input raster. Click the first cell under New values and enter 20. Click and enter 40, 45, and 50 in the next three cells. Then enter NoData in the next two cells. Click OK to dismiss the dialog. *Reclass of landuse_gd* is added to the map.

2. To calculate the index value from *soil_gd*, *reclass of landuse_gd*, and *depwater_gd*, select Raster Calculator from the Spatial Analyst dropdown list. Enter the following expression in the Raster Calculator dialog: [soil_gd] * 3 + [Reclass of landuse_gd] + [depwater_gd]. Click Evaluate.

3. *Calculation* appears in the map. Like the vector-based model, the index (cell) values of *Calculation* range from 145.2 to 250. The value range can be easily converted into a range from 0 to 1. To convert the value range, select Raster Calculator from the Spatial Analyst dropdown list. Enter the following expression in the Raster Calculator dialog: ([Calculation] − 145.2) / (250 − 145.2). The output called *Calculation 2* shows the value range from 0 to 1.

REFERENCES

Aller, L., T. Bennett, J. H. Lehr, R. J. Petty, and G. Hackett 1987. *DRASTIC: A Standardized System for Evaluating Groundwater Pollution Potential Using Hydrogeologic Settings.* U.S. Environmental Protection Agency, EPA/600/2-87/035, pp. 622

Banai-Kashani, R. 1989. A New Method for Site Suitability Analysis: The Analytic Hierarchy Process. *Environmental Management* 13: 685–93.

Basnet, B. B., A. A. Apan, and S. R. Raine. 2001. Selecting Suitable Sites for Animal Waste Application Using a Raster GIS. *Environmental Management* 28: 519–31.

Beck, M. B., A. J. Jakeman, and M. J. McAleer. 1993. Construction and Evaluation of Models of Environmental Systems. In A. J. Jakeman, M. B. Beck, and M. J. McAleer, eds., *Modelling Change in Environmental Systems,* pp. 3–35. Chichester, England: Wiley.

Brooks, R. P. 1997. Improving Habitat Suitability Index Models. *Wildlife Society Bulletin* 25: 163–67.

Busacca, A. J., C. A. Cook, and D. J. Mulla. 1993. Comparing Landscape-Scale Estimation of Soil Erosion in the Palouse Using Cs-137 and RUSLE. *Journal of Soil and Water Conservation* 48: 361–67.

Chang, K., Z. Khatib, and Y. Ou. 2002. Effects of Zoning Structure and Network Detail on Traffic Demand Modeling.

Environment and Planning B 29: 37–52.

Chang, K., and Z. Li. 2000. Modeling Snow Accumulation with a Geographic Information System. *International Journal of Geographical Information Science* 14: 693–707.

Chrisman, N. 2001. *Exploring Geographic Information Systems,* 2d ed. New York: Wiley.

Chuvieco, E., and R. G. Congalton. 1989. Application of Remote Sensing and Geographic Information Systems to Forest Fire Hazard Mapping. *Remote Sensing of the Environment* 29: 147–59.

Chuvieco, E., and J. Salas. 1996. Mapping the Spatial Distribution of Forest Fire Danger Using GIS. *International Journal of Geographical Information Systems* 10: 333–45.

Cochrane, T. A., and D. C. Flanagan. 1999. Assessing Water Erosion in Small Watersheds Using WEPP with GIS and Digital Elevation Models. *Journal of Soil and Water Conservation* 54: 678–85.

Corwin, D. L., P. J. Vaughan, and K. Loague. 1997. Modeling Nonpoint Source Pollutants in the Vadose Zone with GIS. *Environmental Science & Technology* 31: 2157–75.

Couclelis, H. 2002. Modeling Frameworks, Paradigms and Approaches. In K. C. Clarke, B. O. Parks, and M. P. Crane, eds., *Geographic Information Systems and Environmental Modeling*, pp. 36–50. Upper Saddle River, NJ: Prentice Hall.

DeMers, M. N. 2002. *GIS Modeling in Raster.* New York: Wiley.

Densham, P. J. 1991. Spatial Decision Support Systems. In D. J. Maguire, M. F. Goodchild, and D. W. Rhind, eds., *Geographical Informaton Systems: Principles and Applications*, Vol 1., pp. 403–12. London: Longman.

Desmet, P. J. J., and G. Govers. 1996a. Comparison of Routing Systems for DEMs and Their Implications for Predicting Ephemeral Gullies. *International Journal of Geographical Information Systems* 10: 311–31.

Desmet, P. J. J., and G. Govers. 1996b. A GIS Procedure for Automatically Calculating the USLE LS Factor on Topographically Complex Landscape Units. *Journal of Soil and Water Conservation* 51: 427–33.

Finco, M. V., and G. F. Hepner. 1999. Investigating U.S.–Mexico Border Community Vulnerability to Industrial Hazards: A Simulation Study in Ambos Nogales. *Cartography and Geographic Information Science* 26: 243–52.

Foster, G. R. 1994. Comments on "Length-Slope Factor for the Revised Universal Soil Loss Equation: Simplified Method of Estimation." *Journal of Soil and Water Conservation* 49: 171–3.

Gertner, G., G. Wang, S. Fang, and A. B. Anderson. 2002. Effect and Uncertainty of Digital Elevation Model Spatial Resolutions on Predicting the Topographic Factor for Soil Loss

Estimation. *Journal of Soil and Water Conservation* 57: 164–74.

Hall, G. B., F. Wang, and Subaryono. 1992. Comparison of Boolean and Fuzzy Classification Methods in Land Suitability Analysis by Using Geographical Information Systems. *Environment and Planning A* 24: 497–516.

Hall, O., and W. Arnberg. 2002. A Method for Landscape Regionalization Based On Fuzzy Membership Signatures. *Landscape and Urban Planning* 59: 227–40.

Hardisty, J., D. M. Taylor, and S. E. Metcalfe. 1993. *Computerized Environmental Modelling.* Chichester, England: Wiley.

Hickey, R., A. Smith, and P. Jankowski. 1994. Slope Length Calculations from a DEM within ARC/INFO GRID. *Computers, Environment, and Urban Systems* 18: 365–80.

Hobbs, B. F., and P. M. Meier. 1994. Multicriteria Methods for Resource Planning: An Experimental Comparison. *IEEE Transactions on Power Systems* 9 (4): 1811–7.

Hopkins, L. D. 1977. Methods for Generating Land Suitability Maps: A Comparative Evaluation. *Journal of the American Institute of Planners* 43: 386–400.

Jankowski, J. 1995. Integrating Geographic Information Systems and Multiple Criteria Decision-Making Methods. *International Journal of Geographical Information Systems* 9: 251–73.

Jiang, H., and J. R. Eastman. 2000. Application of Fuzzy Measures in Multi-Criteria Evaluation in GIS. *International Journal of*

Geographical Information Science 14: 173–84.

Kliskey, A. D., E. C. Lofroth, W. A. Thompson, S. Brown, and H. Schreier. 1999. Simulating and Evaluating Alternative Resource-Use Strategies Using GIS-Based Habitat Suitability Indices. *Landscape and Urban Planning* 45: 163–75.

Laflen, J. M., L. J. Lane, and G. R. Foster. 1991. WEPP: A New Generation of Erosion Prediction Technology. *Journal of Soil and Water Conservation* 46: 34–38.

Lathrop, R. G., Jr., and J. A. Bognar. 1998. Applying GIS and Landscape Ecologic Principles to Evaluate Land Conservation Alternatives. *Landscape and Urban Planning* 41: 27–41.

Loague, K., and D. L. Corwin. 1998. Regional-Scale Assessment of Non-Point Source Groundwater Contamination. *Hydrologic Processes* 12: 957–65.

Malczewski, J. 1999. *GIS and Multicriteria Decision Analysis.* New York: Wiley.

Malczewski, J. 2000. On the Use of Weighted Linear Combination Method in GIS: Common and Best Practice Approaches. *Transactions in GIS* 4: 5–22.

Menard, S. 2002. *Applied Logistic Regression Analysis,* 2d ed. Thousand Oaks, CA: Sage.

Merchant, J. W. 1994. GIS-Based Groundwater Pollution Hazard Assessment: A Critical Review of the DRASTIC Model. *Photogrammetric Engineering & Remote Sensing* 60: 1117–27.

Miles, J., and Shevlin, M. 2001. *Applying Regression & Correlation: A Guide for*

Students and Researchers. London: Sage.

Mladenoff, D. J., and T. A. Sickley. 1998. Assessing Potential Gray Wolf Restoration in the Northeastern United States: A Spatial Prediction of Favorable Habitat and Potential Population Levels. *Journal of Wildlife Management* 62: 1–10.

Mladenoff, D. J., T. A. Sickley, R. G. Haight, and A. P. Wydeven. 1995. A Regional Landscape Analysis and Prediction of Favorable Gray Wolf Habitat in the Northern Great Lakes Regions. *Conservation Biology* 9: 279–94.

Montgomery, D. R., K. Sullivan, and H. M. Greenberg. 1998. Regional Test of a Model for Shallow Landsliding. *Hydrologic Processes* 12: 943–55.

Moore, I. D., and G. J. Burch. 1986. Physical Basis of the Length-Slope Factor in the Universal Soil Loss Equation. *Soil Science Society of America Journal* 50: 1294–98.

Moore, I. D., A. K. Turner, J. P. Wilson, S. K. Jenson, and L. E. Band. 1993. GIS and Land-Surface-Subsurface Process Modelling. In M. F. Goodchild, B. O. Park, and L. T. Styaert, eds., *Environmental Modelling with GIS,* pp. 213–30. Oxford, England: Oxford University Press.

Moore, I. D., and J. P. Wilson. 1992. Length-Slope Factor for the Revised Universal Soil Loss Equation: Simplified Method of Estimation. *Journal of Soil and Water Conservation* 47:423–28.

Moore, I. D., and J. P. Wilson. 1994. Reply to Comments by Foster on "Length-Slope Factor for the

Revised Universal Soil Loss Equation: Simplified Method of Estimation." *Journal of Soil and Water Conservation* 49: 174–80.

Osborne, C. T. 1993. CRP Status, Future and Policy Options. *Journal of Soil and Water Conservation* 48: 272–78.

Pereira, J. M. C., and L. Duckstein. 1993. A Multiple Criteria Decision-Making Approach to GIS-Based Land Suitability Evaluation. *International Journal of Geographical Information Systems* 7: 407–24.

Pereira, J.M.C., and R.M. Itami. 1991. GIS-Based Habitat Modeling Using Logistic Multiple Regression: A Study of the Mt. Graham Red Squirrel. *Photogrammetric Engineering & Remote Sensing* 57: 1475-86.

Peuquet, D. J. 1994. It's About Time: A Conceptual Framework for the Representation of Temporal Dynamics in Geographic Information Systems. *Annals of the Association of American Geographers* 84: 441–61.

Renard, K. G., G. R. Foster, G. A. Weesies, D. K. McCool, and D. C. Yoder, coordinators. 1997. Predicting Soil Erosion by Water: A Guide to Conservation Planning with the Revised Universal Soil Loss Equation (RUSLE). *Agricultural Handbook 703*. Washington, DC: U.S. Department of Agriculture.

Rogerson, P. A. 2001. *Statistical Methods for Geography*. London: Sage.

Rogowski, A., and J. Goyne. 2002. Dynamic Systems Modeling and Four Dimensional Geographic Information Systems. In K. C.

Clarke, B. O. Parks, and M. P. Crane, eds., *Geographic Information Systems and Environmental Modeling*, pp. 122–59. Upper Saddle River, NJ: Prentice Hall.

Saaty, T. L. 1980. *The Analytic Hierarchy Process*. New York: McGraw-Hill.

Srinivasan, R., and J. G. Arnold. 1994. Integration of a Basin-Scale Water Quality Model with GIS. *Water Resources Bulletin* 30: 453–62.

Steiner, F. 1983. Resource Suitability: Methods for Analyses. *Environmental Management* 7: 401–20.

Stoms, D. M., J. M. McDonald, and F. W. Davis. 2002. Environmental Assessment: Fuzzy Assessment of Land Suitability for Scientific Research Reserves. *Environmental Management* 29: 545–58.

Wischmeier, W. H., and D. D. Smith. 1965. Predicting Rainfall-Erosion Losses from Cropland East of the Rocky Mountains: Guide for Selection of Practices for Soil and Water Conservation. *Agricultural Handbook 282*. Washington, DC: U.S. Department of Agriculture.

Wischmeier, W. H., and D. D. Smith. 1978. Predicting Rainfall Erosion Losses: A Guide to Conservation Planning. *Agricultural Handbook 537*. Washington, DC: U.S. Department of Agriculture.

Wright, L.E., W. Zitzmann, K. Young, and R. Googins. 1983. LESA—agricultural Land Evaluation and Site Assessment.

Journal of Soil and Water Conservation 38: 82-89.

Wu, J., M. D. Random, M. D. Nellis, G. J. Kluitenberg, H. L. Seyler, and B. C. Rundquist. 2002. Using GIS to Assess and Manage the Conservation Reserve Program in Finney County, Kansas. *Photogrammetric Survey and Remote Sensing* 68: 735–44.

Xiang, W. 2001. Weighting-by-Choosing: A Weight Elicitation Method for Map Overlays. *Landscape and Urban Planning* 56: 61–73.

Yager, R. 1988. On Ordered Weighted Averaging Aggregation Operators in Multicriteria Decision Making. *IEEE Transactions on Systems, Man, and Cybernetics* 18: 183–90.

Young, R. A., C. A. Onstad, D. D. Bosch, and W. P. Anderson. 1987. AGNPS, Agricultural Non-point-Source Pollution Model: A Large Watershed Analysis Tool. *Conservation Research Report 35*. Washington, DC: Agricultural Research Service, U.S. Department of Agriculture.

Young, R. A., C. A. Onstad, D. D. Bosch, and W. P. Anderson. 1989. AGNPS: A Nonpoint-Source Pollution Model for Evaluating Agricultural Watersheds. *Journal of Soil and Water Conservation* 44: 168–73.

Zhang, Z., and D. A. Griffith. 2000. Integrating GIS Components and Spatial Statistical Analysis in DBMSs. *International Journal of Geographical Information Science* 14: 543–66.

15

REGIONS

CHAPTER OUTLINE

15.1 INTRODUCTION

Regions are higher-level objects built on top of simple arcs and polygons. The regions data model organizes polygons into regions and groups regions into region subclasses. This data model differs from the simple polygon model in several important aspects: (1) a region consists of polygons that may or may not be connected spatially; (2) regions may overlap or cover the same area; and (3) each region subclass can have its own attributes and can function independently for data display, query, and analysis.

The properties of the regions data model are similar to those of polygon shapefiles and the polygon feature classes of the geodatabase model. We can use ArcGIS Desktop to display, query, and analyze regions by treating region subclasses as separate layers in a polygon coverage. We can also convert a region subclass to a polygon shapefile or a geodatabase polygon feature class. But before topology is added to the geodatabase model, ArcGIS users will continue to use ArcInfo Workstation to create regions and run regions-based analysis.

This chapter is divided into the following five sections. Section 15.2 discusses two types of geographic regions, uniform regions and hierarchical regions, and relates the regions data model to geographic regions. Section 15.3 examines the potential of incorporating regions in real-world applications. Section 15.4 reviews methods for creating regions, dropping regions, and managing attribute data of regions. Section 15.5 discusses regions-based overlay and query. Section 15.6 describes other regions-based tools.

15.2 GEOGRAPHIC REGIONS

Perhaps the most important aspect of the regions data model is its capacity to implement the concept of geographic regions from such disciplines as geography, landscape ecology, and forestry (Berry 1968; Bailey 1976; Forman and Godron 1986). A geographic region is an area with certain attributes that give the area a particular character. Examples of geographic regions include ethnic regions, linguistic regions, hydrologic units, and ecological units. Like the regions data model, geographic regions may overlap and each region may include spatially joint or disjoint areas. Therefore, the regions data model is a natural choice for studies of geographic regions.

15.2.1 Uniform Regions

A **uniform region** is a geographic area with similar characteristics. Uniform regions may be simple or complex. A simple uniform region is based on a single variable, whereas a complex uniform region is based on multiple variables.

Simple uniform regions are common in maps and atlases. We are used to seeing regional classification maps of landforms, natural vegetation, soils, land uses, ethnic groups, religions, and languages. Other examples include trade areas, which are often defined by travel time, and traffic analysis zones, which represent the origins and destinations of trips in travel-demand modeling (Miller 1999). Complex uniform regions are composites of simple regions. An example is Omernik's ecoregions, which are defined by land use, potential natural vegetation, land surface form, soils, and other supporting data (Omernik 1987).

Whether simple or complex, uniform regions can overlap. For example, the trade areas of two department stores can overlap. A uniform region can also have spatially disjoint areas. For instance, Omernik's ecoregion called Arizona/New Mexico Mountains consists of several disjoint areas in Arizona and New Mexico.

15.2.2 Hierarchical Regions

Hierarchical regions represent different spatial scales of a hierarchy. A hierarchical structure can divide the Earth's surface into progressively smaller regions of increasingly uniform characteristics. Two well-known examples of hierarchical regions in the United States are census units and hydrologic units.

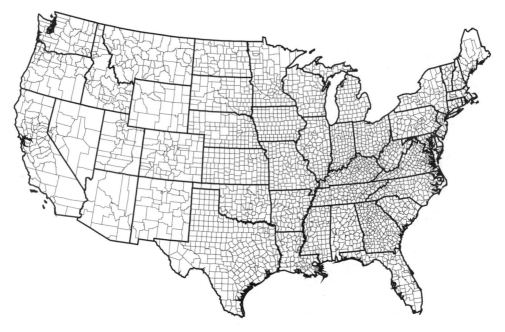

Figure 15.1
A nested hierarchy of counties and states in the conterminous United States.

The U.S. Census Bureau compiles and distributes census data by state, county, census tract, block group, and block. These area units form a nested hierarchy: blocks are within a block group, block groups are within a census tract, census tracts are within a county, and counties are within a state (Figure 15.1). Thus, census units represent different levels of spatial aggregation in a hierarchy.

The U.S. Geological Survey organizes hydrologic units into a nested hierarchy. The system divides the United States at four spatial levels: regions, subregions, accounting units, and cataloging units (**http://water.usgs.gov/GIS/huc. html**). A two-digit code is given at each level. Each hydrologic unit therefore has a unique code of two to eight digits, depending on the unit's classification within the hierarchy.

Ecological units are more complex examples of hierarchical regions (Bailey 1976, 1983; Cleland et al. 1997; McMahon et al. 2001). The National Hierarchical Framework of Ecological Units, adopted by the U.S. Forest Service in 1993, systematically maps areas of the Earth's surface at the scales of ecoregion, subregion, landscape, and land unit (**http://svinet2.fs.fed.us/land/pubs/ecoregions/**). The delineation of ecological units is based on the association of ecological factors. Climate and landform define ecoregion and subregion boundaries. Relief, geologic parent materials, and potential natural communities define landscape boundaries. And topography, soil characteristics, and plant associations define land unit boundaries. Canada has a similar national ecological framework, which is based on the four hierarchical levels of ecozone, ecoprovince, ecoregion, and ecodistrict (**http://sis. agr.gc.ca/cansis/nsdb/ecostrat/intro.html**).

15.2.3 The Regions Approach to Implementing Geographic Regions

Because geographic regions have similar characteristics as those of the regions data model, it is

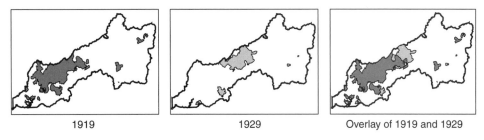

Figure 15.2
This illustration shows burned areas in the Lochsa River Basin in Idaho for 1919, 1929, and both years. Notice that burned areas for 1919 overlaps with burned areas for 1929.

easier to implement geographic regions as regions than as polygons.

Consider the example in Figure 15.2, which shows burned areas in the Lochsa River Basin in Idaho for 1919, 1929, and both years. There are 18 polygons in the 1919 map and 5 polygons in the 1929 map. Some overlap exists between the burned areas of the two maps. The map showing burned areas for both years was created by traditional map overlay using the simple polygon model (Chapter 10). The overlay map has 35 polygons. Although many of these 35 polygons have the same attribute (e.g., burned in 1919), they must be treated as separate polygons according to the simple polygon model.

Now consider two options of using regions to store and manage the same fire data. The first option is to store the 1919 and 1929 data in a single region subclass (Figure 15.3). The region subclass has three regions: the first region consists of all burned areas for 1919, the second region consists of all burned areas for 1929, and the third region consists of void areas (areas not burned). Because the regions data model allows a region to have spatially disjoint components, the burned areas for 1919 can be aggregated into one region. The same is true for the 1929 burned areas and void areas. And, because the regions data model allows regions to overlap, the 1919 region can overlap the 1929 region without having to create a separate region for the overlapped area. Having only three regions, this option is certainly more efficient in managing the fire data than the traditional method.

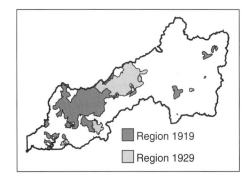

Figure 15.3
This illustration shows a single region subclass that contains three regions: 1919 burned areas, 1929 burned areas, and areas not burned. The 1919 region overlaps with the 1929 region, as allowed in the regions data model.

The second option is to store the burned areas for 1919, 1929, and both years in three region subclasses (Figure 15.4). The 1919 subclass has 18 regions, the 1929 subclass has 5 regions, and the third subclass has 1 region that is made of areas burned in both 1919 and 1929. Although this option creates as many regions as polygons in the 1919 and 1929 subclasses, it has two advantages over the simple polygon model. One advantage relates to the overlay operation for creating a new regions subclass (i.e., the subclass for both years). It can overlay many maps (32 using ArcInfo Workstation) at one time, as opposed to only two maps with the traditional overlay method. The other

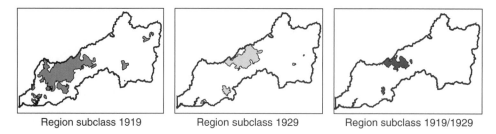

Region subclass 1919   Region subclass 1929   Region subclass 1919/1929

Figure 15.4
This illustration shows three region subclasses: 1919 burned areas, 1929 burned areas, and burned areas for both 1919 and 1929.

advantage is the storage of both the input and output in the same coverage instead of separate coverages. This simplifies the data management task.

15.2.4 Integrated Coverage

An **integrated coverage** stores the geometries of region subclasses and can provide the geometries for new region subclasses. Using burned areas in the Lochsa River Basin as an example, the integrated coverage includes the boundaries of all burned areas for 1919 and 1929 and can provide the boundaries of burned areas for both years. An integrated coverage is therefore a required coverage, which must reside in the same workspace as the region subclasses.

An integrated coverage typically requires large computer memory and extra processing time. Additionally, one must manage an integrated coverage carefully: its hierarchical data structure of arc, polygon, and region must be constantly maintained, and its topology must be rebuilt whenever the inputs are altered. For example, if burned areas in the Lochsa River Basin are expanded to include 1931, it will require a new integrated coverage.

Is there any relationship between an integrated coverage and the integrate command in ArcGIS Desktop (Chapter 5)? Given a feature dataset with two or more feature classes, the integrate command uses a user-specified cluster tolerance to snap all points or lines that fall within the tolerance. Therefore, one can consider using the integrate command to consolidate the boundaries among the input maps before creating an integrated coverage. This would reduce the number of slivers caused by digitizing errors in the integrated coverage.

15.3 APPLICATIONS OF THE REGIONS DATA MODEL: INCORPORATING SPATIAL SCALE IN GIS ANALYSIS

The regions data model provides a more efficient data structure than the simple polygon model for storing and managing geographic regions that overlap or have spatially disjoint areas. The regions data model is also useful for including spatial scale in GIS analysis. Here we will examine the potential application of the regions data model in two research areas.

15.3.1 Studies of Ecosystems

Ecosystems are places where life forms and environment interact. This interaction takes place at different spatial scales (Box 15.1). Hierarchy theory in ecology, for example, has emphasized that ecosystems and their underlying biophysical environments occur in a spatial hierarchy, with smaller ecosystems embedded in larger ones (Allen and Starr 1982; O'Neill et al. 1986). The literature has

Box **15.1** **Map Scale and Spatial Scale**

Scale is a confusing term. Cartographers define the map scale as the ratio between a map distance and the ground distance it represents. A map scale of 1:24,000 means that an inch on the map represents 24,000 inches (2000 feet) on the ground. At a scale of 1:100,000, a 1-inch map distance represents 100,000 inches (about 1.58 miles) on the ground. To cartographers, 1:24,000 is a larger scale than 1:100,000, and a large-scale map covers a smaller area than a small-scale map.

Ecologists use the term spatial scale. Unlike map scale, spatial scale is not rigidly defined. Spatial scale refers to the size of area or extent. A large spatial scale therefore covers a larger area than a small spatial scale. To put it another way, a large spatial scale to an ecologist is a small map scale to a cartographer.

also shown that ecological patterns and processes are scale dependent and scale is an important variable in ecological studies (Meentemeyer and Box 1987; Urban et al. 1987; Wiens 1989; Costanza and Maxwell 1994).

According to hierarchy theory, the conditions and processes of larger ecosystems can affect or constrain those of smaller ecosystems. For example, logging operations on upper slopes of an ecological unit will likely affect streams and riparian habitats in downslope smaller units. At the same time, local conditions and processes can have cumulative effects on the next higher level. For instance, pines in a snow-pine forest landscape convert solar radiation into sensible heat, which moves to the snow cover and melts it faster than would happen in either a wholly snow-covered or a wholly forested basin (Bailey 1983).

Spatial scaling in ecology is a complex topic. A GIS using the regions data model can assist research in this area in two ways. First, it can incorporate spatial scales into region subclasses and store large and small ecosystems in the same workspace. In other words, the regions data model provides a means to build a framework similar to a hierarchical GIS proposed by Walker and Walker (1991) for ecological studies. Second, the GIS can provide the analytical functions for manipulating and analyzing the attributes of ecosystems at different spatial scales. Regions-based query and overlay are covered later in Section 15.5.

15.3.2 Modifiable Areal Unit Problem

The **modifiable areal unit problem (MAUP)** addresses the "modifiable" nature of area units used in spatial analysis and the influence it has on the analysis and modeling results. The modifiable nature of area units appears in two forms: (1) the level of spatial data aggregation and (2) the definition or partitioning of area units for which data are collected. The former is referred to as the scale effect and the latter as the zoning effect in the MAUP literature (Openshaw and Taylor 1979; Wong and Amrhein 1996). Users of census data must be aware of the scale effect because they can choose block groups, census tracts, or counties as area units for analysis. But users of census data do not need to worry about the zoning effect because the boundaries of census units are predefined. The zoning effect does exist if a new procedure is used to aggregate census data spatially (Fotheringham and Wong 1991).

An early study of the MAUP found that the correlation strengthened with spatial aggregation (Gehlke and Biehl 1934). Robinson (1950) reported a similar relationship between estimated correlation coefficients and the spatial scale of census data used in the computation. Other studies have reported MAUP effects in bivariate regression (Clark and Avery 1976; Amrhein, 1995), multivariate regression (Fotheringham and Wong 1991), spatial interaction (Openshaw 1977; Batty and Sikdar 1982a, 1982b), and location-allocation (Goodchild 1979). The MAUP has also been studied in landscape ecology (Jelinski and Wu 1996), wildlife management (Svancara et al. 2002), and transportation planning (Chang et al. 2002).

MAUP effects appear to be consistent for bivariate regression: as the level of spatial aggregation increases, one can expect a general increase in the correlation coefficient, a likely result from the smoothing effect of data aggregation. But MAUP effects are essentially unpredictable in multivariate analysis (both linear regression and logistic regression) according to Fotheringham and Wong (1991). Their regression models relating mean family income to demographic variables in Buffalo, New York, initially showed a coefficient of determination of 0.81 at the census tract level and 0.37 at the block group level. But subsequent experimentation with scaling and zoning showed a wide range of results in terms of estimated regression parameters and the coefficient of determination.

Given the uncertainty of MAUP effects, it is wise to incorporate spatial scale as a variable in studies that are based on modifiable area units. Fotheringham and Wong (1991), for example, recommended the reporting of results at different levels of aggregation and with different zoning systems at the same scale. The regions data model can be useful for studying the MAUP by including each level of spatial aggregation to be studied as a region subclass.

15.4 CREATE REGIONS

There are several methods for creating regions. An understanding of the hierarchical data structure of arcs, polygons, and regions (Chapter 3) is useful in choosing an efficient method. Whenever the choice for making regions is between arcs and polygons, polygons should be used because regions are built directly on polygons. If arcs are

chosen, polygons must be built first from arcs before regions can be created.

One option in creating regions is the choice between noncontiguous and contiguous regions. A **noncontiguous region** may have spatially disjoint components, as allowed in the regions data model. In contrast, a **contiguous region** must be spatially connected. A region subclass made of contiguous regions is like a simple polygon coverage except that it resides with other region subclasses and shares the geometry with the integrated coverage. In most cases, one would opt for the noncontiguous option to simplify the task of data management.

15.4.1 Create Regions Interactively

Regions can be digitized directly but the process is slow because of overlapping regions and their spatially disjoint components. The preferred method is to select arcs or polygons that have already been digitized to make up regions. Figure 15.5 shows a

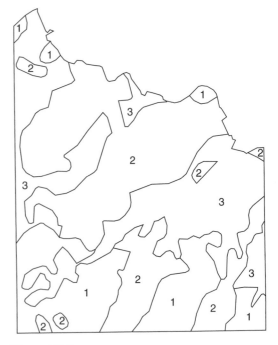

Figure 15.5

A soil map with the suitability values of 1, 2, and 3.

soil suitability map in three class values. Suppose we want to build a region subclass from this map with three different regions, one for each of the three suitability values. The following shows how the region subclass can be created interactively from either existing polygons or existing arcs.

It is relatively simple to create regions from existing polygons. To make up the region with the suitability value of 1 in Figure 15.6, for example, we can select polygons with the suitability value of 1 and use ArcInfo Workstation's MAKEREGION to make the region from the selected polygons. In this case, the output uses the noncontiguous option. Alternatively, the contiguous option can make a region from each polygon and the region subclass will look exactly like the simple polygon coverage.

To create regions from existing arcs, we must select all arcs that define polygons with the suit-

ability value of 1 to make the first region and repeat the same process for the second and third regions (Figure 15.7). The output consists of noncontiguous regions. If two adjacent polygons belong to the same region, the arcs making up each polygon must be selected and processed separately to support the hierarchical data structure of arcs, polygons, and regions (Figure 15.8). Otherwise, a region may not form a closed loop.

15.4.2 Create Regions by Data Conversion

Regions can be converted from polygons or arcs, but data conversion is not a straightforward method. ArcInfo Workstation's POLYREGION, for example, can convert each polygon of a polygon coverage into a region. But the region subclass created by POLYREGION uses the contiguous

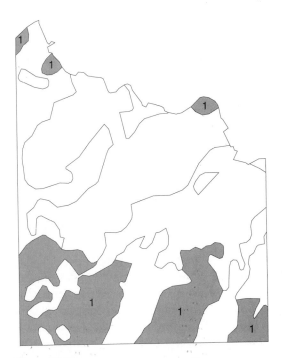

Figure 15.6
MAKEREGION can create a noncontiguous region from the selected polygons that have the suitability value of 1.

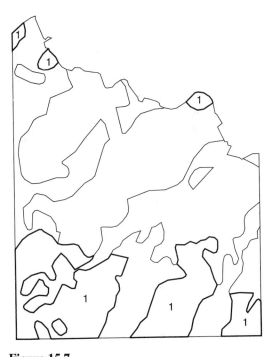

Figure 15.7
To create a region from arcs that make up the polygons with the value of 1, first select the arcs (thicker lines) and then use the MAKEREGION command.

Box 15.2   Convert a Text File with *x*-, *y*-Coordinates to a Region Subclass

The following shows the steps of using ArcInfo Workstation's REGIONCLASS to convert a text file with *x*-, *y*-coordinates into a region subclass.

1. Code all arcs that make up a region component with the same ID value. For example, all arcs that make up the first region in the above suitability example must have the same ID value of 1, and so on.
2. Convert the data file into a line coverage by using the GENERATE command.
3. Build the line coverage's topology, and make sure that the ID item in the attribute table separates arcs by region so that all arcs that

make up region 1 have the ID value of 1, and so on.

4. Use the ID item as the input item and the REGIONCLASS command to create preliminary regions from the arcs in the line coverage.
5. Use the CLEAN command to build the topology of preliminary regions and to complete the conversion process.

If the arcs that make up a region do not close properly and the resulting overshoot or undershoot is not taken care of by cleaning the coverage, the region will have topological errors.

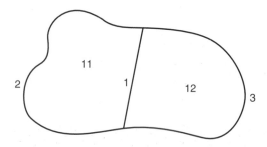

Figure 15.8
Polygon 11 shares arc 1 with polygon 12. Arc 1 must be selected and processed twice to complete the hierarchical data structure of arcs, polygons, and regions used by the regions data model.

option. To change it to the noncontiguous option, one must use another command such as RE-GIONDISSOLVE (see Section 15.6).

REGIONCLASS can convert a line coverage generated from a text file with *x*-, *y*-coordinates into a region subclass (Box 15.2). Although the process is tedious, it clearly illustrates how the higher-level regions are built on polygons and arcs. The process involves compiling arcs for each region first and then relating arcs to regions.

So far, the discussion on data conversion has been limited to ArcInfo coverages. But data con-

version to regions can extend to other data formats. The vector product format (VPF) is the standard format used by the National Imagery and Mapping Agency (NIMA) for large geographic databases (Chapter 4). VPFIMPORT is an ArcInfo Workstation command that can convert VPF files into coverages. If a VPF file contains multiple area features, such as a political file containing states, counties, and other area units, VPFIMPORT converts these area features into region subclasses. VPF is one more reason for GIS users to be familiar with the regions data model.

15.4.3 Create Regions by Using a Related Table

ArcInfo Workstation's REGIONJOIN can create new regions by joining a related table to a region subclass. Figure 15.9 illustrates how REGION-JOIN works. Figure 15.9a shows the attribute table of an existing region subclass called *soils1*. First we must establish a relationship between *soils1* and the suitability table (Figure 15.9b) by using soils1-ID as the key. A one-to-one relationship exists between *soils1*'s attribute table and the suitability table. This one-to-one relationship makes it possible for REGIONJOIN, with the

(a)

Area	Soils1#	Soils1-ID
508,309	1	2
14,022	2	3
19,000	3	4
23,931	4	5
1,283,917	5	6
58,753	6	7
22,764	7	8
833,006	8	9
8,438	9	10
18,385	10	11
188,282	11	12
303,357	12	13
86,361	13	14
516,076	14	16
301,925	15	17
90,328	16	18
11,328	17	19
13,428	18	20

(b)

Soils1-ID	Suit
2	3
3	1
4	1
5	2
6	2
7	3
8	1
9	3
10	2
11	2
12	2
13	2
14	3
16	1
17	1
18	1
19	2
20	2

(c)

Area	Soils2#	Soils2-ID	Suit
1,486,428	1	1	3
964,115	2	2	1
1,851,068	3	3	2

Figure 15.9
REGIONJOIN creates regions from a related table. Soils1-ID relates the region subclass's attribute table (a) to the suitability table (b). Table (c) is the attribute table of the new region subclass.

noncontiguous option, to group regions with the same suit value in *soils1* into a single region in a new region subclass called *soils2*. *Soils2* has three regions, one for each suit value (Figure 15.9c). The geometry of *soils2* is supplied by *soils1*.

If a one-to-many relationship exists between a region and the related table, REGIONJOIN stacks duplicates of the region in the output and lists a different attribute value for each duplicate. Figure 15.10 is the same as Figure 15.9 except that (1) the related table is a crop table and (2) one of the regions in *soils1* (soils1-ID = 10) has three values (2, 3, and 1). *Soils3*, a new region subclass created

by REGIONJOIN, has five regions (Figure 15.10c). The first two are duplicates of soils1-ID = 10 and have the soils1_crop values of 3 and 1, respectively. The other duplicate of soils1-ID = 10 with the soils1_crop value of 2 is merged with other region components of the same value in *soils3*.

15.4.4 Create Regions by Using Regions-Based Commands

Regions-based commands can create regions by querying and analyzing existing coverages or

(a)

Area	Soils1#	Soils1-ID
508,309	1	2
14,022	2	3
19,000	3	4
23,931	4	5
1,283,917	5	6
58,753	6	7
22,764	7	8
833,006	8	9
8,438	9	10
18,385	10	11
188,282	11	12
303,357	12	13
86,361	13	14
516,076	14	16
301,925	15	17
90,328	16	18
113,288	17	19
134,288	18	20

(b)

Soils1-ID	Crop
2	3
3	1
4	1
5	2
6	2
7	3
8	1
9	3
10	3
10	1
10	2
11	2
12	2
13	2
14	3
16	1
17	1
18	1
19	2
20	2

(c)

Area	Soils3#	Soils3-ID	Crop	Soils1_Crop
8,438	1	1	2	3
8,438	2	2	2	1
1,486,428	3	3	3	3
964,115	4	4	1	1
1,851,068	5	5	2	2

Figure 15.10
A one-to-many relationship exists between the region subclass's attribute table (a) and the crop table (b) for soils1-ID = 10. REGIONJOIN creates a stack of three regions for soils1-ID = 10: two are shown in the first two records of table (c), and the third is merged with the region for crop = 2.

region subclasses. These commands are covered in Section 15.5.

15.4.5 Drop or Convert Regions

ArcInfo Workstation's DROPFEATURES can drop both the geometry and attributes of a region subclass from an integrated coverage.

15.4.6 Attribute Data Management with Regions

Region subclasses reside in the same workspace with the integrated polygon coverage. The integrated coverage has its own attribute table and so does each region subclass. The naming convention for a region subclass's attribute table is to have the coverage name as the prefix and the concatenated

all.pat — attribute table for the polygon coverage *all*

all.patveg — attribute table for the region subclass *veg* that resides in the polygon coverage *all*

all.patsoil — attribute table for the region subclass *soil* that resides in the polygon coverage *all*

all.patelev — attribute table for the region subclass *elev* that resides in the polygon coverage *all*

Figure 15.11
The PATs of the integrated polygon coverage *all* and its region subclasses, *veg*, *soil*, and *elev*.

string of PAT (polygon attribute table) and the region subclass name as the extension. For example, the file name *all.patveg* refers to the attribute table for a region subclass named *veg* in an integrated polygon coverage called *all*. The coverage *all* may have other region subclasses such as *soil* and *elev*, each having its own PAT file (Figure 15.11).

The PAT for a region subclass is maintained and managed separately from the integrated coverage and other region subclasses. Each region subclass's PAT can have its own attributes that are different from other PATs and can function as a relation in a relational database. In short, each region subclass can be treated as a separate layer in GIS operations.

15.5 REGIONS-BASED QUERY AND OVERLAY

ArcInfo Workstation has two regions-based query commands, AREAQUERY and REGIONQUERY. These two commands can do more than data query: They can overlay polygons or regions; query a subset from the overlay output; and save the selection into a new region subclass.

AREAQUERY uses a set of polygon coverages as the input and processes the input in two steps. First, it converts each coverage into a region subclass and stores it with the integrated coverage. Second, AREAQUERY evaluates a query statement operated on the region subclasses and saves

the result into a new region subclass. AREAQUERY therefore combines map overlay and attribute data query into a single command. As explained earlier, this alternative to traditional map overlay has its advantages in terms of data management and the number of maps that can be overlaid at one time.

REGIONQUERY can perform query and create regions from a polygon coverage. Suppose a polygon coverage records past fires with three items: fire1 lists the year of fire for areas burned once, fire2 lists the year of the second fire for areas burned twice, and fire3 lists the year of the third fire for areas burned three times. The query statement, fire1 > 0 AND fire2 > 0, used with REGIONQUERY can create a region subclass for areas burned twice.

REGIONQUERY can also use an integrated coverage with region subclasses, which may have been created previously by AREAQUERY, as the input. Suppose an integrated coverage called *all-cov* has three region subclasses of *elev* (elevation), *strmbuf* (stream buffer), and *slope*. Attribute data in each region subclass attribute table are listed in Table 15.1. The following shows two examples of using REGIONQUERY to create new region subclasses. The first example creates a new region subclass containing areas that are in elevation zone 2 and are within the stream buffer:

elev.zone = 2 AND strmbuf.inside = 100

The above statement looks the same as any query statement except for the operands of elev.zone and strmbuf.inside. Each operand consists of two parts: the first part refers to the region subclass, and the second part refers to a field in the region subclass's attribute table.

The second example creates a new region subclass containing areas that are in elevation zone 2, have the slope code of 1, and are outside the stream buffer:

elev.zone = 2 AND slope.slope-code = 1 NOT $strmbuf

$strmbuf in the above statement means that all region features in the *strmbuf* subclass are used in

Box 15.3 **Boolean Operations to Select Regions**

Figure 15.12a shows two region subclasses: elevation zones (*elev*) in thin lines and stream buffer zones (*strmbuf*) in thick lines. Notice that the area extent of the stream buffer goes slightly beyond that of the elevation zone. The Boolean operation, $elev OR $strmbuf, duplicates a UNION operation, with the result shown in Figure 15.12b. Figure 15.12c shows the output if the operation is changed to $elev AND $strmbuf, or an INTERSECT operation. And Figure 15.12d shows the operation, ($elev AND $strmbuf) OR $strmbuf, or an IDENTITY operation using *strmbuf* as the input coverage.

XOR and NOT add more varieties to regions-based operations. Figure 15.13 shows the output if the Boolean operation, $elev XOR $strmbuf, is carried out. The output retains map features of the input region subclasses except those in the overlapped area of the two subclasses. Figure 15.14 shows the output from the operation, $elev NOT $strmbuf. Region features of the subclass *elev* that fall within *strmbuf* are removed from the output.

TABLE 15.1	Attributes and Attribute Descriptions of Region Subclasses, *elev*, *strmbuf*, and *slope*

Subclass	Attribute	Attribute Description
elev	Zone	1—less than 3000ft
		2—3000 to 4000ft
		3—greater than 4000ft
strmbuf	Inside	100—inside buffer zone
		1—outside buffer zone
slope	Slope-code	1—less than 20%
		2—20–40%
		3—40–60%
		4—greater than 60%

the selection. The Boolean connector NOT negates $strmbuf, meaning that the selection excludes areas covered by the *strmbuf* subclass.

Although regions-based query operations can be alternatives to traditional map overlay in many instances, there are three important differences between them. First, regions-based operations are limited to polygon coverages. Traditional map overlay can work with point, line, and polygon coverages. Second, regions-based operations use logical expressions and the proper Boolean connectors of AND, OR, XOR, and NOT to duplicate the overlay methods of UNION, INTERSECT, and IDENTITY. Box 15.3 shows additional examples of using the Boolean connectors in selecting regions. Third, AREAQUERY lets the user assign a relative weight to an input coverage to influence its role in line snapping. Line snapping takes place when arcs in the overlay output are within a specified fuzzy tolerance. Lines on a coverage with a greater weight are less likely to be snapped to lines on a coverage with a smaller weight.

15.6 REGIONS-BASED TOOLS

Besides overlay and query, ArcInfo Workstation has other regions-based tools for dissolve, buffering, and area tabulation between region subclasses.

REGIONDISSOLVE aggregates regions and polygons with the same value for a specified item and saves the output into a new region subclass.

REGIONBUFFER creates buffers around points, lines, polygons, or regions, and saves the output into a region subclass. REGIONBUFFER

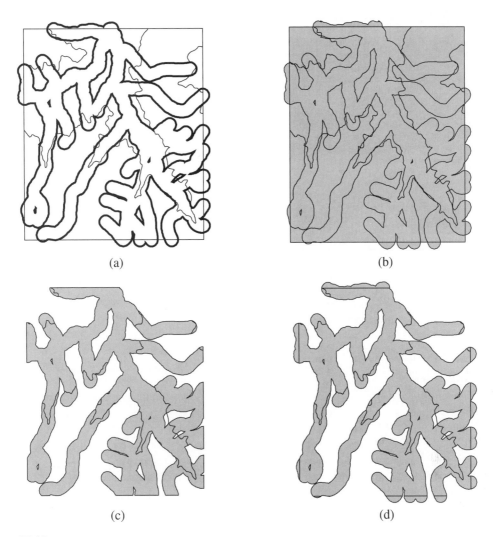

(a) (b) (c) (d)

Figure 15.12
Logical expressions are used on the two region subclasses in (a) to simulate the overlay methods of UNION (b), INTERSECT (c), and IDENTITY (d). The shaded areas represent the output. See Box 15.3 for explanation.

allows either constant buffer distance or different buffer distances. Buffer zones created around individual map features by REGIONBUFFER may be noncontiguous or contiguous. A noncontiguous buffer zone may have spatially disjoint components, whereas a contiguous buffer zone is spatially connected. REGIONBUFFER combines map features from both the input and output (Fig-

ure 15.15). To retain the input coverage, we must specify a new coverage for the output.

REGIONXAREA tabulates all overlap combinations of region subclasses in an integrated coverage and, for each combination, lists the area of overlap and the percent of the first region subclass that is overlapping with the second. Table 15.2 shows the output from REGIONXAREA for a

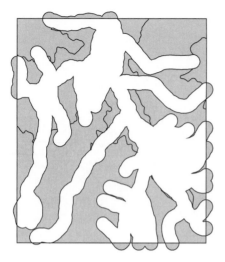

Figure 15.13
The XOR connector retains map features of the two region subclasses in Figure 15.12a except features in the overlapped area. The shaded areas represent the output.

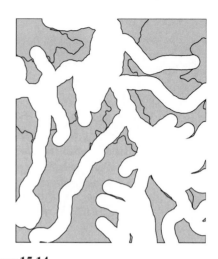

Figure 15.14
The NOT connector excludes map features that fall within *strmbuf* in Figure 15.12a. The shaded areas represent the output.

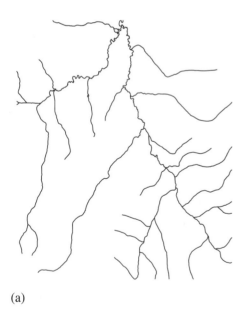

(a)

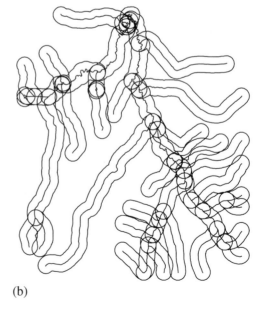

(b)

Figure 15.15
REGIONBUFFER combines map features from the input coverage (a) into the output polygon coverage (b).

TABLE 15.2   A Sample Output from REGIONXAREA

Subclass1	Subclass1#	Subclass2	Subclass2#	Area	Percent
fire1	1	fire12	1	6,325,234.000	43.15
fire1	1	fire12	2	2,709,613.500	18.48
fire1	1	fire123	1	963,893.500	6.58
fire12	1	fire1	1	6,325,234.000	100.00
fire12	2	fire1	1	2,709,613.500	100.00
fire12	2	fire123	1	963,893.500	35.57
fire123	1	fire1	1	963,893.500	100.00
fire123	1	fire12	2	963,893.500	100.00

small coverage with three region subclasses. The region subclass of *fire1* covers the entire area. The region subclass of *fire12* has two regions. And the region subclass of *fire123* has one region. The first record in the table shows that 43.15 percent of

fire1 is overlapping with the first region of *fire12*. The second record shows that 18.48 percent of *fire1* is overlapping with the second region of *fire12*, and so on.

KEY CONCEPTS AND TERMS

Contiguous region: A region with spatially joint components.

Hierarchical regions: Regions representing different spatial scales of a hierarchy.

Integrated coverage: A polygon coverage that contains the geometries of the built-in region layers.

Modifiable areal unit problem (MAUP):   The problem related to the influence that the

definition of units for which data are collected and the level of spatial data aggregation have on results of spatial analysis.

Noncontiguous region: A region with spatially disjoint components.

Uniform region: A geographic area with similar characteristics.

APPLICATIONS: REGIONS

This applications section covers six tasks. Task 1 uses ArcMap to display and query regions. Task 2 uses ArcMap to create regions from existing polygons. The remaining four tasks use ArcInfo Workstation. Tasks 3 and 4 focus on creating regions, and Tasks 5 and 6 tabulate region overlaps and perform regions-based overlay, respectively.

Task 1: Display and Query Regions in ArcMap

What you need: a regions-based coverage called *intcov*.

The integrated polygon coverge *intcov* has six region subclasses: *land* from a land-use coverage, *soil* from a soil coverage, *stream* from a stream buffer coverage, *sewer* from a sewer buffer coverage, and *potsites* and *finalsites* from results of regions-based queries. The purpose of this task is to display and query regions in ArcMap.

1. Start ArcCatalog, and make connection to the Chapter 15 database. Launch ArcMap, and rename Layers Task 1. Click the Add Data button. Navigate to the Chapter 15 database. Double-click on *intcov*. Ten data sets appear. Select *polygon*, *region.finalsites*, *region.land*, *region.potsites*, *region.sewer*, *region.soil*, and *region.stream*, and add them to Task 1. Examine the integrated polygon coverage and each of the six region subclasses.

2. Select Open Attribute Table from the context menu of *intcov region.soil*. Click the Options dropdown arrow and choose Select by Attributes. Enter the following SQL statement in the expression box: "SUIT" = 2. Click Apply. All polygons in *region.soil* that have the suit value of 2 are highlighted. A region subclass behaves like a regular layer for data display and query.

3. Select Open Attribute Table from the context menu of *intcov region.potsites*. Click the Options dropdown arrow and choose Select by Attributes. Enter the following SQL statement in the expression box: "AREA" > 2000. Click Apply. Seven out of 10 polygons in *region.potsites* are selected. This selection can be saved into a new shapefile. Right-click *intcov region.potsites*, point to Data, and select Export Data. Specify *sel_sites.shp* as the output shapefile in the Export Data dialog. *Sel_sites* contains the same polygons as *intcov region.finalsites*.

Task 2: Create Regions from Existing Polygons in ArcMap

What you need: *landuse*, a polygon coverage.

If you are using ArcInfo of ArcGIS, you can use ArcMap to create regions from existing polygons and store them in a region subclass in Task 2. The region subclass, however, must be created first

as an empty subclass in ArcInfo Workstation using the ArcEdit command of CREATEFEATURE:

```
Arcedit: mapextent landuse
Arcedit: editcov landuse
Arcedit: createfeature region.urban
Arcedit: quit /*save changes
Arc: quit
```

1. Select Data Frame from the Insert menu in ArcMap. Rename the new data frame Task 2. Click the Add Data button. Navigate to the Chapter 15 database. Double-click *landuse*. Select *polygon* and *region.urban* and add them to Task 2. *Landuse region.urban* currently has no features.

2. The first step is to select existing polygons to be grouped into the urban region subclass. Select Open Attribute Table from the context menu of *landuse polygon*. Click the Options dropdown arrow and choose Select by Attributes. Enter the following SQL statement in the expression box: "LUCODE" = 100. Click Apply. The selected records are highlighted in both the attribute table and the map.

3. This step is to group urban polygons into a region subclass. Click the Editor Toolbar. Select Starting Editing from the Editor dropdown list. Make sure that the Task is to Create New Feature and the Target is *landuse region.urban*. Click the Editor dropdown arrow, and select Union. The seven urban polygons are now components of the region subclass *urban*. Select Open Attribute Table from the context menu of *landuse region.urban*. A (region) record should appear in the table. Select Stop Editing from the Editor dropdown list. Save the edits.

Task 3: Create Regions from Existing Polygons in ArcInfo Workstation

What you need: *landuse*, same as in Task 2.

This task asks you to create a region from polygons that have the lucode value of 700 in the coverage *landuse*. Go to ArcEdit and set up the editing environment:

> Arcedit: display 9999
> Arcedit: mapextent landuse
> Arcedit: editcov landuse
> Arcedit: drawenv poly label
> Arcedit: draw /* draw polygons and labels of *landuse*
> Arcedit: editfeature label
> Arcedit: select lucode = 700 /* select labels that have lucode = 700
> Arcedit: drawselect /* highlight selected labels
> Arcedit: editfeature polygon /* change the edit feature to polygons
> Arcedit: select many /* click the labels previously highlighted and press 9 to exit
> Arcedit: makeregion marsh /* create a region called *marsh* from the selected polygons
> Arcedit: quit /* save the changes made in the session
> Arc: describe landuse

The DESCRIBE command provides a description of the coverage *landuse* and its contents. The feature class should include regions, and the subclass should list *marsh*. Type 'list landuse.patmarsh' in Arc to view the *marsh* subclass's PAT. Only one record is listed in the PAT because the three polygons are grouped into a single region using the noncontiguous option.

Task 4: Create Regions by Using Regions-Based Commands

What you need: a polygon coverage called *fire*.

The coverage *fire* shows past fires in a national forest. Three attribute items (fire1, fire2, and fire3) record up to three fire events in a given polygon. For example, a polygon with recorded fires in 1909, 1923, and 1970 has the values of 1970, 1923, and 1909 under fire1, fire2, and fire3 respectively. First, you will use the command REGIONDISSOLVE to create a region subclass.

> Arc: regiondissolve

> Usage: REGIONDISSOLVE <in_cover> {out_cover} <out_subclass> <dissolve_item | #ALL> {POLY | REGION.Subclass} {out_item_name} {selection_file}
> Arc: regiondissolve fire # fire2 fire2

REGIONDISSOLVE uses the dissolve item fire2 to create a region subclass called *fire2* in the coverage *fire*. You can use ArcTools, a graphical user interface to ArcInfo Workstation, to display the region subclass *fire2*:

1. Type arctools at the Arc prompt. Select Map Tools from the ArcTools menu. Then select New from the View menu in the Map Tools dialog.

2. In the Add New Theme dialog, select Coverage for Categories and Region for Theme Classes.

3. In the Region Theme Properties dialog, enter *fire2* as the Identifier, *fire* as the Data Source, and *fire2* as the Region. Select red as the symbol to draw regions. Click Preview to view the region subclass of *fire2*.

Exit ArcTools and go to Tables. Use the dir command in Tables to list all the INFO files in your workspace. You should see fire.pat and fire.patfire2.

> Tables: select fire.patfire2 /* select the PAT of *fire2*
> Tables: list /* list records in the PAT

You can remove the *fire2* region subclass from the *fire* coverage by using the command DROPFEATURES:

> Arc: dropfeatures
> Usage: DROPFEATURES <cover> <feature_class> {ATTRIBUTE | GEOMETRY}
> Arc: dropfeatures fire region.fire2

REGIONQUERY is another regions-based command that can create regions. REGIONQUERY uses logical expressions to select polygons to be grouped into regions. In the following, REGIONQUERY creates nested region subclasses of *fire1*, *fire12*, and *fire123* from the coverage *fire*. The *fire1* region layer represents areas burned once, *fire12* burned twice, and *fire123* burned three times. The three layers are superimposed on one another, forming a set of nested regions.

```
Arc: regionquery
Usage: REGIONQUERY <in_cover>
{out_cover} <out_subclass> {selection_file}
{NONCONTIGUOUS | CONTIGUOUS}
{out_item...out_item}
Arc: regionquery fire # fire1 # # fire1
>: reselect fire1 > 0
>: [enter a blank line] /* exit regionquery
Arc: regionquery fire # fire12 # # fire1 fire2
>: reselect fire1 > 0 and fire2 > 0
>: [enter a blank line] /* exit regionquery
Arc: regionquery fire # fire123 # # fire1 fire2
fire3
>: reselect fire1 > 0 and fire2 > 0 and fire3 > 0
>: [enter a blank line] /* exit regionquery
```

Now use ArcTools to display the nested regions of *fire1*, *fire12*, and *fire123*:

1. Type arctools at the Arc prompt. Select Map Tools from the ArcTools menu. Then select New from the View menu in the Map Tools dialog.

2. Select Coverage for Categories and Region for Theme Classes in the Add New Theme dialog.

3. First, set up the draw environment for *fire1*. In the Region Theme Properties dialog, enter *fire1* as the Identifier, *fire* as the Data Source, and *fire1* as the Region. Select red as the symbol to draw regions, and click OK.

4. Repeat Steps 2 and 3 to set up the draw environment for *fire12*. Enter *fire12* as the Identifier, *fire* as the Data Source, and *fire12* as the Region in the Region Theme Properties dialog. Select green for the Symbol, and click OK.

5. Repeat Steps 2 and 3 to set up the draw environment for *fire123*. Enter *fire123* as the Identifier, *fire* as the Data Source, and *fire123*

as the Region in the Region Theme Properties dialog. Select blue for the Symbol, and click OK.

6. *Fire1, fire12,* and *fire123* are now listed under Themes in the Theme Manager dialog. Use the arrow to move the three region layers to the Draw list. Then click on the Draw button to draw the region layers in three different colors. Exit ArcTools.

You can list the attributes contained in the PAT of each region subclass in either Arc or Tables.

```
Arc: list fire.patfire1
Arc: list fire.patfire12
Arc: list fire.patfire123
```

Task 5: Tabulate Region Overlaps

What you need: a polygon coverage called *sfire*.

The Arc command REGIONXAREA runs an AML macro to tabulate all possible region overlaps and their areas and percentage areas. The result is saved into an INFO file. Depending on the number of region subclasses and the number of regions in each subclass, the INFO file can be massive. For this task, you will use a small coverage called *sfire* so that you can better understand how REGIONXAREA works.

1. Use ArcTools to display the nested region subclasses in *sfire*.

2. Now, run REGIONXAREA in Arc. Ignore messages from the execution of REGIONXAREA.

```
Arc: regionxarea
Usage: REGIONXAREA <in_cover>
<out_info_file> {$ALL |
subclass...subclass}
Arc: regionxarea sfire sfire.dat
```

3. List the INFO file created by REGIONXAREA.

```
Arc: list sfire.dat
```

Each record in *sfire.dat* represents an overlap of regions. Subclass1 and subclass2 are the two region subclasses in the overlap. Subclass1# and subclass2# list the regions in the overlap. Area is the size of the overlap, and percent shows the area percentage of subclass1# overlapping with subclass2#. As examples, the first record in *sfire.dat* shows that 43.15 percent of *fire1* is overlapping with the first region of *fire12* and the second record shows that 18.48 percent of *fire1* is overlapping with the second region of *fire12*.

Task 6: Perform Regions-Based Overlay

What you need: coverages of *landuse*, *soils*, *strmbuf*, and *sewerbuf*.

As described in this chapter, AREAQUERY can perform both map overlay and data query in a single operation. AREAQUERY is an alternative to the traditional overlay method for GIS analysis. This task simulates a suitability analysis with the following selection criteria:

- Preferred land use is brushland (lucode = 300 in *landuse*)
- Choose soil types suitable for development (suit >= 2 in *soils*)
- Site must be within 300 meters of existing sewer lines (within *sewerbuf*)
- Site must be beyond 20 meters of existing streams (beyond *strmbuf*)
- Site must contain an area at least 2000 square meters

Arc: areaquery
Usage: AREAQUERY <out_cover>
<out_subclass> {tolerance}
{NONCONTIGUOUS|CONTIGUOUS}
{out_item...out_item}

Arc: areaquery allcov potsites # contiguous
/* initiate AREAQUERY

A dialog command, AREAQUERY first asks for input coverages and their region subclass names.

Enter the 1st coverage: landuse land /* land is the subclass name for *landuse*
Enter the 2nd coverage: soils soil /* soil is the subclass name for *soils*
Enter the 3rd coverage: sewerbuf sewer /*sewer is the subclass name for *sewerbuf*
Enter the 4th coverage: strmbuf stream /*stream is the subclass name for *strmbuf*
Enter the 5th coverage: end

AREAQUERY then asks for a logical expression. Each operand in the logical expression consists of two parts: the first part refers to the region subclass, and the second part refers to the item in the region subclass. The logical expression can include the Boolean connectors of AND, OR, XOR, and NOT.

>: resel land.lucode = 300 and soil.suit >= 2 and $sewer and not $stream
>: [enter a blank line] /* exit areaquery
Arc: list allcov.patpotsites

The output from AREAQUERY is saved into the region subclass *potsites*, which includes sites that satisfy the first four criteria. To complete the task, use REGIONQUERY to select sites that are larger than 2000 m^2 and save the output into *finalsites*.

Arc: regionquery allcov # finalsites # contiguous
>: resel potsites.area > 2000
>: [enter a blank line] /* exit regionquery

REFERENCES

Allen, T. H .F., and T. B. Starr. 1982. *Hierarchy: Perspectives for Ecological Complexity*. Chicago: The University of Chicago Press.

Amrhein, C. G. 1995. Searching for the Elusive Aggregation Effect:

Evidence from Statistical Simulations. *Environment and Planning A* 27: 105–19.

Bailey, R. G. 1976. *Ecoregions of the United States*. Map (scale 1:7,500,000). Ogden, Utah: U.S. Department of Agriculture,

Forest Service, Intermountain Region.

Bailey, R. G. 1983. Delineation of Ecosystem Regions. *Environmental Management* 7: 365–73.

Batty, M., and P. K. Sikdar. 1982a. Spatial Aggregation in Gravity

Models. 1. An Information-Theoretic Framework. *Environment and Planning A* 14: 377–405.

Batty, M., and P. K. Sikdar. 1982b. Spatial Aggregation in Spatial Interaction Models. 2. One-Dimensional Population Density Models. *Environment and Planning A* 14: 524–53.

Berry, B. J. L. 1968. Approaches to Regional Analysis: A Synthesis. In B. J. L. Berry and D. F. Marble, eds., pp. 24–34. *Spatial Analysis: A Reader in Statistical Geography*. Englewood Cliffs, NJ: Prentice-Hall.

Chang, K., Z. Khatib, and Y. Ou. 2002. Effects of Zoning Structure and Network Detail on Traffic Demand Modeling. *Environment and Planning B* 29: 37–52.

Clark, W. A. V., and K. L. Avery. 1976. The Effect of Data Aggregation in Statistical Analysis. *Geographical Analysis* 8: 428–37.

Cleland, D. T., R. E. Avers, W. H. McNab, M. E. Jensen, R. G. Bailey, T. King, and W. E. Russell. 1997. National Hierarchical Framework of Ecological Units. In M. S. Boyce and A. Haney, eds., pp. 181–200. *Ecosystem Management Applications for Sustainable Forest and Wildlife Resources*, New Haven, CT: Yale University Press.

Costanza, R., and T. Maxwell. 1994. Resolution and Predictability: An Approach to the Scaling Problem. *Landscape Ecology* 9: 47–57.

Forman, R. T. T., and M. Godron. 1986. *Landscape Ecology*. New York: Wiley.

Fotheringham, A. S., and D. W. S. Wong. 1991. The Modifiable Areal Unit Problem in Multivariate Statistical Analysis. *Environment and Planning A* 23: 1025–44.

Gehlke, C. E., and K. Biehl. 1934. Certain Effects of Grouping upon the Size of the Correlation Coefficient in Census Tract Material. *Journal of the American Statistical Association* 24:169–70.

Goodchild, M. F. 1979. The Aggregation Problem in Location-Allocation. *Geographical Analysis* 11: 240–55.

Jelinski, D. E., and J. Wu. 1996. The Modifiable Areal Unit Problem and Implications for Landscape Ecology. *Landscape Ecology* 11:129–40.

McMahon, G., S. M. Gregonis, S. W. Waltman, J. M. Omernik, T. D. Thoson, J. A. Freeouf, A. H. Rorick, and J. E. Keys. 2001. Developing a Spatial Framework of Common Ecological Regions for the Conterminous United States. *Environmental Management* 28: 293–316.

Meentemeyer, V., and E. O. Box. 1987. Scale Effects in Landscape Studies. In M. G. Turner, ed., *Landscape Heterogeneity and Disturbance*, pp. 15–34. New York: Springer-Verlag.

Miller, H. J. 1999. Potential Contributions of Spatial Analysis to Geographic Information Systems for Transportation. *Geographical Analysis* 31: 373–99.

Omernik, J. M., 1987. Ecoregions of the Conterminous United States. *Annals of the Association of American Geographers* 77: 118–25.

O'Neill, R. V., D. L. DeAngelis, J. B. Waide, and T. F. H. Allen. 1986. *A Hierarchical Concept of Ecosystems*. Princeton, NJ: Princeton University Press.

Openshaw, S. 1977. Optimal Zoning Systems for Spatial Interaction Models. *Environment and Planning A* 9:169–84.

Openshaw, S., and P. J. Taylor. 1979. A Million or So Correlation Coefficients: Three Experiments on the Modifiable Areal Unit Problem. In N. Wrigley, ed., *Statistical Applications in the Spatial Sciences*, pp. 127–44. London: Pion.

Robinson, W. S. 1950. Ecological Correlation and the Behavior of Individuals. *American Sociological Review* 15: 351–57.

Svancara, L. K., E. O. Garton, K. Chang, J. M. Scott, P. Zager, and M. Gratson. 2002. The Inherent Aggravation of Aggregation: An Example with Elk Aerial Survey Data. *Journal of Wildlife Management* 66: 776–87.

Urban, D. L., D. O'Neill, and H. H. Shugart. 1987. Landscape Ecology. *BioScience* 37: 119–27.

Walker, D. A., and M. D. Walker. 1991. History and Pattern of Disturbance in Alaskan Arctic Terrestrial Ecosystems: A Hierarchical Approach to Analysing Landscape Change. *Journal of Applied Ecology* 28: 244–76.

Wiens, J. A. 1989. Spatial Scaling in Ecology. *Functional Ecology* 3: 385–97.

Wong, D., and C. Amrhein. 1996. Research on the MAUP: Old Wine in a New Bottle or Real Breakthrough. *Geographical Systems* 3: 73–76.

NETWORK AND DYNAMIC SEGMENTATION

16.1 INTRODUCTION

A network consists of connected linear features. A familiar network is a road system. Other networks include railways, public transit lines, bicycle paths, streams, and shorelines. Dynamic segmentation is a data model that is built on lines of a network and allows the use of plane coordinates with linear measures such as mileposts. By using the dynamic segmentation model, one can link linearly referenced data such as accidents and pavement conditions to a geographically referenced road network.

Both the network and dynamic segmentation models must have the appropriate attributes for real-world applications. To use a road network for path finding or distribution of resources, one must add attributes such as travel time, turn impedance, and one-way streets to the arcs and nodes of the network. Likewise, one must have linear measures added to the arcs in order to link data that are measured in mileposts directly to road segments.

Network analysis and dynamic segmentation are available in some GIS packages. MGE Network, PAMAP NETWORKER, and TransCAD can perform network analysis. MGE Segment Manager and TransCAD have dynamic segmentation functions. As of this writing, ArcGIS users can use ArcInfo Workstation for building and analyzing both the network and dynamic segmentation

models. ArcGIS Desktop has a new geometric network model but the model is used primarily for utility management (Box 16.1). The Network Analyst extension to ArcView 3.x, which covers network applications, is not yet available to ArcGIS 8.x users. ESRI Inc. plans to add the Network Analyst extension to ArcGIS 9.

This chapter includes the following five main sections. Section 16.2 covers the preparation of a road network with an example from Moscow, Idaho. Section 16.3 considers network analysis and applications. Section 16.4 discusses the building of a dynamic segmentation model, and Section 16.5 discusses the linking of linearly referenced data to the dynamic segmentation model. Section 16.6 reviews applications of dynamic segmentation.

16.2 NETWORK

A **network** is a line coverage, which is topology-based, and has the appropriate attributes for the flow of objects such as traffic. The geometry of a network can be either digitized or imported from existing data sources. Regardless of its origin, a network must have the appropriate attributes for real-world applications. Because most network applications involve road systems, the following discussion concentrates on feature and attribute data

---

Box 16.1 **Geometric Network in ArcGIS**

The geometric network model has two basic components: edges and junctions. Compared to a line coverage, edges are like arcs and junctions are like nodes. The data structure of a geometric network is based on the geodatabase model: the network is a feature data set, edges and junctions are feature classes, and different connections between edges and junctions are built as topology rules. The attribute tables of edge and junction features can have attributes such as diameter, impedance, and subtypes for network analysis.

As of this writing, the geometric network is designed primarily for management of water and electricity utilities. ArcMap has a Utility Network Analyst toolbar, which can perform various trace operations for utility management. Utility Network Analyst, however, cannot be applied to road networks, which are the main topic of this chapter.

for road networks, including impedance values assigned to network links, turns, one-way streets, and overpasses and underpasses. Later an example from Moscow, Idaho, shows how these data can be put together to form a street network.

16.2.1 Link Impedance

A **link** refers to a segment separated by two nodes in a road network. **Link impedance** is the cost of traversing a link. A simple measure of the cost is the physical length of the link. But the length may not be a reliable measure of cost, especially in cities where speed limits and traffic conditions vary significantly along different streets. A better measure of link impedance is the travel time estimated from the length and the speed limit of a link. For example, if the speed limit is 30 miles per hour and the length is 2 miles, then the link travel time is 4 minutes (2/30 x 60 minutes).

There are variations in measuring the link travel time. The travel time may be directional: the travel time in one direction may be different from the other direction. In that case, the travel time can be entered separately as from-to and to-from, depending on the arc direction. The travel time may also vary by the time of the day and by the day of the week, thus requiring the setup of different network attribute data for different applications.

16.2.2 Turn Impedance

A turn is a transition from one arc to another in a network. **Turn impedance** is the time it takes to complete a turn, which is significant in a congested street network (Ziliaskopoulos and Mahmassani 1996). Because a network typically has many turns with different conditions, a **turn table** is used to assign the turn impedance values. A turn table has three fields: the node number for the intersection, the arc numbers involved in a turn, and the turn impedance value.

A turn table must list all possible turns by arc numbers at each intersection. A driver approaching a street intersection usually has three options: go straight, turn right, or turn left. In some cases the driver has the fourth option of making a U turn.

Assuming the intersection involves four street segments, as most intersections do, this means at least 12 possible turns at the intersection excluding U turns (Figure 16.1).

Depending on the level of detail for a study, GIS users may not have to include every intersection and every possible turn in a turn table. A partial turn table listing only turns at intersections with stoplights may be sufficient for a network application.

Turn impedance is usually directional. For example, it may take 5 seconds to go straight, 10 seconds to make a right turn, and 30 seconds to make a left turn at a stoplight. A negative impedance value for a turn means a prohibited turn, such as turning the wrong way onto a one-way street.

16.2.3 One-Way or Closed Streets

The value of a designated field in a network's attribute table can show one-way or closed streets. FT, for instance, means that travel is permitted in the direction from the from-node to the to-node of an arc (a street segment). TF means that travel is permitted from the to-node to the from-node. And N means that travel is not allowed in either direction.

16.2.4 Overpasses and Underpasses

There are two methods for including overpasses and underpasses in a network. The first method is

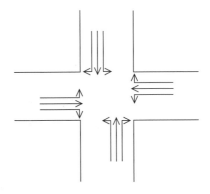

Figure 16.1
Possible turns at an intersection with four street segments. No U turns are allowed in this example.

to use nonplanar features: an overpass and the street underneath it are both represented as continuous lines without a node at their intersection (Figure 16.2). The second method treats overpasses and underpasses as planar features: the two arcs representing the overpass meet at one node, and the two arcs representing the street underneath the overpass meet at another (Figure 16.3). Elevation fields are assigned to the arcs: T-elev is the elevation of the arc entering the crossing and F-elev is the elevation of the arc leaving the crossing. To differentiate the overpass from the street, one can give the node used by the overpass a higher elevation value (e.g., 1) than the node used by the street (e.g., 0).

16.2.5 Putting Together a Street Network

Putting together a road network with the appropriate attributes for real-world applications is not an easy task. The example of Moscow, Idaho, shows how to accomplish the task. A university town of 20,000 people, Moscow has more or less the same network features as other larger cities except for overpasses or underpasses.

TIGER/Line files from the U.S. Census Bureau are a common data source for street networks in the United States. Step 1 in this example is to make a preliminary street coverage from the 1995 TIGER/Line files. TIGER/Line files use longitude and latitude values as measurement units and NAD83 (North American Datum of 1983) as the datum. Therefore, the Moscow street coverage must be converted to plane coordinates.

Step 2 is to edit and update the preliminary street coverage. TIGER/Line files include a long list of attribute data, many of which are not relevant to network applications and can be deleted. Mistakes on the street coverage, which are transferred from TIGER/Line files, must be corrected, and new streets built in the 1990s must be added. Also, pseudo nodes (i.e., nodes that are not required topologically) must be removed so that a street segment between two intersections is not unnecessarily broken up. But a pseudo node is needed at the location where one street changes into another along a continuous arc. Without the pseudo node, the continuous arc will be treated as the same street.

Step 3 is to attribute the street network with link impedance values. Speed limits can be assigned

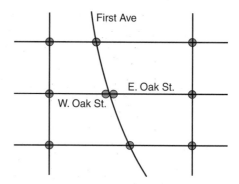

Street name	F-elev	T-elev
First Ave	0	1
First Ave	1	0
W. Oak St.	0	0
E. Oak St.	0	0

Figure 16.3
First Ave. crosses Oak St. with an overpass. A planar representation with two nodes is used at the intersection: one for First Ave., and the other for Oak St. The elevation value of 1 shows that the overpass is along First Ave.

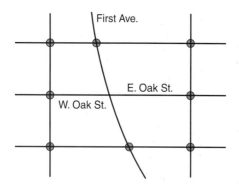

Figure 16.2
First Ave. crosses Oak St. with an overpass. A nonplanar representation with no nodes is used at the intersection of Oak St. and First Ave.

to streets by functional classification. Moscow, Idaho, has three speed limits: 35 miles/hour for principal arterials, 30 miles/hour for minor arterials, and 25 miles/hour for all other city streets. With speed limits in place, the travel time for each street segment can be computed from the segment's length and speed limit.

Step 4 is to attribute one-way streets. Moscow, Idaho, has two one-way streets serving as the northbound and the southbound lanes of a state highway. Those one-way streets are denoted by the value of 1 for a field called *direction*. The arc direction of all street segments that make up a one-way street must be consistent and pointing in the right direction. Street segments that are incorrectly oriented must be flipped.

Step 5 is to prepare a turn table. ArcInfo Workstation has the command TURNTABLE that can generate a turntable with all intersections and possible turns in a network. One can then add the turn impedance values to the turntable using a field, such as minutes or seconds. Attributing turn impedance requires selecting each intersection or node, determining different types of turns at the intersection, and assigning an estimated impedance value to each turn. In this example, a partial turntable is made for street intersections with stop lights only.

Figure 16.4 shows a street intersection at node 341 with no restrictions for all possible turns except U turns. An angle value of 90 means a left turn, for example, turning left from arc 503 to arc 467 (Box 16.2). An angle value of −90 means a right turn. And an angle value of 0 means going straight. This example uses two turn impedance

values: 30 seconds or 0.5 minute for a left turn, and 15 seconds or 0.25 minute for either a right turn or going straight.

Some street intersections do not allow certain types of turns. Figure 16.5 shows a street intersection at node 265 with stop signs posted only for the

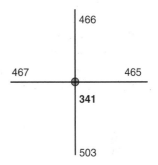

Node#	Arc1#	Arc2#	Angle	Minutes
341	503	467	90	0.500
341	503	466	0	0.250
341	503	465	−90	0.250
341	467	503	−90	0.250
341	467	466	90	0.500
341	467	465	0	0.250
341	466	503	0	0.250
341	466	467	−90	0.250
341	466	465	90	0.500
341	465	503	90	0.500
341	465	467	0	0.250
341	465	466	−90	0.250

Figure 16.4
Possible turns at node 341.

Box 16.2 Turn Angle

The turn angle is the angle between the "from" arc and the "to" arc of a turn. In reality, the angle is often 90° for a right or left turn and 0° for going straight. As measured from the coverage prepared from TIGER/Line files, the angle is likely to deviate from 90° or 0° due to digitizing errors. Because the purpose of the angle is simply to determine the type of turn, we can ignore the deviation.

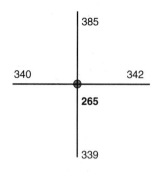

Node#	Arc1#	Arc2#	Angle	Minutes
265	339	342	−87.412	0.000
265	339	340	92.065	0.000
265	339	385	7.899	0.000
265	342	339	87.412	0.500
265	342	340	−0.523	0.250
265	342	385	−84.689	0.250
265	340	339	−92.065	0.250
265	340	342	0.523	0.250
265	340	385	95.834	0.500
265	385	339	−7.899	0.000
265	385	342	84.689	0.000
265	385	340	−95.834	0.000

Figure 16.5

Turn impedance values for turns at node 265 that has stop signs for the east–west traffic.

Node#	Arc1#	Arc2#	Angle	Minutes
339	467	501	90.190	0.500
339	467	462	1.152	0.250
339	467	461	−92.197	−1.000
339	462	501	−90.962	0.250
339	462	467	−1.152	0.250
339	462	461	86.651	−1.000
339	461	501	2.386	0.250
339	461	467	92.197	0.500
339	461	462	−86.651	0.250

Figure 16.6

Turn impedance values for turns at node 339. Node 339 is an intersection between a southbound one-way street and an east–west two-way street.

east–west traffic. Therefore, the turn impedance values are assigned only to turns to be made from arc 342 and arc 340. Figure 16.6 shows a street intersection with stoplights between a southbound one-way street and an east-west two-way street. An impedance value of −1 means a prohibited turn, such as a right turn from arc 467 to arc 461 or a left turn from arc 462 to arc 461.

16.3 NETWORK APPLICATIONS

This section discusses applications that use a network directly for path finding and accessibility measures as well as applications that link a network to a broader topic, such as location–allocation and transportation planning modeling. Direct network applications are usually accessible through commands or menu choices in GIS packages. Applications that involve a network as input often link a GIS package with a discipline-specific software package, such as a transportation planning software package.

16.3.1 Shortest-Path Analysis

Shortest-path analysis finds the path with the minimum cumulative impedance between nodes on a network. The path may connect just two nodes—the origin and destination—or have specific stops between the nodes. Shortest-path analysis can help a traveler plan a trip, a van driver set up a schedule for dozens of deliveries, or an emergency service connect a dispatch station, accident location, and hospital.

The idea behind shortest-path analysis is similar to that of cost distance measure operations using raster data (Chapter 11). The main difference is the data model. Shortest-path analysis is vector-based and uses an existing network. The cost distance measure operation is raster-based and uses an input grid and a cost grid to search for the least accumulative cost path for proposed facilities such as water lines and pipelines.

The shortest-path problem has been studied extensively in operations research, computer science, spatial analysis, and transportation engineering. A new application of shortest-path analysis is in the development of in-vehicle route guidance systems (RGS). Important to any RGS is an algorithm that can find the optimal (shortest) route from the origin to the destination (Fu and Rilett 1998).

Shortest-path analysis typically begins with an impedance matrix, in which a value represents the impedance of a direct link between two nodes on a network and a 0 or infinity means no direct connection. The analysis then follows an iterative process of finding the shortest distance from node 1 to all other nodes (Dijkstra 1959). Different computing algorithms for solving the shortest path have been proposed in the literature (Dreyfus 1969).

Here we will look at an example of six cities on a road network (Figure 16.7). Table 16.1 shows the impedance matrix of travel time measured in minutes. The value of 0 above and below the principal diagonal in the impedance matrix means no direct path between two nodes. Suppose we want to find the shortest path from node 1 to all other nodes in Figure 16.7. We can solve the problem by using an iterative procedure (Lowe and Moryadas 1975). At each step, we choose the shortest path from a list of candidate paths and place the node of the shortest path in the solution list.

The first step is to choose the minimum among the three paths from node 1 to nodes 2, 3, and 4, respectively:

$$\min (p_{12}, p_{13}, p_{14}) = \min (20, 53, 58)$$

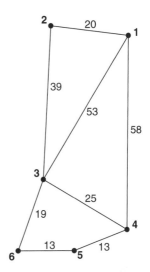

Figure 16.7

Six cities on a road network for shortest-path analysis.

	(1)	(2)	(3)	(4)	(5)	(6)
(1)	0	20	53	58	0	0
(2)	20	0	39	0	0	0
(3)	53	39	0	25	0	19
(4)	58	0	25	0	13	0
(5)	0	0	0	13	0	13
(6)	0	0	19	0	13	0

TABLE 16.1 The Impedance Matrix among Six Nodes in Figure 16.7

We choose p_{12} because it has the minimum impedance value among the three candidate paths. We then place node 2 in the solution list with node 1.

The second step is to prepare a new candidate list of paths that are directly or indirectly connected to nodes in the solution list (nodes 1 and 2):

$$\min (p_{13}, p_{14}, p_{12} + p_{23}) = \min (53, 58, 59)$$

TABLE 16.2		Shortest Paths from Node 1 to All Other Nodes in Figure 16.7	
From-Node	To-Node	Shortest Path	Minimum Cumulative Impedance
1	2	p_{12}	20
1	3	p_{13}	53
1	4	p_{14}	58
1	5	$p_{14} + p_{45}$	71
1	6	$p_{13} + p_{36}$	72

We choose p_{13} and add node 3 to the solution list. To complete the solution list with other nodes on the network we need to go through the following steps:

$$\min (p_{14}, p_{13} + p_{34}, p_{13} + p_{36}) =$$
$$\min (58, 78, 72)$$

$$\min (p_{13} + p_{36}, p_{14} + p_{45}) = \min (72, 71)$$

$$\min (p_{13} + p_{36}, p_{14} + p_{45} + p_{56}) =$$
$$\min (72, 84)$$

Table 16.2 summarizes the solution to the shortest-path problem from node1 to all other nodes.

The **traveling salesman problem** is a more complex form of shortest-path analysis because it has two constraints: (1) the salesman must visit each of the select stops only once and (2) the salesman may start from any stop but must return to the original stop. The objective is to determine which route, or tour, the salesman can take to minimize the total impedance value. A common solution to the traveling salesman problem uses a heuristic method (Lin 1965). Beginning with an initial random tour, the method runs a series of locally optimal solutions by swapping stops that yield reductions in the cumulative impedance. The iterative process ends when no improvement can be found by swapping stops. This heuristic approach

Figure 16.8

Shortest path from a street address in Moscow, Idaho, to its closest fire station (shown by the square symbol).

can usually create a tour with a minimum, or near minimum, cumulative impedance.

16.3.2 Closest Facility

One type of shortest-path analysis is to find the closest facility, such as a hospital, fire station, or ATM, to any location on a network. A **closest-facility** algorithm first computes the shortest paths from the select location to all candidate facilities, and then chooses the closest facility among the candidates. Figure 16.8 shows, for example, the closest fire station to a street address in Moscow, Idaho.

Location-based services (LBS), a new application heavily promoted by GIS companies in recent years, depend on closest-facility analysis. An LBS user can use a Web-enabled mobile phone to be located and to access, through mobile location services, information that is relevant to the user's location. For example, if the user wants to find the closest ATM, the LBS provider can run a closest facility analysis and send the information to the user.

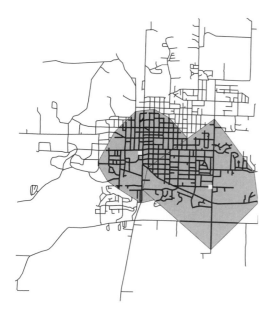

Figure 16.9
Service areas of two fire stations within a 2-minute response time in Moscow, Idaho.

Figure 16.10
Service areas of two fire stations within a 5-minute response time in Moscow, Idaho.

16.3.3 Allocation

Allocation is the study of the spatial distribution
of resources through a network. Resources in allocation studies often refer to public facilities, such
as fire stations or schools. Because the distribution
of the resources defines the extent of the service
area, the main objective of spatial allocation analysis is to measure the efficiency of these resources.

A common measure of efficiency in the case
of emergency services is the response time—the
time it takes for a fire truck or an ambulance to
reach an incident. Figure 16.9, for example, shows
areas in Moscow, Idaho, covered by two existing
fire stations within a 2-minute response time. The
map shows a large portion of the city is outside the
2-minute response zone. The response time to the
city's outer zone is about 5 minutes (Figure 16.10).

If residents of Moscow, Idaho, demand that
the response time to any part of the city be 2 minutes or less, then the options are either to relocate
the fire stations or, more likely, to build new fire

stations. A new fire station should be located to
cover the largest portion of the city unreachable in
2 minutes by the existing fire stations. The problem then becomes a location and allocation problem, which is covered in the next section.

16.3.4 Location–Allocation

An important topic in operations research and spatial analysis, **location–allocation** solves problems
of matching the supply and demand by using sets
of objectives and constraints. Location–allocation
algorithms used to be available in stand-alone
computer programs, but GIS packages, such as
ArcInfo Workstation and MGE Network, have incorporated some algorithms as application tools.

The private sector offers many location–
allocation examples. Suppose a company operates
soft-drink distribution facilities to serve supermarkets. The objective in this example is to minimize
the total distance traveled, and a constraint, such
as a 2-hour drive distance, may be imposed on the

problem. A location–allocation analysis is to match the distribution facilities and the supermarkets while meeting both the objective and the constraint.

Location–allocation is also important in the public sector. For instance, a local school board may decide that (1) all school-age children should be within 1 mile of their schools and (2) the total distance traveled by all children should be minimized. In this case, schools represent the supply, and school-age children represent the demand. The objective of this location–allocation analysis is to provide equitable service to a population while maximizing efficiency in the total distance traveled.

The setup of a location–allocation problem requires inputs in supply, demand, and distance measures. The supply consists of facilities or centers at point locations. The demand may consist of points, lines, or polygons, depending on the data source and the level of spatial data aggregation. For example, the locations of school-age children may be represented as individual points, aggregate points along streets, or aggregate points (centroids) in unit areas such as census block groups. Use of aggregate demand points is likely to introduce errors in location–allocation analysis (Francis et al. 1999). Distance measures between the supply and demand are often included in a distance matrix or a distance list. Distances may be measured along the shortest path between two points on a road network or along the straight line connecting two points. In location–allocation analysis shortest-path distances are likely to yield more accurate results than straight-line distances.

Perhaps the most important input to a location–allocation problem is the model or algorithm for solving the problem. Two most common models are minimum distance and maximum covering (Ghosh et al. 1995; Church 1999). The **minimum distance model**, also called the *p-median location model*, minimizes the total distance traveled from all demand points to their nearest supply centers (Hakimi 1964). The *p*-median model has been applied to a variety of facility location problems, including food distribution facilities, public libraries,

and health facilities. The **maximum covering model** maximizes the demand covered within a specified time or distance (Church and ReVelle 1974). Public sector location problems, such as the location of emergency medical and fire services, are ideally suited for the maximum covering model. The model is also useful for many convenience-oriented retail facilities, such as movie theaters, banks, and fast-food restaurants.

Both the minimum distance and maximum covering models may have added constraints or options. A maximum distance constraint may be imposed on the minimum distance model so that the solution, while minimizing the total distance traveled, ensures that no demand is beyond the specified maximum distance. A desired distance option may be used with the maximum covering model to cover all demand points within the desired distance.

Here we will use public libraries in Latah County, Idaho, as an example to examine three location–allocation scenarios. The demand for public libraries in Latah County consists of 33 demand points, each representing the total number of households in a census block group (Figure 16.11). Distances are measured along straight lines between the demand points and public library locations.

The first scenario assumes that (1) every block group is a possible candidate for a public library and (2) the county government wants to set up three libraries using the maximum covering model with a desired distance of 10 miles. Figure 16.12 shows the three selected candidates. The solution leaves one demand point with 344 households uncovered. To cover every demand point, the county government must either increase the number of public libraries or relax the desired distance.

The second scenario has the same setup as the first except that the county government wants to use the minimum distance model. Figure 16.13 shows the three selected candidates, one of which is not included in Figure 16.12 from the maximum covering model.

The third scenario assumes that one of the three library locations is fixed or has been previously chosen. Figure 16.14 shows the two selected new sites using the minimum distance model with

Figure 16.11
Demand points in Latah County, Idaho, for a public library location problem.

Figure 16.12
Three candidates selected by the maximum covering model and the demand points served by the candidates.

the maximum distance constraint of 10 miles. The solution leaves three demand points with 1124 households uncovered.

The location–allocation algorithms and other network analysis functions available in ArcInfo Workstation are summarized in Box 16.3.

16.3.5 Urban Transportation Planning Model

An urban transportation planning (UTP) model typically uses the four-step process of trip generation, trip distribution, modal choice, and trip assignment (Nyerges 1995; Khatib et al. 2001). Trip generation estimates the number of trips to and from each traffic analysis zone (TAZ) in the model. Trip distribution uses the number of trips generated in each TAZ and the distance or travel time between TAZs to produce a trip interchange matrix between TAZs. Modal choice splits trips by travel mode. And trip assignment loads estimated traffic volumes from the model on the network.

UTP modeling is a complex process. Each step in the four-step process involves selecting algorithms or models and requires a good understanding of transportation planning from the software developer and the user (Waters 1999). This is why specialized software packages are normally used for UTP modeling.

One of the necessary inputs to UTP modeling is a transportation network. The network is used to measure the travel impedance between TAZs. Both trip distribution and modal choice in the four-step process need the travel impedance data. The network is also used in trip assignment because at this step the estimated traffic volume is assigned to every link of the network.

To be used for UTP modeling a network must have the arc–node data structure. Nodes are road intersections and critical turns. Arcs or links are road segments between nodes. Link attributes needed for transportation planning usually include distance, speed limit, capacity, and functional classification.

Figure 16.13
Three candidates selected by the minimum distance model and the demand points served by the candidates.

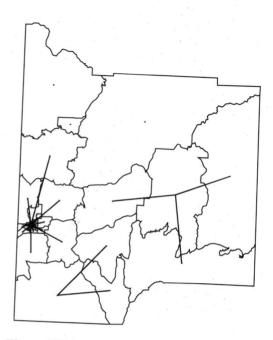

Figure 16.14
One fixed and two selected candidates using the minimum distance model with the maximum distance constraint of 10 miles. The fixed candidate is the one to the west.

16.4 DYNAMIC SEGMENTATION

Built upon arcs of a network, the dynamic segmentation data model has the basic elements of routes, sections, and events. A route is a collection of sections and resides in a line coverage as a subclass. A section refers directly to arcs in the line coverage and provides measures to a route system. Events are occurrences, such as accidents or pavement condition, and are related to a route system by linear measures of their locations (Box 16.4). The process of developing a dynamic segmentation model therefore involves creating routes, computing measures for routes, and building event tables.

16.4.1 Create Routes on New Arcs

One option is to digitize arcs that make up a new route and use ArcInfo Workstation's REMEASURE to compute the route measures. When the topology of the line coverage including routes is built, a section table and a route table will be automatically created. The section table shows measures of each section by relating sections to arcs, and the route table uses the route ID to link to the section table.

16.4.2 Create Routes on Existing Arcs

Creating routes on the existing arcs of a line coverage is an efficient option. This option offers the interactive and data conversion methods. The interactive method consists of two steps: (1) select arcs that make up a route and (2) use ArcInfo Workstation's MAKEROUTE to convert the arcs to a route and to compute the route measures.

The data conversion method applies to a line coverage rather than a select group of arcs as in the interactive method and can create many routes at once by referencing a field in the line coverage's attribute table. For example, if the field is highway number, the conversion creates a route system for each numbered highway.

Box 16.3 **Network Analysis Using ArcInfo Workstation**

The Network Analyst extension to ArcView 3.x can solve shortest-path, closest-facilities, and allocation problems. But the extension is not available to ArcGIS 8.x users. Before ESRI Inc. adds the Network Analyst extension to ArcGIS, network analysis functions are only available through ArcInfo Workstation commands.

PATH and TOUR are two commands for solving the shortest-path problem. PATH uses a user-defined order of stops to solve the problem, whereas TOUR determines the order in which stops are visited to minimize the cumulative impedance. ArcInfo Workstation saves the shortest-path solution as a route, which can be displayed on top of a network. Both PATH and TOUR offer the option of providing directions for the shortest path with street names, turns, and distances.

ALLOCATE works with the allocation problem by assigning arcs and/or nodes on a network to centers (e.g., fire stations). LOCATEALLOCATE works

with the location–allocation problem. ArcInfo Workstation offers six location–allocation models including the minimum distance model, the maximum covering model, and their variations. ArcInfo Workstation also offers two location–allocation computing algorithms: GRIA (global regional interchange algorithm) (Densham and Rushton 1992), and TEITZBART (Teitz and Bart 1968). Both algorithms are heuristics. These algorithms can solve location–allocation problems reasonably fast but may not produce optimal solutions, especially with many supply and demand points. The maximum covering model in ArcInfo Workstation has a required parameter of maximum distance and an optional parameter of desired distance. Although the objective is maximum covering, the model can actually leave some demand points uncovered, as seen in Figure 16.12. This apparent discrepancy is a result of using a minimum distance solution technique to solve a maximum covering model (Church 1999).

Box 16.4 **Linear Location Referencing System**

Transportation agencies normally use a linear location referencing system to locate events such as accidents and potholes, and facilities such as bridges and culverts, along roads and transit routes. To locate events a linear referencing system uses distance measures from known points, like the beginning of a route, a milepost, or a road intersection. The address of an accident location, for example, contains a route name and a distance offset from a milepost. The ad-

dress for a linear event, on the other hand, uses distance offsets from two reference posts.

A vector-based GIS uses *x*-, *y*-coordinates to locate points, lines, polygons, and higher-level objects. A plane coordinate system is fundamentally different from a linear location referencing system. The dynamic segmentation model uses routes, measures, and events to bring the two systems together.

ArcInfo Workstation has ARCROUTE and MEASUREROUTE that can perform data conversion. ARCROUTE works with topologically continuous routes, whereas MEASUREROUTE

works with either continuous or discontinuous routes. A topologically discontinuous route refers to a route that is divided by a canal or river into two or more parts.

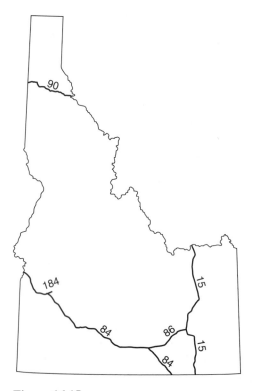

Figure 16.15
Interstate highways in Idaho.

TABLE 16.3	The Route Table for the Interstates in Idaho	
Inter#	Inter-ID	High_Number
1	1	15
2	2	84
3	3	86
4	4	90
5	5	184

Figure 16.15 shows a line coverage with the following interstate highways in Idaho: 15, 84, 86, 90, and 184. Suppose we want to create a route system for each interstate highway. We can use ARCROUTE to perform the conversion. Table 16.3 is the output route table, which has a route for each interstate highway. Table 16.4 is the section table, which uses routelink# to link to the route system and arclink# to link to arcs in the line coverage. Each route system is measured separately as shown by the fields of *f*-meas and *t*-meas in the section table.

16.4.3 Create Different Types of Routes

Routes may be grouped into the following four types:

- **Simple route**: a route follows one direction and does not loop or branch.

- **Combined route**: a route is joined with another route.
- **Split route**: a route subdivides into two routes.
- **Looping route**: a route intersects itself.

Simple routes are easy to create, but the other three types require special handling.

An example of combined routes is an interstate highway, which has different traffic conditions depending on the traffic direction. In this case, two routes can be built for the interstate highway, one for each direction. An example of split routes is highway ramps. Similar to combined routes, split routes can be separated into two or more routes: one route for the continuous path of the highway and individual routes for the split sections.

A looping route requires breaking up the route into sections. Unless the route is dissected into sections, the route measures will be off. To explain the procedure, we will use a bus route as an example (Figure 16.16). The bus route has two loops. Therefore, we can dissect the bus route into three sections: (1) from the origin on the west side of the city to the first crossing of the route, (2) between the first and the second crossing, and (3) from the second crossing back to the origin (Figure 16.17). After selecting arcs that make up the first section, we can use MAKEROUTE to create the initial bus route. Next, we select arcs of the second section, append them to the route, and use REMEASURE

TABLE 16.4	The Section Table for the Interstates in Idaho						
Routelink#	Arclink#	*f*-meas	*t*-meas	*f*-pos	*t*-pos	Inter#	Inter-ID
1	5	0	44,700	100	0	1	1
1	4	44,700	123,648	100	0	2	2
1	3	123,648	199,791	100	0	3	3
1	16	199,791	239,375	100	0	4	4
1	15	239,375	315,194	100	0	5	5
2	6	0	74,024	0	100	6	6
2	7	74,024	78,964	0	100	7	7
2	9	78,964	154,873	0	100	8	8
2	10	154,873	226,153	0	100	9	9
2	11	226,153	303,050	0	100	10	10
2	12	303,050	356,992	0	100	11	11
2	17	356,992	433,769	0	100	12	12
2	18	433,769	443,570	0	100	13	13
3	13	0	78,065	0	100	14	14
3	14	78,065	101,154	0	100	15	15
4	1	0	72,033	0	100	16	16
4	2	72,033	117,974	0	100	17	17
5	8	0	6,348	100	0	18	18

to compute measures of the second section. We can complete the bus route by appending and remeasuring arcs of the third section.

16.4.4 Create Routes with Measured Polylines

Measured polylines are line features with measured values. Usually entered manually, these measured values can be treated as route measures in creating a route.

16.5 EVENT TABLES

Events are attribute data measured on a linear reference system. Using the route ID and location measures of events, the dynamic segmentation model can link events to a geographically referenced route system. Events may be point, continuous, or linear events:

• **Point events**, such as accidents and stop signs, occur at point locations. To relate point

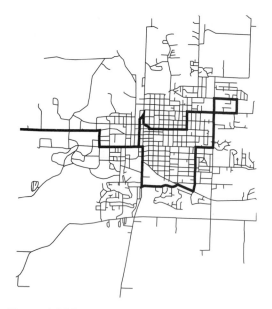

Figure 16.16
A bus route with two loops.

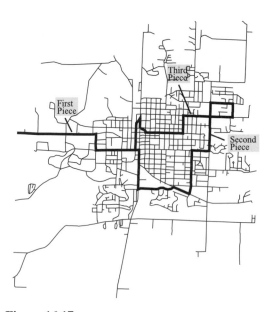

Figure 16.17
Because the bus route is a looping route, the route system is dissected into three sections.

events to a route system, a point event table must have the route ID, the location measures of the events, and attributes describing the events.

- **Continuous events**, such as speed limits, cover the entire route with no gaps. A continuous event table relates to a route system by the route ID and the To measure of the event.
- **Linear events**, such as pavement conditions, are discontinuous events along a route system. To relate linear events to a route system, a linear event table must have the route ID and the From and To measures. Because gaps may be present between linear events, both the From and To measures are required to mark the location of a linear event.

Except for the required route ID and measures, event tables are like regular attribute data tables in preparation and use.

16.5.1 Use Point or Polygon Coverages to Prepare Event Tables

An event table can be prepared from a point or polygon coverage so long as the coverage is based on the same coordinate system as the route system. The procedure is similar to map overlay except that the result is an event table. By placing a point coverage on a route system, the location of points in relation to the route system can be measured with the help of a fuzzy tolerance. If the distance between a point and the route system is within the fuzzy tolerance, then the point is a point event and its location is measured. By placing a polygon coverage on a route system, each segment of the route system is assigned with the attribute data of the polygon it crosses. The following shows two examples of creating event tables from coverages.

Figure 16.18 shows the bus stops along a bus route. To display these bus stops on the route system and the attributes at the stops, such as numbers

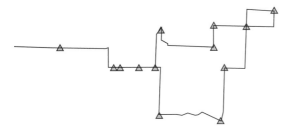

Figure 16.18
Bus stops along a bus route.

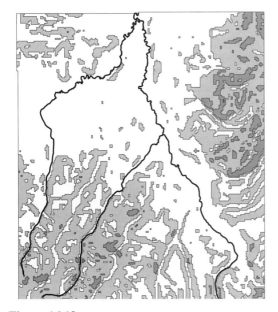

Figure 16.19
A stream route system and a slope coverage with four slope classes.

TABLE 16.5	A Point Event Table Showing Bus Stops along a Bus Route	
Busstations#	**Bus#**	**Measure**
1	1	899.930
2	1	2,359.145
3	1	2,476.239
4	1	2,849.655
5	1	3,163.485
6	1	4,173.557
7	1	5,446.844
8	1	6,451.580
9	1	9,368.944
10	1	8,509.497
11	1	10,002.686
12	1	10,412.696
13	1	11,728.987

of passengers getting on and off, we can treat the bus stops as point events. First, we prepare a point coverage containing the bus stops. Next, we use ArcInfo Workstation's ADD-ROUTEMEASURE to measure the locations of the bus stops along the bus route and to write the information to a point event table (Table 16.5). Because the point event table is an INFO file, we can easily add to the file attribute data, such as numbers of passengers getting on and off.

The second example was a stream route system (Emida) and a slope coverage with four slope classes to create a linear event table (Figure 16.19). ArcInfo Workstation's POLYGON-EVENTS first computes the intersection between the route system and the slope coverage. It assigns a slope-code value to each segment of the route system and writes the information to a linear event table. Table 16.6 shows a portion of the table.

16.5.2 Prepare Event Tables as INFO Files

Although using point and polygon coverages to create event tables is relatively easy, the method cannot be applied to event data in tabular format. The alternative is to prepare event tables directly as INFO files or text files, as shown in the following two examples.

TABLE 16.6	A Linear Event Table Showing Slope-Code along a Stream Route (Emida–ID = 1)		
Emida-ID	From	To	Slope-Code
1	0	7638	1
1	7638	7798	2
1	7798	7823	1
1	7823	7832	2
1	7832	8487	1
1	8487	8561	1
1	8561	8586	2
1	8586	8639	1
1	8639	8643	2

TABLE 16.7	A Linear Event Table Showing Year of Pavement Resurfacing on the Interstate Highway Route System		
Inter-ID	From	To	Year
1	44,700	90,000	1995
1	123,648	180,000	1989
1	239,375	270,000	1992
2	74,024	78,000	1988
2	154,873	180,000	1993
2	356,992	400,000	1987
3	78,065	90,000	1988
4	40,000	72,033	1986

The first example is a linear event (pavement resurfacing) table for the interstate highway route system in Idaho. The event table or INFO file has four fields: Inter-ID (interstate ID), From (from measure), To (to measure), and Year (year of pavement resurfacing) (Table 16.7). The first three fields relate the event table to the route system: (1) Inter-ID links the event table to the route table, and (2) From and To together relate the event table to the section table (Figure 16.20). Using the INFO file, we can query and display Year on the route system (Figure 16.21).

The second example also applies to the interstate highway route system in Idaho but uses a continuous event table. This example assumes that each interstate highway must be divided into 50-mile segments (the last segment may be shorter than 50 miles) for the purpose of sampling. Table 16.8 is a continuous event table. Inter-ID, or the interstate highway number, links the event table to the route table. To, or the to measure, relates the event table to the section table for continuous

events. (Because the highway coverage is measured in meters, the To values are also measured in meters rather than miles.) Unit is the unit number of 50-mile (80,465-meter) segments, starting with 1 for each interstate highway. Figure 16.22 plots the first segment of each interstate in thicker line.

16.6 APPLICATIONS OF DYNAMIC SEGMENTATION

Applications of the dynamic segmentation model include the display, query, and analysis of routes, measures, and events. Similar to regions-based commands, ArcInfo Workstation has specific commands to work with dynamic segmentation. For example, EVENTSOURCE or EVENTMENU can link an event source and a route system as required in many applications. EVENT-SOURCE uses a command statement to set up the input arguments for the linkage, whereas EVENTMENU invokes a menu for the inputs.

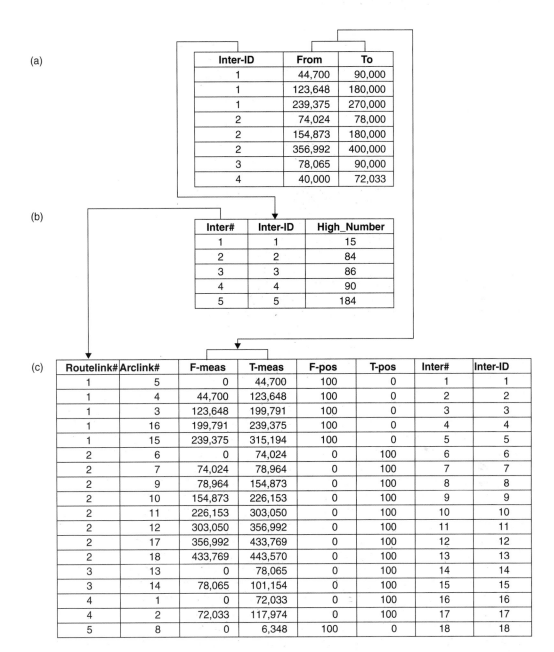

(a)

Inter-ID	From	To
1	44,700	90,000
1	123,648	180,000
1	239,375	270,000
2	74,024	78,000
2	154,873	180,000
2	356,992	400,000
3	78,065	90,000
4	40,000	72,033

(b)

Inter#	Inter-ID	High_Number
1	1	15
2	2	84
3	3	86
4	4	90
5	5	184

(c)

Routelink#	Arclink#	F-meas	T-meas	F-pos	T-pos	Inter#	Inter-ID
1	5	0	44,700	100	0	1	1
1	4	44,700	123,648	100	0	2	2
1	3	123,648	199,791	100	0	3	3
1	16	199,791	239,375	100	0	4	4
1	15	239,375	315,194	100	0	5	5
2	6	0	74,024	0	100	6	6
2	7	74,024	78,964	0	100	7	7
2	9	78,964	154,873	0	100	8	8
2	10	154,873	226,153	0	100	9	9
2	11	226,153	303,050	0	100	10	10
2	12	303,050	356,992	0	100	11	11
2	17	356,992	433,769	0	100	12	12
2	18	433,769	443,570	0	100	13	13
3	13	0	78,065	0	100	14	14
3	14	78,065	101,154	0	100	15	15
4	1	0	72,033	0	100	16	16
4	2	72,033	117,974	0	100	17	17
5	8	0	6,348	100	0	18	18

Figure 16.20

The linking of an event table (a), a route table (b), and a section table (c). See text for explanation.

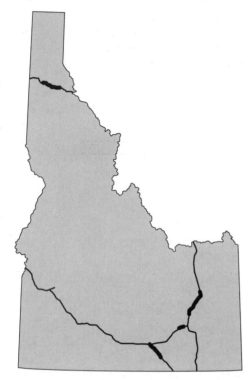

Figure 16.21

This map displays the result from querying the event table (Year <= 1990).

TABLE 16.8	A Continuous Event Table Dividing Each Interstate Highway into 50-Mile Segments	
Inter-ID	To	Unit
1	80,465	1
1	160,930	2
1	241,395	3
1	315,194	4
2	80,465	1
2	160,930	2
2	241,395	3
2	321,860	4
2	402,325	5
2	443,570	6
3	80,465	1
3	101,155	2
4	80,465	1
4	117,974	2

16.6.1 Data Query with Events

Like attribute data tables, event tables can be queried by logical expressions. Suppose a linear event table has a field called Year for year of pavement resurfacing. We can query the event table and select highway sections resurfaced in different years.

Querying event data spatially, such as querying past accidents within 10 miles of a recent accident, requires that we first convert event data into a coverage. ArcInfo Workstation's commands, EVENTARC and EVENTPOINT, can perform the conversion. For instance, EVENTPOINT can convert the point event table of accidents into a point coverage, which can then be queried graphically.

16.6.2 Data Analysis with Routes and Events

To buffer a route in ArcInfo Workstation, the route must be converted into arcs first by using the command ROUTEARC or SECTIONARC. Like buffering streams, buffering routes can be a useful management tool. For example, to study the efficiency or equity of public transit routes, we can buffer the routes with a distance of one-half of a mile and overlay the buffer zone with census blocks or block groups. Demographic data from the buffer zone (e.g., population density, income levels, number of vehicles owned per family, and commuting time to work) can provide useful information for the planning and operation of public transit.

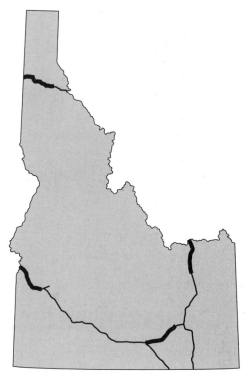

ArcInfo Workstation's OVERLAYEVENTS can overlay event data with the following options: line-on-line or point-on-line, and UNION or INTERSECT. Suppose a point event table shows accidents and a linear event table shows the pavement condition. To see if a relationship exists between accidents and the pavement condition, we can overlay the two event tables by using the point-on-line option and the INTERSECT option. The result can determine whether accidents are more likely to happen on stretches of the route with poor pavement conditions.

Figure 16.22
This map displays the first 50-mile segment of Interstates 15, 84, 86, and 90.

KEY CONCEPTS AND TERMS

Allocation: A study of the spatial distribution of resources on a network.

Closest facility: A network analysis, which computes the shortest paths from the select location to all candidate facilities and then finds the closest facility among the candidates.

Combined route: A route that is joined with another route.

Continuous events: Events that cover the entire route with no gaps on a network, such as speed limits.

Linear events: Discontinuous events along a route system, such as pavement conditions.

Link: A segment separated by two nodes in a road network.

Link impedance: The cost of traversing a link, which may be measured by the physical length or the travel time.

Location–allocation: A spatial analysis, which matches the supply and demand by using sets of objectives and constraints.

Looping route: A route that intersects itself.

Maximum covering model: An algorithm for solving the location–allocation problem by maximizing the demand covered within a specified time or distance.

Minimum distance model: An algorithm for solving the location–allocation problem by minimizing the total distance traveled from all demand points to their nearest supply centers. The model is also called the *p-median location model.*

Network: A line coverage, which is based on the arc–node model and has the appropriate attributes for the flow of objects such as traffic.

Point events: Events that occur at point locations on a network, such as accidents and stop signs.

Shortest-path analysis: A network analysis, which finds the path with the minimum cumulative impedance between nodes on a network.

Simple route: A route that follows one direction and does not loop or branch.

Split route: A route that subdivides into two routes.

Traveling salesman problem: A network analysis, which finds the best route with the conditions of visiting each stop only once, and returning to the original stop where the journey starts.

Turn impedance: The cost of completing a turn on a road network, which is usually measured by the time delay.

Turntable: A table that can be used to assign the turn impedance values on a road network.

APPLICATIONS: NETWORK AND DYNAMIC SEGMENTATION

This applications section has five tasks. Task 1 uses ArcMap and the dynamic segmentation model to display a bus route, and point events along the route. Task 2 lets you use ArcMap to create a bike route from the existing arcs of a street coverage. Tasks 3 and 4 use ArcInfo Workstation to run a shortest-path analysis and an allocation problem, respectively. Because ArcGIS does not yet offer the Network Analyst extension as of this writing, ArcInfo Workstation is used for network applications. Task 5 shows you how to create a bike route from the existing arcs of a street coverage in ArcInfo Workstation.

Task 1. Display Point Events on a Route System in ArcMap

What you need: *moscowst,* a street coverage, and a point event table called *stations.txt.*

Moscowst is a street coverage of Moscow, Idaho. A bus route system has been built on the arcs of *moscowst.* The point event table *stations.txt* has the fields of busstation, bus_id, measure, and adp. The first three fields refer to bus station ID, route ID, and measure of bus station, respectively. Adp refers to the average daily number of passen-

gers. The purpose of this task is for you to become familiar with use of the dynamic segmentation data model in ArcMap.

1. Start ArcCatalog, and make connection to the Chapter 16 database. Launch ArcMap, and rename Layers Tasks 1&2. Click the Add Data button. Navigate to the Chapter 16 database, double-click on *moscowst,* and select *arc* and *route.bus* to add to Tasks 1&2.

2. This step is to open a point event table and to convert the table to a shapefile. Right-click *stations.txt* in the Catalog tree, point to Create Feature Class, and select From Route Event Table. Make sure that *stations.txt* is the input route event table in the first panel. The second panel asks you for the route event type and associated attributes. Check that the events are point events, select bus_id from the event key field dropdown list, and verify that measure is the location field. Click Next. Click the browse button in the third panel, navigate to the Chapter 16 database, double-click on *moscowst,* and select *route.bus* for the Route feature class. In the fourth panel, select Shapefile for the output type. Then

enter *stations.shp* for the name of the output shapefile. In the next panel, opt to convert all events. Click Finish in the summary sheet.

3. Add *stations.shp* to Tasks 1&2. Select Open Attribute Table from the context menu of *stations*. Click the Options dropdown arrow and choose Select by Attributes. Enter the following SQL statement in the expression box: "adp" > 30. Click Apply. The selection shows bus stations that have the average daily number of passengers greater than 30. Those selected stations are highlighted in both the table and the map. You can also display the average daily number of passengers in graduated symbols.

Task 2: Create a Bike Route in ArcMap

What you need: *mosst*, a network coverage.

If you are using ArcInfo of ArcGIS, Task 2 lets you build a bike route on the arcs and nodes of *mosst* in ArcMap. But you need to use ArcInfo Workstation commands at the beginning and the end of the task. *Mosst* is a street coverage of Moscow, Idaho, same as *moscowst* in Task 1. First, create the route subclass as an empty subclass in ArcInfo Workstation by using the ArcEdit command of CREATEFEATURE:

Arcedit: mapextent mosst
Arcedit: editcov mosst
Arcedit: createfeature route.bike2
Arcedit: quit /*save changes

1. Click the Add Data button in ArcMap. Navigate to the Chapter 16 database. Double-click on *mosst*. Select *arc* and *route.bike2* to add to Tasks 1&2. *Route.bike2* currently has no features.

2. Click the Editor Toolbar button. Click the Editor dropdown arrow and select Start Editing. In the next dialog, select ArcInfo Workspace to be the folder or database to edit data from and click OK. Make sure that the Task is to Create New Feature and the Target is *mosst route.bike2*.

3. This step is to select streets from *mosst arc* to be included in the bike route subclass. Select Open Attribute Table from the context menu of *mosst arc*. Click the Options dropdown arrow and choose Select by Attributes. Double-click FENAME in the Fields menu. Click on the Complete List button, and enter the following SQL statement in the expression box: "FENAME" = 'Sixth'. Click Apply. The selected street segments are highlighted in both the table and the map.

4. Click the Editor dropdown arrow and select Union. The selected street segments become the components of *mosst route.bike2*. Click the Editor dropdown arrow and select Stop Editing. Save the edits.

5. Run the Arc command of MEASUREROUTE in ArcInfo Workstation to convert the selected arcs to a route:

Arc: measureroute route mosst bike2
Arc: describe mosst /* check *bike2* is created

Task 3: Run Shortest-Path Analysis in ArcInfo Workstation

What you need: *uscities.shp,* a point shapefile, and *interstates.shp,* a line shapefile.

Uscities.shp contains cities in the conterminous United States, and *interstates.shp* is a network of the interstate highways. Use the Shapefile to Coverage conversion tool in ArcToolbox to convert *uscities.shp* into a point coverage and name the coverage *uscities* and to convert *interstates.shp* into a line coverage and name the coverage *interstates*. This conversion is necessary because ArcInfo Workstation does not work with shapefiles.

The objective of this task is to find the best (shortest) route between any two nodes or cities. The shortest-path analysis in ArcInfo Workstation cannot use *uscities* as input but must first find nodes on *interstates* that are closest to the points of interest in *uscities*. Identification of nodes actually occupies a major portion of the task.

This task uses travel time to derive the shortest path. The speed limit for calculating the travel time is 65 miles/hour, and the travel time considers only link impedance. The cities to be queried are Helena, Montana, and Raleigh, North Carolina.

1. First build a node attribute table (NAT) for *interstates*.

   Arc: build interstates nodes

2. Go to ArcPlot and set up the draw environment.

   Arcplot: display 9999
   Arcplot: mapex interstates /* set the map extent as that of *interstates*
   Arcplot: arcs interstates /* plot *interstates*
   Arcplot: nodes interstates /*plot nodes of *interstates*

3. Next plot Helena and Raleigh.

   Arcplot: reselect uscities point city_name = 'Helena' or city_name = 'Raleigh'
   /* select Helena and Raleigh from *uscities*
   Arcplot: pointmarkers uscities 3 /* plot Helena and Raleigh in green

4. Find the closest node on *interstates* to Helena.

   Arcplot: mapex * /* zoom in the area around Helena
   Arcplot: clear /* clear the display window
   Arcplot: arcs interstates /* plot *interstates* around Helena
   Arcplot: nodes interstates /*plot nodes of *interstates* around Helena
   Arcplot: pointmarkers uscities 3 /* plot Helena in green
   Arcplot: identify interstates node * /* click node closest to Helena and record the INTERSTATES-ID of the node (33)

5. Clear the display window and repeat the same procedure to find the INTERSTATES-ID of the node closest to Raleigh (328). After the nodes representing Helena and Raleigh are identified, use the following commands to run a shortest-path analysis and to plot the result.

Arcplot: mapex interstates /* reset the map extent as that of *interstates*
Arcplot: arcs interstates /* plot *interstates*
Arcplot: nodes interstates /* plot nodes of *interstates*
Arcplot: pointmarkers uscities 3 /* plot Helena and Raleigh in green
Arcplot: netcover interstates trip /* specify *interstates* for the network coverage and *trip* for the route system
Arcplot: impedance minutes /* specify minutes as the link impedance
Arcplot: path 33 328 end 1 /* specify the stops of 33 and 328 and the route-ID of 1
Arcplot: routelines interstates trip 4 /* plot the shortest path in blue
Arcplot: directions 1 route minute /* get the travel directions and the total travel time

Task 4: Run Allocation in ArcInfo Workstation

What you need: a network coverage called *moscowst* and a point shapefile called *firestat.shp*.

Moscowst contains a street network and *firestat.shp* has two fire stations in Moscow, Idaho. Use the Shapefile to Coverage conversion tool in ArcToolbox to convert *firestat.shp* into a point coverage and name the coverage *firestat*. Task 4 deals with an allocation problem. The objective is to measure the efficiency of the two fire stations. Similar to Task 3, you cannot use *firestat* directly as input. Instead, the nodes of *moscowst* closest to the two fire stations are used as input to ALLO-CATE. These two nodes have been identified as having the moscow-ID values of 603 and 848. The efficiency or the service area of the two fire stations can be measured in physical distance or response time. For this task, you will use a response time of 3 minutes.

1. Go to ArcPlot and set up the draw environment.

   Arcplot: display 9999
   Arcplot: mapex moscowst /* set the extent as that of *moscowst*

Arcplot: arcs moscowst /* plot *moscowst*
Arcplot: nodes moscowst /* plot nodes of
moscowst
Arcplot: pointmarkers firestat 3 /* plot
firestat in green

2. Specify the network and impedance before
running ALLOCATE, and plot the results.

Arcplot: netcover moscowst fire /* specify
moscowst for the network coverage and *fire*
for the route system
Arcplot: impedance minutes /* specify
minutes or response time as impedance
Arcplot: allocate 603 848 end 3 out /*
specify 603 and 848 for the stations and 3
minutes for the response time
Arcplot: routelines moscowst fire 2 /* plot
the streets within the service area in red

Task 5: Create a Bike Route in ArcInfo Workstation

What you need: a network coverage called *mosst*.

In Task 5, you will build a bike route on the
arcs and nodes of *mosst* by using the ArcEdit command MAKEROUTE. The bike route is a simple
route along Sixth St. from Perimeter Rd. to Washington St.

1. Go to ArcEdit and set up the draw and edit
environment.

Arcedit: display 9999
Arcedit: mapex mosst
Arcedit: editcov mosst
Arcedit: drawenv arc
Arcedit: draw
Arcedit: editfeature arc

2. Next, select and display Sixth St, Perimeter
Rd, and Washington St.

Arcedit: select fename = 'Sixth' /* fename
contains the street name
Arcedit: drawselect /* to highlight Sixth St
Arcedit: aselect fename = 'Perimeter' /* add
Perimeter Rd to the subset
Arcedit: drawselect

Arcedit: aselect fename = 'Washington' /*
add Washington St to the subset
Arcedit: drawselect

3. The bike route follows Sixth St. between
Perimeter Rd. and Washington St. Zoom in
the area of the bike route. To create the bike
route on the existing arcs of *mosst*, first select
the arcs that make up the route and then run
the command MAKEROUTE.

Arcedit: select many /* select arcs of the bike
route and press 9 to exit
Arcedit: makeroute
Usage: makeroute <subclass> {route-id}
{measure_item} {UL | UR | LL | LR | * | xy}
{START <start_measure>} {NOGAP | GAP}
{CONNECT <connect_distance>}
Arcedit: makeroute bike /* use selected arcs
to make a route called *bike*

4. Exit ArcEdit and save edits. Go to Tables.

Tables: select mosst.ratbike /* select the *bike*
route table
Tables: list /* the route table shows a single
route
Tables: select mosst.secbike /* select the *bike*
section table
Tables: list

5. The section table has the following items:
ROUTELINK#, ARCLINK#, F-MEAS, T-
MEAS, F-POS, T-POS, BIKE#, and BIKE-
ID. ROUTELINK# links the sections to the
bike route, while ARCLINK# links the
sections to the arcs in *mosst*. F-MEAS and T-
MEAS provide the cumulative measures of
the sections. F-POS and T-POS measure the
beginning and ending position of each
section relative to the underlying arc. BIKE#
and BIKE-ID are the machine ID and label
ID of the sections.

6. You can use the Arc command
DROPFEATURES to drop routes and
sections from a coverage.

Arc: dropfeatures
Usage: DROPFEATURES <cover>
<feature_class> {ATTRIBUTES |
GEOMETRY}

Arc: dropfeatures mosst route.bike /* remove
the *bike* route
Arc: dropfeatures mosst section.bike /*
remove the *bike* sections

REFERENCES

Church, R. L. 1999. Location Modelling and GIS. In P. A. Longley, M. F. Goodchild, D. J. MaGuire, and D. W. Rhind, eds., *Geographical Information Systems*, 2d ed., pp. 293–303. New York: Wiley.

Church, R. L., and C. S. ReVelle. 1974. The Maximal Covering Location Problem. *Papers of the Regional Science Association* 32: 101–18.

Densham, P. J., and G. Rushton. 1992. A More Efficient Heuristic for Solving Large *p*-Median Problems. *Papers in Regional Science* 71: 307–29.

Dijkstra, E. W. 1959. A Note on Two Problems in Connexion with Graphs. *Numerische Mathematik* 1: 269–71.

Dreyfus, S. 1969. An Appraisal of Some Shortest Path Algorithms. *Operations Research* 17: 395–412.

Francis, R. L., T. J. Lowe, G. Rushton, and M. B. Rayco. 1999. A Synthesis of Aggregation Methods for Multifacility Location Problems: Strategies for Containing Error.

Geographical Analysis 31: 67–87.

Fu, L., and L. R. Rilett. 1998. Expected Shortest Paths in Dynamic and Stochastic Traffic Networks. *Transportation Research B* 32: 499–516.

Ghosh, A., S. McLafferty, and C. S. Craig. 1995. Multifacility Retail Networks. In Z. Drezner, ed., *Facility Location: A Survey of Applications and Methods*, pp. 301–30. New York: Springer.

Hakimi, S. L. 1964. Optimum Locations of Switching Centers and the Absolute Centers and Medians of a Graph. *Operations Research* 12: 450–59.

Khatib, Z., K. Chang, and Y. Ou. 2001. Impacts of Analysis Zone Structures on Modeled Statewide Traffic. *Journal of Transportation Engineering* 127: 31–38.

Lin, S. 1965. Computer Solutions of the Travelling Salesman Problem. *Bell System Technical Journal* 44: 2245–69.

Lowe, J. C., and S. Moryadas. 1975. *The Geography of Movement*. Boston: Houghton Mifflin.

Nyerges, T. L. 1995. Geographical Information System Support for Urban/Regional Transportation Analysis. In S. Hanson, ed., *The Geography of Urban Transportation*, 2d ed., pp. 240–65. New York: Guilford Press.

Teitz, M. B., and P. Bart. 1968. Heuristic Methods for Estimating the Generalised Vertex Median of a Weighted Graph. *Operations Research* 16: 953–61.

Waters, N. 1999. Transportation GIS: GIS-T. In P. A. Longley, M. F. Goodchild, D. J. MaGuire, and D. W. Rhind, eds., *Geographical Information Systems*, 2d ed., pp. 827–44. New York: Wiley.

Ziliaskopoulos, A. K., and H. S. Mahmassani. 1996. A Note on Least Time Path Computation Considering Delays and Prohibitions for Intersection Movements. *Transportation Research B* 30: 359–67.

List of GIS Books

APPENDIX A

Aronoff, S. 1989. *Geographic Information Systems: A Management Perspective.* Ottawa: WDL Publications.

Bernhardsen, T. 2002. *Geographic Information Systems: An Introduction,* 3d ed. New York: Wiley.

Bolstad, P. 2002. *GIS Fundamentals.* Eider Press.

Bonham-Carter, G. F. 1994. *Geographic Information Systems for Geoscientists: Modeling with GIS.* New York: Pergamon Press.

Booth B. 1999. *Getting Started with ArcInfo.* Redlands, CA: ESRI Press.

Burrough, P. A., and R. A. McDonnell. 1998. *Principles of Geographical Information Systems.* Oxford, England: Oxford University Press.

Chrisman, N. R. 2001. *Exploring Geographic Information Systems,* 2d ed. New York: Wiley.

Clarke, K. C. 2000. *Getting Started with Geographic Information Systems,* 3d ed. Upper Saddle River, NJ: Prentice Hall.

DeMers, M. N. 2000. *Fundamentals of Geographic Information Systems,* 2d ed. New York: Wiley.

ESRI Press. 1998. *Getting to Know ArcView GIS,* 3d ed. Redlands, CA.

Heywood, I., S. Cornelius, and S. Carver. 1998. *An Introduction to Geographical Information Systems.* Upper Saddle River, NJ: Prentice Hall.

Huxhold, W. E. 1991. *An Introduction to Urban Geographic Information Systems.* New York: Oxford University Press.

Laurini, R., and D. Thompson. 1992. *Fundamentals of Spatial Information Systems.* London: Academic Press.

Lee, J., and D. W. S. Wong. 2001. *Statistical Analysis with ArcView GIS.* New York: Wiley.

Lo, C. P., and A. K. W. Yeung. 2002. *Concepts and Techniques in Geographic Information Systems.* Upper Saddle River, NJ: Prentice Hall.

Longley, P. A., M. D. Goodchild, D. J. Maguire, and D. W. Rhind. 1999. *Geographical Information Systems,* 2d ed. New York: Wiley.

Longley, P. A., M. D. Goodchild, D. J. Maguire, and D. W. Rhind. 2001. *Geographical Information Systems and Science.* New York: Wiley; and Redland, CA: ESRI Press.

MacDonald, A. 1999. *Building a Geodatabase.* Redlands, CA: ESRI Press.

Maguire, D. J., M. F. Goodchild, and D. W. Rhind. 1991. *Geographical Information Systems: Principles and Applications.* Harlow, Essex, England: Longman.

Malczewski, J. 1999. *GIS and Multicriteria Decision Analysis.* New York: Wiley.

Martin, D. S. 1996. *Geographic Information Systems: Socioeconomic Applications,* 2d ed. London: Routledge.

Mitchell, A. 1999. *The ESRI Guide to GIS Analysis,* Vol. 1: *Geographic Patterns & Relationships.* Redland, CA: ESRI Press.

Ormsby, T., and Alvi, J. 1999. *Extending ArcView GIS.* Redland, CA: ESRI Press.

Ormsby, T., E. Napoleon, R. Burke, C. Groessl, and L. Feaster. 2001. *Getting to Know ArcGIS Desktop.* Redland, CA: ESRI Press.

Verbyla, D. L. 2002. *Practical GIS Analysis.* London: Taylor and Francis.

Worboys, M. F. 1995. *GIS: A Computing Perspective.* London: Taylor and Francis.

Zeiler, M. 1999. *Modeling our World: The ESRI Guide to Geodatabase Design.* Redlands, CA: ESRI Press.

APPENDIX B

List of GIS Magazines and Journals

ONLINE MAGAZINES

GEO World, http://www.geoplace.com/gw/

GeoSpatial Solutions, http://www.geospatial-online.com/geospatialsolutions/

JOURNALS

International Journal of Geographical Information Science

Cartography and Geographic Information Science

Transactions in GIS

Journal of Geographic Information and Decision Analysis, http://www.geodec.org/

Geographical Systems

Photogrammetric Engineering and Remote Sensing

Remote Sensing of the Environment

The Geologic Survey of Alabama
http://www.gsa.state.al.us/gsa/GIS/clearinghouse.html

Alaska State Geospatial Data Clearinghouse
http://www.asgdc.state.ak.us/

The University of Arizona Library
http://dizzy.library.arizona.edu/library/teams/sst/geo/guide/gis/

The Arkansas Geolibrary Catalog
http://cast.uark.edu/local/isite/GeoLibraryCatalog.htm

The California Spatial Information Library
http://www.gis.ca.gov/

Colorado Office of Emergency Management, Cartography/GIS Section
http://www.dola.state.co.us/oem/cartography/cartog.htm

Map and Geographic Information Center,
University of Connecticut
http://magic.lib.uconn.edu/

Delaware Geospatial Clearinghouse Node
http://gis.smith.udel.edu/fgdc/gateway/

Florida Geographic Data Library, University of Florida
http://www.fgdl.org/

Georgia GIS Data Clearinghouse
http://gis.state.ga.us/

Hawaii Statewide GIS Program
http://www.state.hi.us/dbedt/gis/

INSIDE Idaho Geospatial Clearinghouse
http://inside.uidaho.edu/

Illinois Natural Resources Geospatial Data Clearinghouse
http://www.isgs.uiuc.edu/nsdihome/

Indiana GIS Initiative
http://www.state.in.us/ingisi/

Iowa Department of Natural Resources, Natural Resources Geographic Information System Library
http://samuel.igsb.uiowa.edu/nrgis/gishome.htm

State of Kansas Geographic Information Systems Initiative's Data Access & Support Center
http://gisdasc.kgs.ukans.edu/dasc.html

Office of Geographic Information, Kentucky
http://ogis.state.ky.us/

Louisiana Geographic Information Center
http://lagic.lsu.edu/

Maine Office of GIS
http://apollo.ogis.state.me.us/

Maryland State and Local Government Metadata Clearinghouse Node
http://www.msgic.state.md.us/

Massachusetts Geographic Information System
http://www.state.ma.us/mgis/

Michigan Information Center
http://www.state.mi.us/dmb/mic/gis/

Minnesota Department of Natural Resources
http://deli.dnr.state.mn.us/

Mississippi Automated Resource Information System
http://www.maris.state.ms.us/

Missouri Spatial Data Information Service
http://msdis.missouri.edu/

Montana Natural Resource Information System
http://nris.state.mt.us/

Nebraska Geospatial Data Clearinghouse
http://geodata.state.ne.us/

University of Nevada–Reno, Keck Earth Sciences and Mining Research Information Center
http://keck.library.unr.edu/

New Hampshire GRANIT website
http://www.granit.sr.unh.edu/

New Jersey Spatial Data Clearinghouse
http://njgeodata.state.nj.us/

New Mexico Resource Geographic Information System
http://rgis.unm.edu/

New York State GIS Clearinghouse
http://www.nysgis.state.ny.us/

North Carolina Geographic Data Clearinghouse
http://cgia.cgia.state.nc.us/ncgdc/

North Dakota Geological Survey, Geographic Information Systems Center
http://www.state.nd.us/ndgs//GIS/ndgs_gis.htm

Ohio Department of Administrative Services, Geographic Information Systems Service Bureau
http://www.state.oh.us/das/dcs/gis/

Spatial and Environmental Information Clearinghouse, Oklahoma State University
http://www.seic.okstate.edu/

Oregon Geospatial Data Clearinghouse
http://www.sscgis.state.or.us/

Pennsylvania Spatial Data Access
http://www.pasda.psu.edu/

Rhode Island Geographic Information System
http://www.edc.uri.edu/rigis/

South Carolina Department of Natural Resources GIS Data Clearinghouse
http://water.dnr.state.sc.us/gisdata/

South Dakota Geological Survey
http://www.sdgs.usd.edu/

The Map Library at the University of Tennessee, Knoxville
http://www.lib.utk.edu/~cic/

Texas Natural Resources Information System
http://www.tnris.state.tx.us/

Utah Automated Geographic Reference Center
http://agrc.its.state.ut.us/

Vermont Geographic Information System
http://www.vcgi.org/

University of Virginia Library, Geospatial & Statistical Data Center
http://fisher.lib.virginia.edu/

Washington State Geospatial Clearinghouse
http://wa-node.gis.washington.edu/

West Virginia GIS Technical Center
http://wvgis.wvu.edu/

Wisconsin Land Information Clearinghouse
http://badger.state.wi.us/agencies/wlib/sco/pages/

Wyoming Natural Resources Data Clearinghouse
http://www.sdvc.uwyo.edu/clearinghouse/

APPENDIX D

The following is a complete example of metadata downloaded from the Interactive Numeric & Spatial Information Data Engine (Inside) website at the University of Idaho (http://inside.uidaho.edu/). "Data about data" are absolutely important to GIS users. For example, the spatial reference information section describing the map's coordinate system and its parameters will be needed to reproject the map onto a different coordinate system.

IDENTIFICATION INFORMATION

Section Index

Citation
    Citation Information
        Originator: University of Idaho
         Library—INSIDE Idaho Project
        Publication Date: January 2000
        Title: Agroclimate Zones in Idaho
        Geospatial Data Presentation Form: Map
        Series Information
          Series Name: INSIDE Idaho GIS
         Database
        Publication Information
          Publication Place: Moscow, Idaho
          Publisher: University of Idaho Library
        Online Linkage: http://inside.uidaho.edu
         /data/statewide/ics/metadata/agclim
         _id_ics.htm
Description
    Abstract
        Multivariate statistical analysis and geographic information systems were used to delineate homogeneous agroclimate zones for Idaho for the purpose of applying successful dryland agricultural research practices and management decision throughout these areas of relative climatic uniformity. Data used to produce the classification are from the para-meter-elevation regressions on independent slopes model (PRISM), developed at Oregon State University. PRISM has produced gridded estimates of mean monthly and annual climatic parameters from point data and a digital elevation model (DEM). Principal components analysis was performed on 55 variables including various temperature and precipitation parameters, the number of growing degree-days, the mean annual number of freeze-free days, the mean annual day of first freeze in the fall, and the mean annual day of last freeze in the spring. Cluster analysis was used to identify 16 agroclimate zones each having similar climatic conditions regardless of its spatial location. As a result, successful dryland agricultural practices and management decisions that are based on new technologies and developed for one part of the state may potentially be applied to other parts of the state that fall within the same agroclimate zone.

Purpose
    1. To demonstrate that a combination of geographic information systems (GIS) and multivariate statistical procedures can be used to map climate using data from the Parameter-elevation Regressions on Independent Slopes Model (PRISM).
    2. To delineate agroclimate zones for the purpose of applying successful dryland agricultural research management practices throughout areas of relative climatic uniformity.

Supplemental Information: Grid cells sizes
  are approximately 4 km by 4 km.
Time Period of Content
  Time Period Information

Range of Dates/Times
  Beginning Date: 1961
  Ending Date: 1990
Currentness Reference: Climatological period
Status
  Progress: Complete
  Maintenance and Update Frequency: None planned
Spatial Domain
  Bounding Coordinates
    West Bounding Coordinate: −117.54391666
    East Bounding Coordinate: −110.62758333
    North Bounding Coordinate: 49.00517893
    South Bounding Coordinate: 41.93336634
Keywords
  Theme
    Theme Keyword Thesaurus: None
    Theme Keyword: Agroclimate
    Theme Keyword: Climate
  Place
    Place Keyword Thesaurus: None
    Place Keyword: Idaho
Access Constraints: none
Use Constraints: This information is not accurate for legal or navigational purposes. This information is for use by Idaho State Climate Services.
Point of Contact
  Contact Information
    Contact Organization Primary
      Contact Organization: Idaho State Climate Services
    Contact Address
      Address Type: Mailing and physical address
      Address
        Department of Biological and Agricultural Engineering
        University of Idaho
        EP425
      City: Moscow
      State or Province: Idaho
      Postal Code: 83844-0904
    Contact Voice Telephone: 208-885-7004
    Contact Facsimile Telephone: 208-885-7908
    Contact Electronic Mail Address: climate@uidaho.edu
    Hours of Service: 0900–1700 Pacific Time

Browse Graphic
  Browse Graphic File Name: http://inside.uidaho.edu/data/statewide/ics/atlas/agclim_id_ics.gif
  Browse Graphic File Description: Agroclimate Zones in Idaho
  Browse Graphic File Type: GIF
Data Set Credit: Idaho State Climate Services—Bruce R. Godfrey, Myron P. Molnau
Cross Reference
  Citation Information
    Originator: University of Idaho Library—INSIDE Idaho Project
    Publication Date: 11/22/99
    Title: Precipitation in Idaho; Average Monthly and Annual (1961–1990) (Raster)
    Geospatial Data Presentation Form: Map
    Series Information:
      Series Name: INSIDE Idaho GIS Database
    Publication Information:
      Publication Place: Moscow, Idaho
      Publisher: University of Idaho Library
      Online Linkage: http://inside.uidaho.edu/data/statewide/ocs/metadata/ppt_id_ocs.htm

DATA QUALITY INFORMATION

Section Index

Lineage
  Source Information
    Source Citation
      Citation Information
        Originator: C. Daly, R. P. Neilson, and D. L. Phillips
        Publication Date: 1994
        Title: A Statistical-Topographic Model for Mapping Climatological Precipitation over Mountainous Terrain
        Publication Information
          Publication Place: Portland, Oregon
          Publisher: Water and Climate Center of the Natural Resources Conservation Service
        Other Citation Details
          Chris Daly of Oregon State University and George Taylor of the Oregon Climate Service at Oregon State University. Natural Resource Conservation Service (NRCS), Water and

Climate Center, NRCS National Cartography and Geospatial Center (NCGC), PRISM Model, and the Oregon Climate Service at Oregon State University.
Online Linkage: *http://www.ocs.orst.edu/ prism/prism_new.html*
Source Citation Abbreviation: PRISM
Source Contribution: PRISM datasets—gridded ASCII data for numerous climate variables
Source Information
  Source Citation
    Citation Information
      Originator: Idaho State Climate Services— Bruce R. Godfrey and Myron P. Molnau
      Publication Date: December 1999
      Title: Delineation of Agroclimate Zones in Idaho
      Geospatial Data Presentation Form: Map
      Publication Information
        Publication Place: Moscow, Idaho
        Publisher: University of Idaho
      Other Citation Details
        This coverage is a product of the thesis (M.S.): "Delineation of Agroclimate Zones in Idaho." Bruce R. Godfrey and Myron P. Molnau—Program in Environmental Sciences Dissertations. University of Idaho Library Call Number: S600.43.M37G63 1999
      Online Linkage:
        http://snow.ag.uidaho.edu/ Clim_map/agroclimate.htm
Source Citation Abbreviation: AGROZONES
Source Contribution: Agroclimate Zones coverage in Arc/Info Grid format
Process Step
  Process Description
    It is beyond the scope of this metadata to describe in detail the procedures used to create this coverage. A full description of the methods used can be found at *http://www.uidaho.edu/ ~climate*. In general, the procedures used were:

1. Downloaded PRISM data from Oregon Climate Services website (*http://www.ocs.orst.edu/prism/ prism_new.html*):

   - Mean temperature
   - Minimum temperature
   - Maximum temperature
   - Precipitation
   - Growing degree-days
   - Day of spring freeze
   - Day of fall freeze
   - Growing season length

2. ARC/INFO GIS was used to extract information needed for agroclimate classification. The ASCII data were converted to ARC/INFO grids and projected to IDTM. The 55 grids used in the classification were created from various seasonal temperature and precipitation parameters, the seasonal number of growing degree-days, the mean annual number of freeze-free days, the mean annual day of first freeze in the fall, and the mean annual day of last freeze in the spring.

3. The data were exported to a statistical software package and PCA and cluster analysis were performed. A georeferenced ASCII file was created that contained the 16 cluster groupings (i.e., the agroclimate zones).

4. The GIS was then used to display the agroclimate zones.

Source Used Citation Abbreviation: PRISM
Process Date: Fall 1999
Process Contact
  Contact Information
    Contact Organization Primary
      Contact Organization: Idaho State Climate Services
    Contact Address
      Address Type: Mailing and physical address
      Address
        Department of Biological and Agricultural Engineering University of Idaho EP425
      City: Moscow
      State or Province: Idaho
      Postal Code: 83844-0904
    Contact Voice Telephone: 208-885-7004
    Contact Facsimile Telephone: 208-885-7908

Contact Electronic Mail Address:
climate@uidaho.edu
Hours of Service: 0900–1700
Pacific Time
Process Step
Process Description
This Grid coverage was acquired from
the Idaho State Climate Services
(*http://www.uidaho.edu/ ~climate*) for
use on the INSIDE Project.
This metadata file was created in SMMS
using metadata acquired with the
coverage.
Source Used Citation Abbreviation:
AGROZONES
Process Date: January 2000
Process Contact
Contact Information
Contact Person Primary
Contact Person: Lily Wai
Contact Organization: University of
Idaho Library
Contact Position: Head of Government
Documents
Contact Address
Address Type: Mailing and physical
address
Address: University of Idaho Library
City: Moscow
State or Province: ID
Postal Code: 83844-2350
Country: USA
Contact Voice Telephone:
208-885-6344
Contact Facsimile Telephone:
208-885-6817
Contact Electronic Mail Address:
lwai@uidaho.edu
Hours of Service: 9:00 A.M.–5:00 P.M.,
Monday–Friday

SPATIAL DATA ORGANIZATION INFORMATION

Section Index

Direct Spatial Reference Method: Raster
Raster Object Information
Raster Object Type: Grid cell
Row Count: 199
Column Count: 128

SPATIAL REFERENCE INFORMATION

Section Index

Horizontal Coordinate System Definition
Planar
Map Projection
Map Projection Name: Transverse
Mercator
Transverse Mercator
Scale Factor at Central Meridian:
0.9996
Longitude of Central Meridian: −114
Latitude of Projection Origin: 42
False Easting: 500000
False Northing: 100000
Planar Coordinate Information
Planar Coordinate Encoding Method:
Row and column
Geodetic Model
Horizontal Datum Name: North American
Datum of 1927
Ellipsoid Name: Clarke 1866
Semimajor Axis: 6378206
Denominator of Flattening Ratio:
294.9786982

ENTITY AND ATTRIBUTE INFORMATION

Section Index

Detailed Description
Entity Type
Entity Type Label: agclim_id_ics.vat
Entity Type Definition: Attribute table of
agclim_id_ics
Entity Type Definition Source: ARC/INFO
Attribute
Attribute Label: Agclim_zone
Attribute Definition: Agroclimate Zone
number
Attribute Domain Values
Range Domain
Range Domain Minimum: 1
Range Domain Maximum: 16
Attribute
Attribute Label: Avg_avg_t
Attribute Definition: Mean average
temperature
Attribute Domain Values
Range Domain
Attribute Units of Measure: C*100

Attribute
    Attribute Label: Avg_gdd
    Attribute Definition: Number of growing
      degree-days
    Attribute Domain Values
      Range Domain
        Attribute Units of Measure: (C-days)
Attribute
    Attribute Label: Avg_max_t
    Attribute Definition: Maximum average
      temperature
    Attribute Domain Values
      Range Domain
        Attribute Units of Measure: C*100
Attribute
    Attribute Label: Avg_min_t
    Attribute Definition: Average minimum
      temperature
    Attribute Domain Values
      Range Domain
        Attribute Units of Measure: C*100
Attribute
    Attribute Label: Avg_ppt
    Attribute Definition: Average precipitation
    Attribute Domain Values
      Range Domain
        Attribute Units of Measure: mm
Attribute
    Attribute Label: Count
    Attribute Definition: Number of GRID cells
      of a VALUE
Attribute
    Attribute Label: Elev
    Attribute Definition: Elevation
    Attribute Domain Values
      Range Domain
        Attribute Units of Measure: Meters
Attribute
    Attribute Label: Frz_fall
    Attribute Definition: Day of first freeze in
      the fall
    Attribute Domain Values
      Range Domain
        Attribute Units of Measure: Day
          of year
Attribute
    Attribute Label: Frz_free
    Attribute Definition: Number of freeze-free
      days
    Attribute Domain Values
      Range Domain

        Attribute Units of Measure: Number
          of days
Attribute
    Attribute Label: Frz_spr
    Attribute Definition: Day of last freeze in
      the spring
    Attribute Domain Values
      Range Domain
        Attribute Units of Measure: Day
          of year
Attribute
    Attribute Label: High_avg_t
    Attribute Definition: Highest average
      temperature
    Attribute Domain Values
      Range Domain
        Attribute Units of Measure: C*100
Attribute
    Attribute Label: High_max_t
    Attribute Definition: Highest maximum
      temperature
    Attribute Domain Values
      Range Domain
        Attribute Units of Measure: C*100
Attribute
    Attribute Label: High_min_t
    Attribute Definition: Highest minimum
      temperature
    Attribute Domain Values
      Range Domain
        Attribute Units of Measure: C*100
Attribute
    Attribute Label: High_ppt
    Attribute Definition: Maximum precipitation
    Attribute Domain Values
      Range Domain
        Attribute Units of Measure: mm
Attribute
    Attribute Label: Low_avg_t
    Attribute Definition: Lowest average
      temperature
    Attribute Domain Values
      Range Domain
        Attribute Units of Measure: C*100
Attribute
    Attribute Label: Low_max_t
    Attribute Definition: Lowest maximum
      temperature
    Attribute Domain Values
      Range Domain
        Attribute Units of Measure: C*100

Attribute
   Attribute Label: Low_min_t
   Attribute Definition: Lowest minimum
     temperature
   Attribute Domain Values
     Range Domain
       Attribute Units of Measure: C*100
Attribute
   Attribute Label: Low_ppt
   Attribute Definition: Minimum
     precipitation
   Attribute Domain Values
     Range Domain
       Attribute Units of Measure: mm
Attribute
   Attribute Label: Value
   Attribute Definition: Internal feature number
     for GRIDs
Detailed Description
   Entity Type
     Entity Type Label: agclim_id_ics.sta
     Entity Type Definition: Statistics table
     Entity Type Definition Source: GRID
   Attribute
     Attribute Label: Max
     Attribute Definition: Maximum GRID cell
       VALUE
   Attribute
     Attribute Label: Mean
     Attribute Definition: Mean GRID cell VALUE
   Attribute
     Attribute Label: Min
     Attribute Definition: Minimum GRID cell
       VALUE
   Attribute
     Attribute Label: stdv
     Attribute Definition: Standard deviation of
       GRID cell VALUEs

DISTRIBUTION INFORMATION

Section Index

Distributor
   Contact Information
     Contact Organization Primary
       Contact Organization: Idaho State
         Climate Services
     Contact Address
       Address Type: Mailing and physical
         address

       Address
         Department of Biological and
          Agricultural Engineering
         University of Idaho
         EP425
       City: Moscow
       State or Province: Idaho
       Postal Code: 83844-0904
     Contact Voice Telephone: 208-885-7004
     Contact Facsimile Telephone:
       208-885-7908
     Contact Electronic Mail Address:
       climate@uidaho.edu
     Hours of Service: 0900–1700 Pacific Time
Standard Order Process
   Digital Form
     Digital Transfer Information
       Format Name: ARCE
       Format Specification: ARC/INFO Grid
       File Decompression Technique: Zip
     Digital Transfer Option
       Online Option
         Computer Contact Information
           Network Address:
             Network Resource Name:
               http://snow.ag.uidaho.edu/
               Clim_Map/agroclimate.htm

DISTRIBUTION INFORMATION

Section Index

Distributor
   Contact Information
     Contact Person Primary
       Contact Person: Lily Wai
       Contact Organization: University of Idaho
         Library
     Contact Position: Head of Government
       Documents
     Contact Address
       Address Type: Mailing and physical ad-
         dress
       Address: University of Idaho Library
       City: Moscow
       State or Province: ID
       Postal Code: 83844-2350
       Country: USA
     Contact Voice Telephone: 208-885-6344
     Contact Facsimile Telephone:
       208-885-6817

Contact Electronic Mail Address:
lwai@uidaho.edu
Hours of Service: 9:00 A.M.–5:00 P.M.,
Monday–Friday
Distribution Liability
The users must assume responsibility to determine the usability of this data for their purposes.

The University of Idaho Library provides these geographic data "as is." UIL makes no guarantee or warranty concerning the accuracy of information contained in the geographic data. UIL further make no warranties, either expressed or implied as to any other matter whatsoever, including, without limitation, the condition of the product, or its fitness for any particular purpose. The burden for determining fitness for use lies entirely with the user. Although these data have been processed successfully on computers of UIL, no warranty, expressed or implied, is made by UIL regarding the use of these data on any other system, nor does the fact of distribution constitute or imply any such warranty.

In no event shall the UIL have any liability whatsoever for payment of any consequential, incidental, indirect, special, or tort damages of any kind, including, but not limited to, any loss of profits arising out of use of or reliance on the geographic data or arising out of the delivery, installation, operation, or support by UIL.

Standard Order Process
Digital Form
Digital Transfer Information
Format Specification: A directory containing the Arc/Info Grid coverage is compressed.
File Decompression Technique: Decompress using software capable of extracting a tgz (.tgz) file.
Digital Transfer Option
Online Option
Computer Contact Information
Network Address:
Network Resource Name:
http://inside.uidaho.edu/data/
statewide/ics/archive/
agclim_id_ics.tgz
Fees: None
Standard Order Process
Nondigital Form: N/A

METADATA REFERENCE INFORMATION
Section Index
Metadata Date: 11/17/1999
Metadata Review Date:
Metadata Future Review Date:
Metadata Contact
Contact Information
Contact Organization Primary
Contact Organization: Idaho State Climate Services
Contact Address
Address Type: Mailing and physical address
Address:
Department of Biological and Agricultural Engineering
University of Idaho
EP425
City: Moscow
State or Province: Idaho
Postal Code: 83844-0904
Contact Voice Telephone:
208-885-7004
Contact Facsimile Telephone:
208-885-7908
Contact Electronic Mail Address:
climate@uidaho.edu
Hours of Service: 0900–1700 Pacific Time
Metadata Standard Name: FGDC Content Standards for Digital Geospatial Metadata
Metadata Standard Version: FGDC-STD-001-1998
Metadata Time Convention: local time
Metadata Access Constraints: None
Metadata Use Constraints: None
Metadata Security Information
Metadata Security Classification: Unclassified

INDEX